Experimental Methods in the Physical Sciences

VOLUME 29C

ATOMIC, MOLECULAR, AND OPTICAL PHYSICS:
ELECTROMAGNETIC RADIATION

EXPERIMENTAL METHODS IN THE PHYSICAL SCIENCES

Robert Celotta and Thomas Lucatorto, *Editors in Chief*

Founding Editors

L. MARTON
C. MARTON

Volume 29C

Atomic, Molecular, and Optical Physics: Electromagnetic Radiation

Edited by

F. B. Dunning
Department of Physics
Rice University, Houston, Texas

and

Randall G. Hulet
Department of Physics
Rice University, Houston, Texas

ACADEMIC PRESS
San Diego   New York   Boston   London   Sydney   Tokyo   Toronto

This book is printed on acid-free paper.

Copyright © 1997 by ACADEMIC PRESS, INC.

All Rights Reserved.
No part of this publication may be reproduced or transmitted in any form or by any means, electronic or mechanical, including photocopy, recording, or any information storage and retrieval system, without permission in writing from the Publisher.

The appearance of the code at the bottom of the first page of a chapter in this book indicates the Publisher's consent that copies of the chapter may be made for personal or internal use of specific clients. This consent is given on the condition, however, that the copier pay the stated per-copy fee through the Copyright Clearance Center, Inc. (222 Rosewood Drive, Danvers, Massachusetts 01923) for copying beyond that permitted by Sections 107 or 108 of the U.S. Copyright Law. This consent does not extend to other kinds of copying, such as copying for general distribution, for advertising or promotional purposes, for creating new collective works, or for resale. Copy fees for pre-1997 chapters are as shown on the title pages. If no fee code appears on the title page, the copy fee is the same as for current chapters.
1079-4042/97 $25.00

Academic Press
a division of Harcourt Brace & Company
525 B Street, Suite 1900, San Diego, CA 92101-4495, USA
1300 Boylston Street, Chestnut Hill, MA 02167, USA
http://www.apnet.com

Academic Press Limited
24-28 Oval Road, London NW1 7DX, UK
http://www.hbuk.co.uk/ap/

International Standard Serial Number: 1079-4042

International Standard Book Number: 0-12-475977-7

PRINTED IN THE UNITED STATES OF AMERICA
97 98 99 00 01 02 BC 9 8 7 6 5 4 3 2 1

CONTENTS

CONTRIBUTORS.. xi
PREFACE.. xiii
VOLUMES IN SERIES... xv

1. Laser-Produced Plasmas as Short-Wavelength Incoherent Optical Sources
   by JAMES F. YOUNG

   1.1. Introduction 1
   1.2. Black-Body Radiators 6
   1.3. Laser-Produced Plasmas 7
   1.4. Practical Considerations 13
        References 18

2. Synchrotron Radiation
   by PETER D. JOHNSON

   2.1. Introduction 23
   2.2. Synchrotron Radiation Characteristics 24
   2.3. Light Monochromatization 33
   2.4. Applications 40
        References 42

3. Continuous Wave Dye Lasers
   by ANDREW DIENES AND DIEGO R. YANKELEVICH

   3.1. Introduction 45
   3.2. Basic Dye Laser Principles 46
   3.3. Simple CW Dye Laser Theory 50
   3.4. Actual CW Dye Lasers 56
   3.5. Alignment of a CW Dye Laser 71
        References 73

4. Semiconductor Diode Lasers
   by R. W. FOX, A. S. ZIBROV, AND L. HOLLBERG

   4.1. Introduction 77
   4.2. General Characteristics of Diode Lasers 77
   4.3. Extended-Cavity Lasers 84
   4.4. Electronics 89
   4.5. Optical Coatings on Laser Facets 95
   4.6. Diode Laser Frequency Noise and Stabilization 97
   4.7. Extending Wavelength Coverage 99
        References 100

5. Frequency Stabilization of Tunable Lasers
   by MIAO ZHU AND JOHN L. HALL

   5.1. Introduction 103
   5.2. Optical Frequency References 106
   5.3. Transducers 121
   5.4. Loop Filter 124
   5.5. Design Examples 126
   5.6. Summary 134
        References 134

6. Pulsed Lasers
by MICHAEL G. LITTMAN and XIAO WANG

    6.1. Introduction 137
    6.2. Pulsed Lasers 138
    6.3. Buyer's Guide 150
    6.4. Builder's Guide 153
    6.5. Summary 169
          References 169

7. Techniques for Modelocking Fiber Lasers
by IRL N. DULING III

    7.1. Introduction 171
    7.2. Cavity Building 171
    7.3. Modelocking 175
    7.4. Diagnostics... 187
          References 190

8. Characterization of Short Laser Pulses
by T. FEURER AND R. SAUERBREY

    8.1. Introduction 193
    8.2. Spatial Characterization and Focusing 196
    8.3. Conventional Detectors for nsec to psec Pulses 198
    8.4. Streak Camera 199
    8.5. Autocorrelation and Cross-Correlation Techniques 203
    8.6. Special Techniques for the VUV and X-Ray Regions 223
          References 227

9. Nonlinear Optical Frequency Conversion Techniques
   by U. SIMON and F. K. TITTEL

   9.1. Introduction 231
   9.2. Second-Harmonic Generation 233
   9.3. Sum- and Difference-Frequency Generation. 247
   9.4. Third-Harmonic Generation and Four-Wave Mixing 252
   9.5. Optical Parametric Amplifiers (OPAs) and
        Oscillators (OPOs)...................................... 255
   9.6. Raman Shifters 266
   9.7. Up-Conversion Lasers 268
        References 270

10. Optical Wavelength Standards
    by JÜRGEN HELMCKE

    10.1. Introduction 279
    10.2. Basic Scheme of an Optical Wavelength Standard 280
    10.3. Iodine-Stabilized Lasers 288
    10.4. Wavelength Standards Utilizing Narrow
          Resonances of Laser-Cooled Absorbers 294
    10.5. Optical Frequency Measurement 303
    10.6. Conclusions 307
          References 307

11. Precise Wavelength Measurement of Tunable Lasers
    by MIAO ZHU AND JOHN L. HALL

    11.1. Introduction 311
    11.2. The λ-Meter (Scanning Michelson Interferometer) 312

11.3. The Fizeau Wavemeter	331
11.4. Plane-Parallel Interferometers with CCD Readout	337
11.5. Summary and Outlook	338
References	339

12. Optical Materials and Devices
by SAMI T. HENDOW

12.1. Introduction	343
12.2. Optical Materials and Performance	343
12.3. Optical Components	347
12.4. Polarization-Controlling Components	353
12.5. Passive Optical Devices	358
References	366

13. Guided-Wave and Integrated Optics
by LEON MCCAUGHAN

13.1. Introduction	369
13.2. Optical Waveguides	369
13.3. Fibers	371
13.4. Guided-Wave Integrated Optics	381
13.5. Concluding Points	392
References	393

INDEX	397

CONTRIBUTORS

Numbers in parentheses indicate the pages on which the authors' contributions begin.

ANDREW DIENES (45), *Department of Electrical and Computer Engineering, University of California, Davis, California 95616*

IRL N. DULING III (171), *Naval Research Laboratories, Washington, DC 20375*

T. FEURER (193), *Institut für Optik und Quantenelektronik, 07743 Jena, Germany*

R. W. FOX (77), *National Institute of Standards and Technology, Boulder, Colorado 80303*

JOHN L. HALL (103, 311), *National Institute of Standards and Technology, Boulder, Colorado 80303*

JÜRGEN HELMCKE (279), *Physikalisch-Technische Bundesanstalt, D-38116 Braunschweig, Germany*

SAMI T. HENDOW (343), *Newport Corporation, Irvine, California 92714*

L. HOLLBERG (77), *National Institute of Standards and Technology, Boulder, Colorado 80303*

MICHAEL G. LITTMAN (137), *Department of Mechanical and Aerospace Engineering, Princeton University, Princeton, New Jersey 08544*

PETER D. JOHNSON (23), *Physics Department, Brookhaven National Laboratory, Upton, New York 11973*

LEON MCCAUGHAN (369), *Department of Electrical and Computer Engineering, The University of Wisconsin, Madison, Wisconsin 53706*

R. SAUERBREY (193), *Institut für Optik und Quantenelektronik, 07743 Jena, Germany*

U. SIMON (231), *Department of Electrical and Computer Engineering, Rice University, Houston, Texas 77005*

F. K. TITTEL (231), *Department of Electrical and Computer Engineering, Rice University, Houston, Texas 77005*

XIAO WANG (137), *Bell Laboratories, Whippany, New Jersey 07981*

DIEGO R. YANKELEVICH (45), *Centro de Investigación Científica y de Educación Superior de Ensenada, Ensenada, B.C., Mexico*

JAMES F. YOUNG (1), *Department of Electrical and Computer Engineering, Rice University, Houston, Texas 77005*

MIAO ZHU (103, 311), *Hewlett-Packard Laboratories, Palo Alto, California 94304*

A. S. ZIBROV (77), *Lebedev Institute of Physics, Moscow 117924, Russia*

PREFACE

Since the publication in 1967 of "Atomic Sources and Detectors," Volumes 4A and 4B of this series, the field of atomic, molecular, and optical physics has seen exciting and explosive growth. Much of this expansion has been tied to the development of new sources, such as the laser, which have revolutionized many aspects of science, technology, and everyday life. This growth can be seen in the dramatic difference in content between the present volumes and the 1967 volumes. Not all techniques have changed, however, and for those such as conventional electron sources, the earlier volumes still provide a useful resource to the research community. By carefully selecting the topics for the present volumes, Barry Dunning and Randy Hulet have provided us with a coherent description of the methods by which atomic, molecular, and optical physics is practiced today. We congratulate them on the completion of an important contribution to the scientific literature.

Beginning with Volume 29A, the series is known as *Experimental Methods in the Physical Sciences* instead of *Methods of Experimental Physics*. The change recognizes the increasing multidisciplinary nature of science and technology. It permits us, for example, to extend the series into interesting areas of applied physics and technology. In that case, we hope such a volume can serve as an important resource to someone embarking on a program of applied research by clearly outlining the experimental methodology employed. We expect that such a volume would appeal to researchers in industry, as well as scientists who have traditionally pursued more academic problems but wish to extend their research program into an applied area. We welcome the challenge of providing an important and useful series of volumes for all of those involved in today's broad research spectrum.

Robert J. Celotta
Thomas B. Lucatorto

VOLUMES IN SERIES

EXPERIMENTAL METHODS IN THE PHYSICAL SCIENCES

(formerly Methods of Experimental Physics)

Editors-in-Chief
Robert Celotta and Thomas Lucatorto

Volume 1. Classical Methods
Edited by Immanuel Estermann

Volume 2. Electronic Methods, Second Edition (in two parts)
Edited by E. Bleuler and R. O. Haxby

Volume 3. Molecular Physics, Second Edition (in two parts)
Edited by Dudley Williams

Volume 4. Atomic and Electron Physics—Part A: Atomic Sources and Detectors; Part B: Free Atoms
Edited by Vernon W. Hughes and Howard L. Schultz

Volume 5. Nuclear Physics (in two parts)
Edited by Luke C. L. Yuan and Chien-Shiung Wu

Volume 6. Solid State Physics—Part A: Preparation, Structure, Mechanical and Thermal Properties; Part B: Electrical, Magnetic and Optical Properties
Edited by K. Lark-Horovitz and Vivian A. Johnson

Volume 7. Atomic and Electron Physics—Atomic Interactions (in two parts)
Edited by Benjamin Bederson and Wade L. Fite

Volume 8. Problems and Solutions for Students
Edited by L. Marton and W. F. Hornyak

Volume 9. Plasma Physics (in two parts)
Edited by Hans R. Griem and Ralph H. Lovberg

Volume 10. Physical Principles of Far-Infrared Radiation
By L. C. Robinson

Volume 11. Solid State Physics
Edited by R. V. Coleman

Volume 12. Astrophysics—Part A: Optical and Infrared Astronomy
Edited by N. Carleton

Part B: Radio Telescopes; Part C: Radio Observations
Edited by M. L. Meeks

Volume 13. Spectroscopy (in two parts)
Edited by Dudley Williams

Volume 14. Vacuum Physics and Technology
Edited by G. L. Weissler and R. W. Carlson

Volume 15. Quantum Electronics (in two parts)
Edited by C. L. Tang

Volume 16. Polymers—Part A: Molecular Structure and Dynamics; Part B: Crystal Structure and Morphology; Part C: Physical Properties
Edited by R. A. Fava

Volume 17. Accelerators in Atomic Physics
Edited by P. Richard

Volume 18. Fluid Dynamics (in two parts)
Edited by R. J. Emrich

Volume 19. Ultrasonics
Edited by Peter D. Edmonds

Volume 20. Biophysics
Edited by Gerald Ehrenstein and Harold Lecar

Volume 21. Solid State: Nuclear Methods
Edited by J. N. Mundy, S. J. Rothman, M. J. Fluss, and L. C. Smedskjaer

Volume 22. Solid State Physics: Surfaces
Edited by Robert L. Park and Max G. Lagally

Volume 23. Neutron Scattering (in three parts)
Edited by K. Sköld and D. L. Price

Volume 24. Geophysics—Part A: Laboratory Measurements; Part B: Field Measurements
Edited by C. G. Sammis and T. L. Henyey

Volume 25. Geometrical and Instrumental Optics
Edited by Daniel Malacara

Volume 26. Physical Optics and Light Measurements
Edited by Daniel Malacara

Volume 27. Scanning Tunneling Microscopy
Edited by Joseph Stroscio and William Kaiser

Volume 28. Statistical Methods for Physical Science
Edited by John L. Stanford and Stephen B. Vardaman

Volume 29. Atomic, Molecular, and Optical Physics—Part A: Charged Particles; Part B: Atoms and Molecules; Part C: Electromagnetic Radiation
Edited by F. B. Dunning and Randall G. Hulet

1. LASER-PRODUCED PLASMAS AS SHORT-WAVELENGTH INCOHERENT OPTICAL SOURCES

James F. Young

*Department of Electrical and Computer Engineering,
Rice University
Houston, Texas*

1.1 Introduction

Incoherent sources are useful experimental tools that have long been used throughout the optical spectrum, subject to availability. Recent technological advances have extended the available range into the vacuum ultraviolet and soft x-ray range through the use of laser-produced plasmas. This chapter will concentrate on the description of these new sources. Traditional vacuum ultraviolet sources and appropriate experimental techniques in this spectral range have been discussed in detail previously [1]. Conventional sources in other spectral regions—from the infrared, through the visible, to the ultraviolet quartz absorption edge at about 160 nm—have also been reviewed [2].

Figure 1 shows a typical configuration of a laser-produced plasma. A pulsed high-power laser is focused to a small spot on a metal target. A small, hot, radiating plasma or spark is formed that is characterized by very high densities of electrons and ions, very large density gradients, and very high temperatures. The parameters can be comparable to those in stellar interiors. Much of the work on laser-produced plasmas was motivated by the laser inertial confinement fusion program, and therefore concentrated on low-atomic-number targets irradiated at extremely high power densities, perhaps 10^{15} W cm^{-2}; interest was focused on shock wave formation, ablation rates, and radiation above 1 keV [3, 4]. Researchers in other fields, however, realized that laser-produced plasmas could be produced at much lower laser fluences, and therefore with smaller, practical lasers, and yet still provide unique sources of short-wavelength radiation [5–18]. Laser-produced plasmas can be used as a point source to illuminate a separate sample, but often, as shown in Fig. 1, a gas of the atoms to be irradiated surrounds the plasma, eliminating collection optics. The combination of a laser-produced plasma and a surrounding gas that can be photoionized also represents a unique pulsed source of hot electrons that can be used to excite optically forbidden transitions [19, 20]. Even a modest pumping laser can produce an effective electron current density of >10^5 A-cm^{-2} with a subnanosecond rise time.

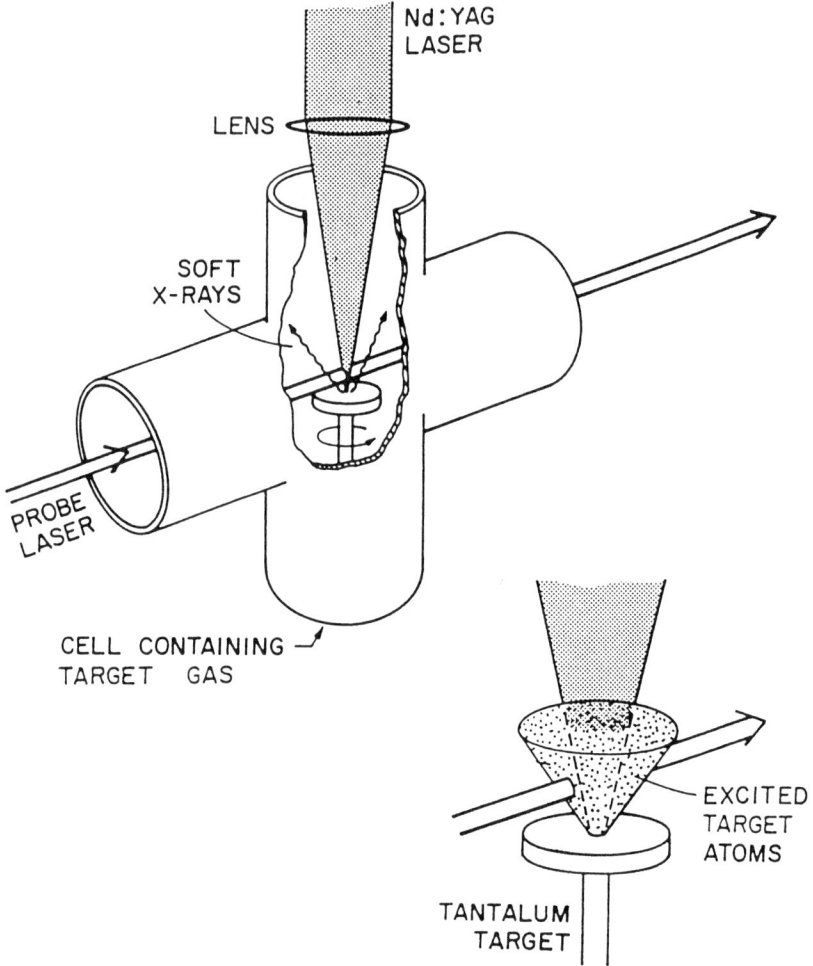

FIG. 1. Schematic of a laser-produced plasma light source. In this geometry the plasma radiation illuminates atoms that are near, but outside, the heated plasma, and the resulting excited levels are measured with a probe laser.

The experimental appeal of laser-produced plasma light sources results from their combination of simplicity and unique characteristics. The physical reality of a typical source is not far different from the schematic in Fig. 1, yet it constitutes a radiation source with a bandwidth and spectral intensity that is difficult to match even at national facilities. The emission from a laser-produced plasma depends on both the laser and the target parameters, but can have a broad smooth spectrum that is often characterized as a black-body radiator with a temperature from 10 eV

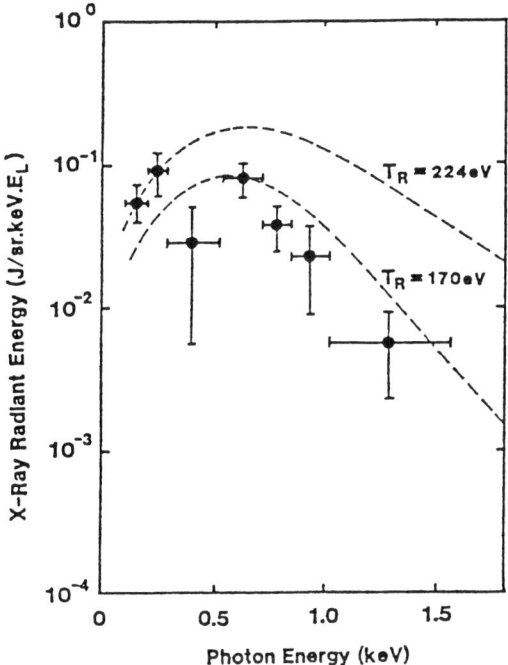

FIG. 2. Radiant energy spectra from an Au target illuminated by a 100-psec 530-nm pulse focused to 4×10^{14} W cm^{-2}. The dashed lines show black-body radiation spectral profiles for the indicated temperatures. From reference [51]. Reprinted with permission from the American Institute of Physics.

to several keV. Figure 2 shows a comparison between measured radiant energy spectra from an Au target and calculated black-body radiation spectral profiles. The agreement is reasonably good, but the black-body temperature of a laser-produced plasma, although useful, should be used only qualitatively. The conversion efficiency from input laser energy to total energy radiated by a plasma can vary from a few percent to more than 60%, as shown in Fig. 3 [21]. These data also show the advantages of short-wavelength pump lasers. The time behavior of the plasma temperature and radiation generally follows that of the driving laser pulse, at least down to about 10-ps time scales, providing a unique tool for time-resolved studies [22–25].

Before reviewing the characteristics of black-body radiators and summarizing the physics of laser-heated plasmas, it is helpful to present some typical numbers to place laser-produced plasmas in a physical context. High focused light intensities are required to initiate and sustain a plasma. While densities of 10^{15} W cm^{-2} and higher have been used to produce high x-ray yields, typical values for laboratory sources are 10^{11} to 10^{13} W cm^{-2}. Thus, for a focal spot of 50 μm diameter, a

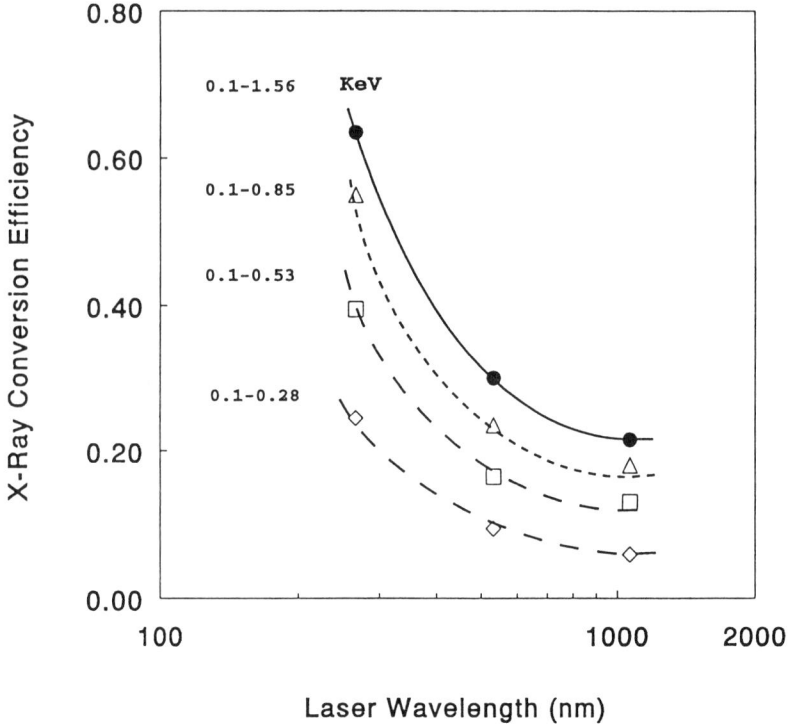

FIG. 3. X-ray conversion efficiency for an Au target illuminated at a power density of 7×10^{13} W cm^{-2} at three wavelengths. The four curves represent measurements in different spectral bands. From reference [21]. Reprinted with permission from the American Institute of Physics.

peak laser power of ~10 MW is required, and pump lasers having a short pulse length must be used to keep the energies practical. The Nd laser fundamental wavelength of 1064 nm plus its harmonics (see Fig. 3) are common choices for the driving radiation because of the commercial availability of the laser. Early fusion work studied CO_2 laser-driven plasmas extensively because of the economy and scalability of CO_2 lasers, but the long wavelength and generally longer pulse lengths reduce plasma heating efficiency. Typically, a plasma heated by 1064 nm radiation will have a spectral radiance 10 to 30 times greater than one heated by 10.6-μm radiation for the same input energy [26, 27]. This difference is illustrated particularly clearly by the data shown in Fig. 4. Excimer lasers are potentially important sources for laser-produced plasmas, but currently the large beam divergence and long pulse lengths of standard commercial lasers are major problems [28].

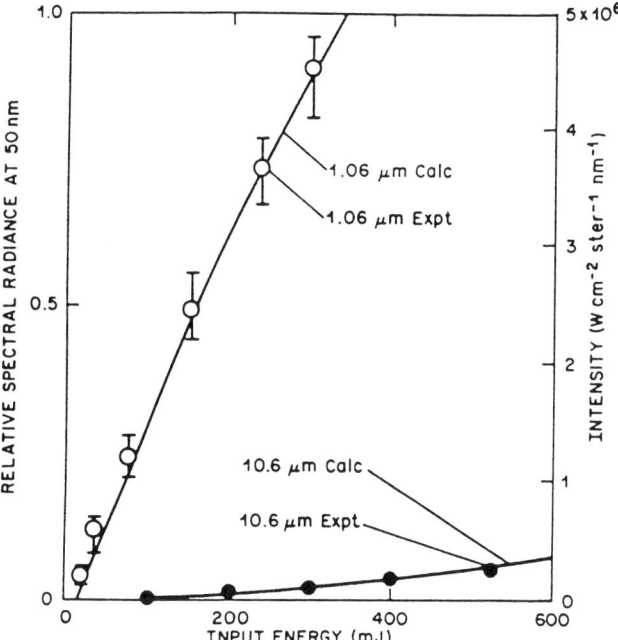

FIG. 4. Spectral radiance at 50 nm as a function of pump input energy for 1064 nm (○) and 10.6 nm (●). The solid curves are theoretical calculations. From reference [27]. Reprinted with permission from the Optical Society of America.

Although Nd Q-switched lasers with pulse lengths up to 10 nsec are used to heat plasmas, Nd mode-locked lasers with pulse lengths of about 100 psec are used more often. Minimum required pulse energies are therefore in the range of 1 to 200 mJ. The recent commercial availability of such sources with high repetition rates has made laser-produced plasmas practical experimental light sources. Ultrashort subpicosecond pulse length laser systems based on dye or Ti:sapphire gain media have been used to produce plasmas with unique characteristics [29–34]. The plasmas produced by very short high-intensity pump pulses are fundamentally different from those produced by more conventional pump sources, which are discussed here. For example, Fig. 5 shows the high-energy spectrum radiated by a plasma heated by a 0.5-TW 120-fsec pulse focused to a density greater than 10^{18} W cm^{-2}. Hard x-ray radiation extending beyond 1 MeV was observed. The spectrum was taken through 19 mm of lead to avoid saturating the detector. The physics of plasma formation and heating under such conditions are quite different from that presented here and are still an active area of research.

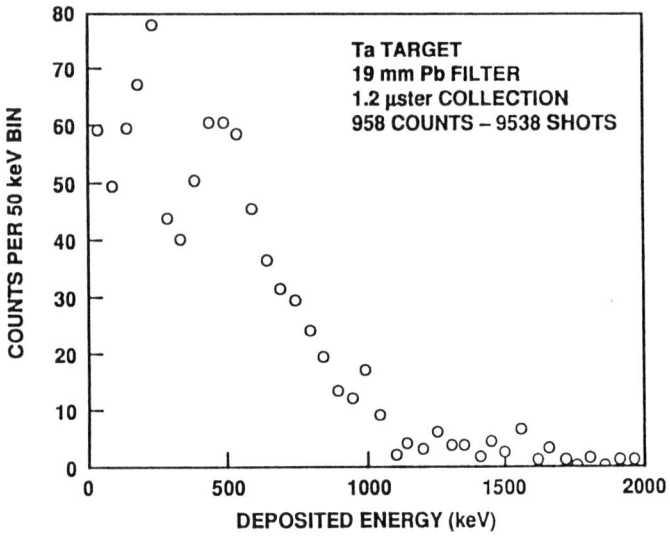

FIG. 5. Pulse height spectrum of radiation from a tantalum target irradiated with a 40-mJ 120-fsec pulse of 807 nm radiation focused to 10^{18} W cm^{-2}. From reference [32]. Reprinted with permission from the IEEE.

1.2 Black-Body Radiators

For many experimental applications, for example, absorption spectroscopy, a source with a broad bandwidth and a flux that is reasonably independent of wavelength is required. The meaning of the vague qualifiers "broad" and "reasonably" vary widely, of course, with the specific application. The prototype incoherent broadband source is the black-body radiator with a spectral radiance given by Planck's law:

$$W(\lambda,T) = \frac{A/\lambda^5}{\exp(B/\lambda T) - 1} \text{W cm}^{-2} \text{ sr}^{-1} \text{ nm}^{-1}, \tag{1}$$

where the constants A and B are given by

$$A = 2hc^2 = 3.97 \times 10^{20} \text{ nm}^5,$$

and

$$B = \frac{ch}{k} = 1240 \text{ nm eV},$$

h is Planck's constant, c is the velocity of light, and k is Boltzmann's constant. The numerical values above and those elsewhere in this chapter assume wavelengths, λ, in nm, and temperatures, T, in units of electron volts, T (eV) = T (K)/11605.

Although the dependence of the black-body spectral radiance on source temperature (Eq. 1) is quite familiar, it is worth repeating that, as the temperature is raised, not only does the peak emission shift to shorter wavelengths, but the emission at *every* wavelength also increases. Thus, for such sources hotter is always better, in terms of spectral radiance, although in some cases it may be advantageous to tune the emission peak to the wavelength of experimental interest in order to minimize the variation of spectral radiance with wavelength.

The wavelength of maximum spectral radiance is

$$\lambda_{max}(T) = 249.9/T \text{ nm}, \qquad (2)$$

and the radiance at the peak is

$$W_{max}(T) = (2.873 \times 10^6)T^5 \text{ W cm}^{-2} \text{ s}^{-1} \text{ nm}^{-1}. \qquad (3)$$

The total output flux of a black-body radiator can be found by integrating Eq. (1):

$$W_{total}(T) = \int_0^\infty W(\lambda,T) \, d\lambda = 32{,}650 \, T^4 \text{ W cm}^{-2} \text{ sr}^{-1}, \qquad (4)$$

which is known as the Stefan–Boltzmann law. Tabulations of these functions, along with integrals of the spectral radiance over finite intervals, are available [35].

It is possible to construct good approximations to low-temperature black-body sources using heated elements, and they are often used in the infrared. But conventional sources with constant temperatures above about 2000 K or $T = 0.17$ eV, corresponding to an emission maximum at 1450 nm, are impractical. Experiments at short wavelengths require a source with an effective temperature at least 100 times higher.

Sources of all types are often described as being effective black-body radiators or as having a particular black-body temperature, but the meaning of these designations is nonstandard and often unclear. An effective black-body temperature may be assigned based on Eq. (2) or (3) using measured values of the source's peak radiance or the wavelength of maximum spectral radiance, even though the wavelength dependence of the source radiance fails to match Eq. (1) very closely. Sometimes it means only that the source has a broad bandwidth relative to the application requirements and a smoothly varying spectral profile. Laser-produced plasma sources are often modeled as a black-body source characterized by an effective temperature. It is a useful qualitative description but is rarely accurate over a broad range. It does, however, have the considerable practical advantages of being easy to explain and easy to use in estimates of flux levels.

1.3 Laser-Produced Plasmas

It is not possible to provide a comprehensive review of laser–plasma interactions here. The subject is far too extensive and complex, and is the focus of much active research. The summary given later is intended to present the basic physics involved, the dependence of such parameters as plasma temperature on experimental variables, and a feeling for typical values. It is necessarily incomplete and incorporates a number of assumptions and simplifications, not all of which are explicitly stated. Laser–plasma interactions have been the subject of several detailed reviews [3, 36–38].

Intense laser light focused on a metal surface penetrates into the surface a skin depth, perhaps 15 nm, and creates hot electrons through multiphoton ionization and collisional absorption. The electrons transfer energy to the metal atoms and ions, creating ever higher stages of ionization. The vaporization temperature is rapidly reached in the local focal region, material flows outward, and an expanding plasma is formed above the surface.

The expansion velocity of the plasma may be taken as the local speed of sound:

$$v_p = \left(\frac{ZkT_e}{m_i}\right)^{1/2} = 10^6 \left(\frac{ZT_e}{A_i}\right) \text{ cm sec}^{-1}, \tag{5}$$

where Z is the ion charge number, T_e is the electron temperature in electron volts, m_i is the ion mass, and A_i is the ion mass number. For the cool ($T_e \approx 20$ eV) plasmas formed from high-atomic-weight materials commonly used for optical sources, v_p is in the range of 10^6 to 10^7 cm sec^{-1}. This result should be treated only as a guide, as various theories and hydrodynamic simulations indicate that the expansion velocity can be subsonic or supersonic in different plasma regions.

The incident laser light propagates into the plasma from regions of low electron density to high electron density, until the local plasma frequency is equal to the laser frequency and the propagation vector becomes imaginary. At this point the light is reflected. For normal incidence, the critical electron density at this surface is

$$n_C = \frac{m\omega^2}{4\pi e^2} = \frac{10^{27}}{\lambda^2} \text{ cm}^{-3}, \tag{6}$$

where m is the electron mass, and ω and λ are the laser frequency and wavelength, respectively. For the common laser wavelength of 1064 nm, $n_C = 8.8 \times 10^{20}$ cm^{-3}. This result is sensitive to geometry. For a wave incident at an angle θ from normal, the critical density for reflection is reduced by a factor of $\cos^2 \theta$, and, as will be shown later, absorption will be reduced. The interaction of the pumping radiation and the plasma depends strongly on the spatial gradient of the plasma density, which is often characterized by a scale length, L_e, the distance over which the density changes from n_C to a low value. For example, if the plasma density is assumed to have an exponential dependence, L_e would be the $1/e$ length.

In the plasma corona, the spatial gradient, or scale length, of the plasma electron density is usually determined by the dynamics of the plasma expansion, since electron and ion temperature gradients are small and have little physical effect. During short laser pulse lengths the plasma expands a distance much less than the focal spot diameter, and the electron density scale length is roughly $L_e = \tau v_p$, where τ is the pulse length. For long pulse lengths or in the limit of very small focal spots, however, the plasma isodensity contours are determined by the geometrical divergence of the plasma flow and are nonplanar, as illustrated in Fig. 6. Thus, if the product τv_p is larger than the focal spot diameter, then the scale length L_e will be reduced to approximately the focal spot radius. There are two reasons to operate in a regime of planar geometry that maximizes L_e. First, the analysis is simpler and the results are likely to be more reasonable; we assume a planar geometry in all the following calculations. Second, as we will show later, the laser energy absorbed by the plasma increases rapidly with L_e. Practically, however, it may be difficult to operate in the planar regime when using long laser pulses, because the required laser energy increases as τ^3 for fixed focal intensity.

The primary heating mechanism in laser-produced plasmas of interest here is inverse Bremsstrahlung, or collisional absorption. Each electron in the optical electric field $\mathcal{E}$ experiences a periodic acceleration that drives an oscillatory motion. Within a few optical cycles, this quiver motion reaches steady state, and the electron acquires an average energy of

$$E_p = \frac{e^2 \mathcal{E}^2}{2m\omega^2} = 10^{-7} I \lambda^2 \text{ eV} \tag{7}$$

in addition to its thermal energy, where I is the laser intensity in watts per square centimeter. If the electron suffers a dephasing collision with an ion while it is oscillating, the associated ponderomotive (or quiver) energy E_p is transferred permanently from the field to the electron. Thus, the average electron heating rate is equal to the product of E_p, the electron density, and the electron–ion collision frequency, which is also proportional to the electron density. Most of the plasma heating therefore takes place in the vicinity of n_C, the highest electron density the field can penetrate. Figure 7 is a schematic of a laser-produced plasma indicating this geometry.

The fraction of laser energy absorbed as the light propagates into the plasma to the critical density surface and back out depends on the shape of the density profile and the angle of incidence. The electron density profile of the coronal plasma is often modeled as $n_e(z) = n_C \exp(-z/L_e)$, in which case the fractional absorption is:

$$F = 1 - \exp\left(\frac{8 v_C L_e}{3c} \cos^3 \theta\right), \tag{8}$$

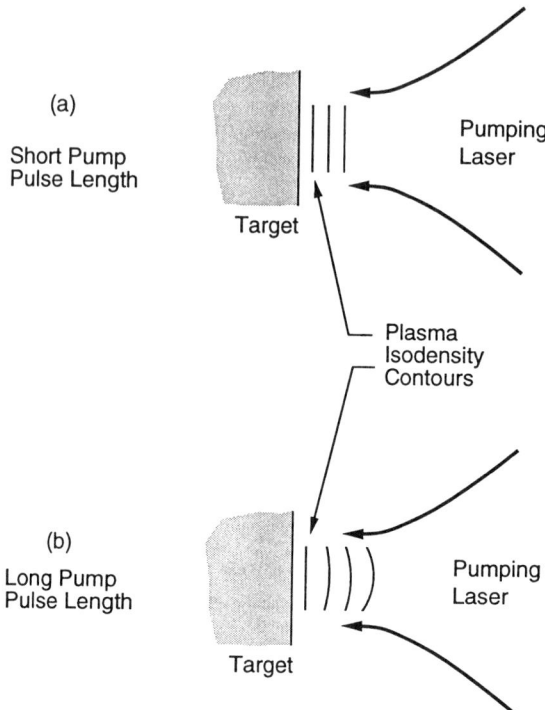

FIG. 6. (a) For short-pulse pumping the plasma isodensity contours are planar and L_e is determined by the pulse length and the plasma expansion velocity. (b) For very short pulses or small focal spots, the isodensity contours are determined by the divergence of the plasma expansion and L_e is approximately equal to the radius of the focal spot.

where v_C is the electron–ion collision frequency at the critical density surface [37]. From basic physical arguments, $v_C \propto Zn_C/T_e^{3/2} \propto Z/T_e^{3/2}\lambda^2$, and is on the order of 10^{15} sec^{-1} for $\lambda = 1064$ nm and the moderate temperature plasmas of laboratory light sources. Thus, the absorption is strongly dependent on pump wavelength (see Fig. 3). In addition, short scale lengths (short laser pulses) or high angles of incidence result in less efficient plasma heating. For a linear density profile, Eq. (8) is still approximately correct for normal incidence, but the dependence on angle is more severe, becoming $\cos^5 \theta$.

Because collisional absorption is a strong function of electron density, most of the absorption occurs close to the critical electron density surface, and it can be seriously reduced by very steep density profiles near n_C [39, 40]. Other absorption processes may then be important. Light obliquely incident on steep plasma density gradients can excite resonant oscillations that transfer significant energy from the

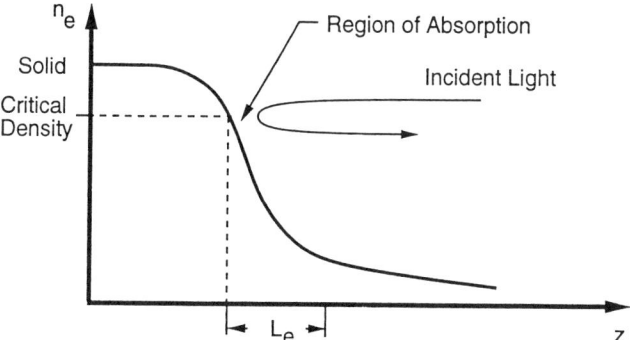

FIG. 7. Electron density versus position for a simple, one-dimensional, laser-heated plasma.

field to the plasma. The conditions under which resonance absorption or other higher-order processes are important are generally present only in laser inertial confinement fusion plasmas and will not be considered here.

The plasma electron temperature is determined by a balance between the absorbed laser energy and the loss of energy from the plasma. The primary loss mechanism is the conduction of energy to the colder, higher-density regions of the plasma near the target, which cannot be heated directly by the laser light. This cooling process is dominated by electron thermal conductivity, since the ion thermal velocity is much lower. Unfortunately, classical calculations of diffusive heat flow fail badly for the strong temperature gradients of laser-heated plasmas. Even with various extensions, they generally predict energy fluxes greater than $n_e E_T v_T$, where E_T and v_T are the electron thermal energy and velocity, respectively [37]. This is inconsistent, since diffusive energy flow must be limited to the product of the electron energy density times the thermal velocity. It is tempting to ascribe this breakdown to a transition to a collisionless energy transfer regime in which electrons free-stream rather than diffuse randomly, but experiments do not support this view. There is a large amount of experimental evidence covering a wide range of conditions that indicates that the heat flux in laser-produced plasmas is in fact well below the classical diffusive limit [41–45]. Despite extensive work, this anomalous inhibition of electron heat conduction is still not well understood, and no theoretical formulation is available. Reasonable fits to the experimental results, however, have been made by postulating a simple flux limit:

$$Q = f n_e E_T v_T, \qquad (9)$$

where f is an empirical parameter that has been found to range from 0.03 to 0.1 over a wide variety of experimental conditions. This simplistic model is physically unsatisfying, yet little else is available. At the least, it is easy to apply.

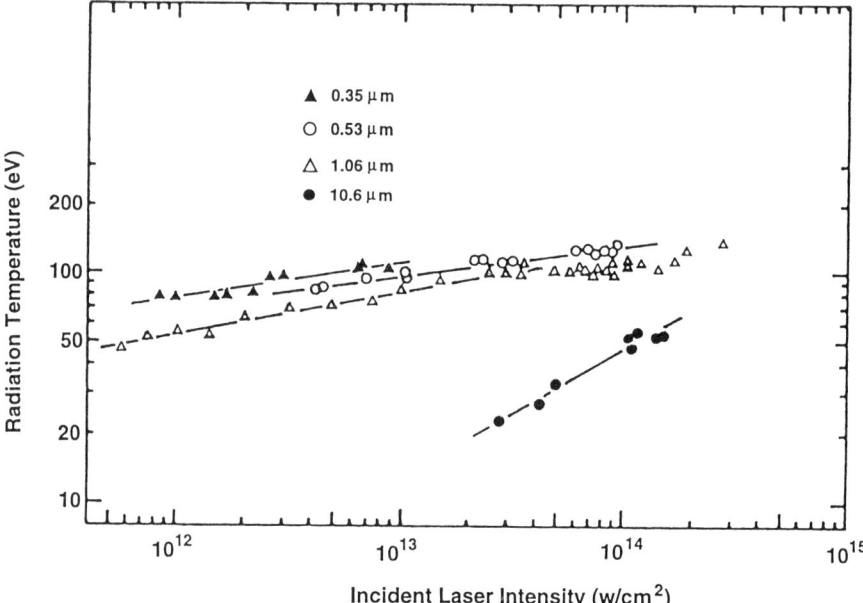

FIG. 8. Radiation temperature of the emission from a gold plasma as a function of incident laser intensity for various pumping wavelengths. The spot sizes were 150 μm; the pump pulse length was 1 nsec for 10.6 μm pumping, and about 0.5 nsec for the other wavelengths. From reference [26]. Reprinted with permission from the American Institute of Physics.

Energy balance using Eq. (9) leads to an electron temperature

$$T_e = (2.8 \times 10^{-11}) \left(\frac{1}{f} I_{abs} \lambda^2\right)^{2/3} \text{ eV,} \qquad (10)$$

where I_{abs} is the absorbed laser power density in watts per square centimeter. Note that the product ($I_{abs} \lambda^2$), and thus T_e, is independent of λ for collisional absorption. Experimental results supporting this conclusion are shown in Fig. 8. The low temperatures for 10.6 μm pumping are the result of a different heat conduction mechanism. Clearly, Eq. (10) should be considered qualitative only. In addition to the problems of the flux-limited model [46], different plasma conditions and different geometries yield different results. For example, low-atomic-number plasmas are essentially completely ionized and optically thin. It is reasonable in that case to assume that energy loss is predominately due to Bremsstrahlung radiation, which is proportional to the square of the density and the square root of the temperature. This model also leads to an electron temperature $T_e \propto I_{abs}^{2/3}$, but with different numerical factors [47]. Surprisingly, a number of different models all predict

$$T_e \propto (I_{abs}\lambda^2)^p, \qquad (11)$$

where the exponent p varies between 1/4 and 2/3 [45].

To this point we have concentrated on electron processes and largely ignored the ions. Laser-produced plasmas are generally dense enough, at least in the hot radiating regions, to be considered in local thermodynamic equilibrium, with T_e determining the electron velocity distribution, the occupancy of excited states, and the ionization stage through the Maxwell velocity distribution, the Boltzmann formula, and the Saha equation, respectively. In addition, the energy equipartition time between electrons and ions is short enough, about a picosecond even for a 10-eV electron temperature, that T_e may be taken as the ion temperature as well. To first order, the distribution of ions will be peaked at the stage that has an ionization energy of a few T_e. Thus, a 15-eV plasma produced from a gold target will be dominated by Au^{+5} ions.

Radiation is emitted from laser-produced plasmas as electrons recombine with ions and decay downward through the various levels and ionization stages. High-atomic-number hot plasmas have many ionization stages with many radiating and absorbing levels that are highly broadened. They are optically dense over a broad spectral region, and the radiation is broad and relatively featureless. They are often characterized or modeled as a black-body radiator of temperature T_e, although the radiation does not rigorously obey Plank's law. Low-atomic-number plasmas necessarily have only a few ionization stages and are often optically thin except for the resonance lines. The resulting spectrum consists primarily of resonance line radiation with a low broad background. There have been many studies of the atomic number dependence of plasma spectra [10, 12, 14, 17, 48–52]. Figure 9 provides a useful summary: the conversion efficiency into several photon energy ranges is shown as a function of target atomic number. Although the curves have structure, the efficiency varies less than a factor of 3 for atomic numbers higher than 20. Extremely hot high-atomic-number plasmas can produce both broadband soft x-ray radiation from cooler regions, and K-shell and higher lines with kilovolt photon energies from hot highly ionized regions [53].

1.4 Practical Considerations

1.4.1 Targets

The simplest target is a bulk piece of the desired material in the form of a plate, disk, or rod. Since the spectrum or effective temperature of the source is only moderately dependent on target atomic number, at least for atomic numbers greater than about 20, the choice of target material is often determined by practical

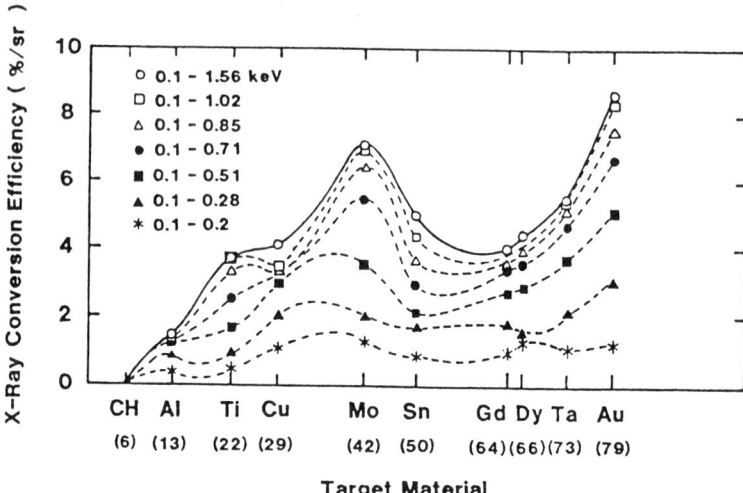

FIG. 9. Target atomic number dependence of x-ray conversion efficiency into various photon energy bands. The pump laser characteristics were: $\lambda = 530$ nm, $\tau \sim 1$ nsec, and a focal intensity of 10^{14} W cm^{-2}. From reference [51]. Reprinted with permission from the American Institute of Physics.

considerations. For example, a laser-produced plasma is commonly formed in a cell containing the species to be illuminated, and in that case, chemical, thermal, or vacuum comparability may dictate the choice. Economy and ease of fabrication are also considerations. Tungsten and tantalum are durable, refractory, high-atomic-number materials but are difficult to machine. Gold is a common prototype target for plasma studies, but few research budgets can accommodate bulk samples on a regular basis. The rare earth elements have been used because of their particularly line-free spectra [10, 14]. Special target compositions have been proposed for specific applications [54], but, in the end, ordinary stainless steel (iron and nickel) is often used, and in most applications appears to work as well as more exotic choices.

The focused laser beam ablates material from the target surface and leaves a crater. If subsequent plasmas are formed at the same point, in the crater, the radiation pattern will be altered by the geometric shielding. The change in geometry from planar to semienclosed and alterations in surface conditions can reduce absorbed power and affect plasma radiation conditions, but no studies have been published. These problems are avoided by moving the target to provide a fresh surface for each laser pulse. The method depends on the experimental geometry, but often a simple rotating disk or cylinder is used. Raster scans of planar targets and moving wires or tapes have also been used. The latter provide for very long experimental runs without requiring target renewal. Reasonable mechanical design must be used so that the target surface remains within the focal region as it moves so as to assure reproducible plasma formation. There is no evidence that special surface treatments improve plasma formation. Common machining techniques to achieve a reasonably flat surface and cleaning for vacuum seem adequate. (One exception to this is mentioned in the later section on short-pulse pumping.)

Other target configurations have been used for special applications. High-pressure gas [18] and gas jets [55, 56] avoid the problem of target replacement and are very clean. Liquid targets have also been used as self-renewing target surfaces [57]. Materials that are expensive or difficult to machine are often used in thin film form, either free standing or evaporated onto a plastic film. The film thickness can be adjusted so that the plasma just ablates through the film, providing radiation from both sides of the film [58]. Microcavities, several millimeters in diameter, have recently been studied as a way of producing true black-body radiators of very high temperature [59]. The laser is focused through a small hole and produces a plasma by heating a spot on the interior wall. Radiation from the plasma heats the entire interior wall, and subsequent reemission and absorption results in an intense thermal field with high uniformity obeying Planck's law. The complexity of the target, however, makes this a very low repetition rate source with low average output.

1.4.2 Focusing

Any focusing geometry that produces the required high power density is appropriate. Small focal spots reduce laser power requirements, but if the spot is very small, a few microns, the plasma density scale length, L_e, will be limited approximately to the spot radius and reduce the absorbed laser energy and plasma temperature. Similarly, for angles of incidence θ from normal, the light penetrates the plasma only to an electron density $n_C \cos^2 \theta$, and absorption and heating efficiency are reduced. If the experimental conditions require a high angle of incidence, special grooved target surfaces can be used to provide a local normal angle of incidence [60]. Single spot, multiple spot, and line foci have been used, formed by both refractive and reflective optics. Limiting the focal geometry to f-numbers greater than 10 reduces aberrations and allows simple uncorrected optics to be used.

1.4.3 Short-Pulse-Length Pumping

There are several advantages to using short-pulsed lasers for plasma generation. Damage and energy storage limitations force laser apertures and system cost to increase with pulse energy, usually much faster than linearly. Short-pulse excitation, therefore, offers a means of achieving the required high focal intensities for

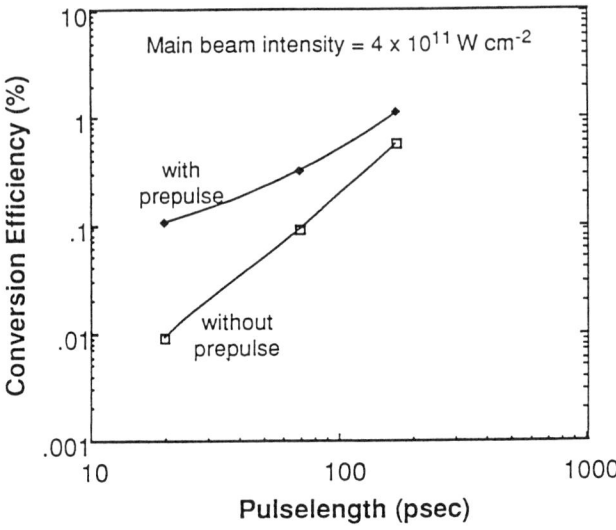

FIG. 10. Conversion efficiency into the 120–285 eV spectral band as a function of pulse length. The prepulse arrives at the target about 1.5 nsec in advance of the main pulse. From reference [63]. Reprinted with permission from the Optical Society of America.

plasma heating with a modest laser investment. In addition, the temperature and radiation of laser-produced plasmas generally track the time behavior of the pump pulse. The rise times are driven by the very high laser heating rates and follow the pump pulse, but the plasma decay time is determined by cooling rates that are independent of the laser pulse and typically limit the minimum radiation pulse length to ~10 psec. Nevertheless, short-pulse excited plasmas offer a unique source for time-resolved studies at very short wavelengths. But the laser–plasma interaction is not independent of time scale. As noted earlier, the plasma scale length decreases directly with pulse length, and that, in turn, decreases the fraction of laser energy absorbed. Figure 10 illustrates this reduction in conversion efficiency for pump pulse lengths below about 200 psec. Physically, it takes time for a plasma to form and expand, and that time is independent of the exciting pulse duration. During a very short laser pulse, there is insufficient time to form a significant volume of low-density plasma where light can be absorbed. Instead, most of the light is reflected from a near-solid density plasma at the target surface.

One approach to reducing this problem is to apply a prepulse to the target that initiates a small plasma [31, 61–63]. This plasma then expands during the interval until the main pulse arrives, resulting in a large plasma volume with a long scale length that can absorb light efficiently. Figure 11 shows the significant improvement that can result by using just a small fraction of the pump energy, about 10%,

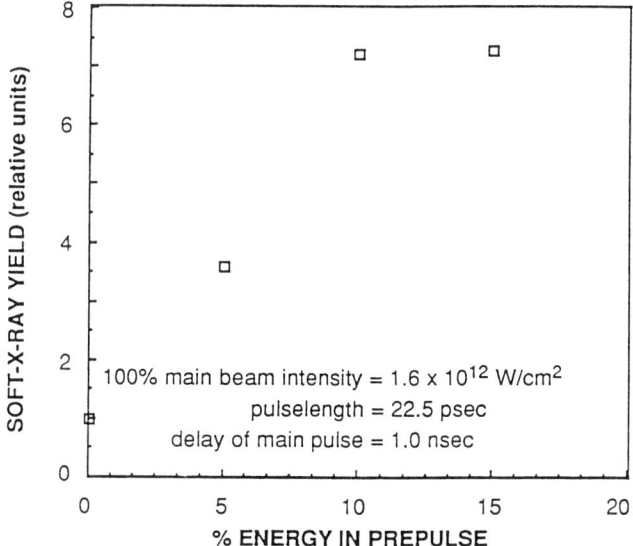

FIG. 11. Enhancement in plasma x-ray yield versus the fraction of main pulse energy used for the prepulse. From reference [63]. Reprinted with permission from the Optical Society of America.

for the prepulse. In these experiments the exact interval between the prepulse and the main pulse was not critical. Delays in the range of 1 to 2 nsec, corresponding to a plasma expansion of ~10 µm, were optimum. Experiments with very-high-intensity, 10^{17} W cm^{-2}, subpicosecond pulses yielded short optimum delays of about 20 psec [29]. In this parameter range, the coupling of plasma expansion, heating, and cooling processes are quite different.

Another approach to increasing efficiency when using ultrashort pulses is to use a target surface that has a built-in density gradient to reduce pump reflection. Etched microgrooves and colloidal porous coatings have been used successfully [64]. This technique increases the absorption of the pump laser energy into a plasma that is near solid density and results in high conversion efficiencies to x-rays above 1 keV energy.

1.4.4 Debris

Target material ablated by a laser-produced plasma ultimately is deposited on nearby surfaces. Laser-produced plasmas are effective evaporation sources and are frequently employed for vacuum deposition. In addition, small particles and droplets can be ejected from the target [65]. Thus, cell walls, windows, collection optics, gratings, etc., in the vicinity of the target will become coated with a thin film of the target material over time [66]. Simple optics, like flat quartz windows, can generally be cleaned to restore transparency, but coated optics, and especially gratings, will eventually be destroyed. The window for the input driving laser is less of a problem than might be expected: the laser beam tends to continuously remove deposits by local heating. After a long experiment, the input window frequently resembles a metal mirror with a clear spot the shape of the input beam [17]. Nevertheless, window damage problems can occur if the pump beam alignment is adjusted in the course of an experiment and the input beam moves into an area coated with absorbing material.

A number of approaches can be used to reduce the debris problem, depending on the wavelength region of interest and experimental requirements. The best approach, if feasible, is to use a sacrificial thin film window [67] to isolate the plasma source from other components. A vacuum tight seal is unnecessary since most of the debris deposit will be limited to the line of sight of the source. Thin film replaceable coatings have also been used to protect x-ray collection optics placed close to plasma sources [68]. A low ambient He pressure can significantly reduce the mean diffusion length of debris. Pressure–length products of only 1.5 torr-cm can reduce debris accumulation by an order of magnitude [69]. Synchronized high-pressure gas jets have been proposed [65] to sweep larger debris particles out of the path of optical components. Thin film targets tend to minimize debris simply by limiting the mass that is available for ablation. One novel approach is to use a target material that is compatible with nearby optics; thus the interest in gold thin film targets. If a gold-coated reflector is used to collect light from a gold target, the effect of target ablation is merely to continuously recoat the reflector with gold, although droplet or particulate debris can increase light scattering. Another approach is to use a cold target consisting of the liquid or frozen phase of a material that is normally a vapor at the ambient cell temperature.

References

1. Samson, J. A. (1967). *Techniques of Vacuum Ultraviolet Spectroscopy*, Pied Publications, Lincoln, NE.
2. Raith, W., Christensen, R. L., and Ames, I. (1967). "Sources of Atomic Particles: Photons," in *Methods of Experimental Physics*, Volume 4, Part A, V. W. Hughes and H. L. Schultz (eds.), Chapter 1.5, pp. 284–318, Academic Press, New York.
3. Max, C. E. (1982). "Physics of the Coronal Plasma in Laser Fusion Targets," in *Laser–Plasma Interaction*, R. Bahan and C. C. Adam (eds.), pp. 302–410, North-Holland, Amsterdam.
4. Yamanaka, C. (1991). "Laser Plasma and Inertial Confinement Fusion," in *Physics of Laser Plasma*, A. Rubenchik and W. Witkowski (eds.), *Handbook of Plasma Physics*, Vol. 3, Chapter 1, pp. 1–62, North-Holland, Amsterdam.

5. Fawcett, B. C., Gabriel, A. H., Irons, F. E., Peacock, N. J., and Saunders, P. A. H. (1966). *Proc. Phys. Soc. (Great Britain)* **88**, 1051.
6. Ehlers, A. W., and Weissler, G. L. (1966). *Appl. Phys. Lett.* **8**, 89.
7. Breton, C., and Popoular, R. (1975). *J. Opt. Soc. Am.* **63**, 1275.
8. Carroll, P. K., Kennedy, E. T., and O'Sullivan, G. (1978). *Opt. Lett.* **2**, 72.
9. Mahajan, C. G., Baker, E. A. M., and Burgess, D. D. (1979). *Opt. Lett.* **4**, 283.
10. Carroll, P. K., Kennedy E. T., and O'Sullivan, G. (1980). *Appl. Opt.* **19**, 1454.
11. O'Sullivan, G., Carroll, P. K., McIlrath, T. J., and Ginter, M. L. (1981). *Appl. Opt.* **20**, 3043.
12. Nicolosi, P., Jannitti, E., and Tondello, G. (1981). *Appl. Phys. B* **26**, 117.
13. Caro, R. G., Wang, J. C., Falcone, R. W., Young, J. F., and Harris, S. E. (1983). *Appl. Phys. Lett.* **42**, 9.
14. Orth, F. B., Ueda, K., McIlrath, T. J., and Ginter, M. L. (1986). *Appl. Opt.* **25**, 2215.
15. Gohil, P., Kaufman, V., and McIlrath, T. J. (1986). *Appl. Opt.* **25**, 2039.
16. Ginter, M. L., and McIlrath, T. J. (1986). *Nucl. Instrum. Methods A* **246**, 779.
17. Bridges, J. M., Cromer, C. L., and McIlrath, T. J. (1986). *Appl. Opt.* **25**, 2208.
18. Laporte, P., Damany, N., and Damany, H. (1987). *Opt. Lett.* **12**, 987.
19. Wang, J. C., Caro, R. G., and Harris, S. E. (1983). *Phys. Rev. Lett.* **51**, 767.
20. Benerofe, S. J., Yin, G. Y., Barty, C. P. J., Young, J. F., and Harris, S. E. (1991). *Phys. Rev. Lett.* **66**, 3136.
21. Kodama, R., Okada, K., Ikeda, N., Mineo, M., Tanaka, K. A., Mochizuki, T., and Yamakoshi, C. (1986). *J. Appl. Phys.* **59**, 3050.
22. Epstein, H. M., Schwerzel, R. E., Mallozzi, P. J., and Campbell, B. E. (1983). *J. Am. Chem. Soc.* **105**, 1466.
23. Lunney, J., Dobson, P. J., Hares, J. D., Tabatabaei, S. D., and Eason, R. W. (1986). *Opt. Comm.* **58**, 269.
24. Murnane, M. M., Kapteyn, H. C., and Falcone, R. W. (1990). *Appl. Phys. Lett.* **56**, 1948.
25. Sher, M. H., Mohideen, U., Tom, H. W. K., Wood II, O. R., and Aumiller, G. D. (1993). *Opt. Lett.* **18**, 646.
26. Nishimura, H., Matsuoka, F., Yagi, M., Yamada, K., Nakai, S., McCall, G. H., and Yamanaka, C. (1983). *Phys. Fluids* **26**, 1688.
27. Wood II, O. R., Silfvast, W. T., Macklin, J. J., and Maloney, P. J. (1986). *Opt. Lett.* **11**, 198.
28. Broughton, J. N., and Fedosejevs, R. (1992). *Appl. Phys. Lett.* **60**, 1818.
29. Külke, K., Herpers, U., and von der Linde, D. (1987). *Appl. Phys. Lett.* **50**, 1785.
30. Wood II, O. R., Silfvast, W. T., Tom, H. W. K., Knox, W. H., Fork, R. L., Brito-Cruz, C. H., Downer, M. C., and Maloney, P. J. (1988). *Appl. Phys. Lett.* **53**, 654.
31. Tom, H. W. K., and Wood II, O. R. (1989). *Appl. Phys. Lett.* **54**, 517.

32. Kmetec, J. D. (1992). *IEEE J. Quantum Electronics* **28**, 2382.
33. Teubner, U., Bergmann, J., van Wonterghem, B., Schafer, F. P., and Sauerbrey, R. (1993). *Phys. Rev. Lett.* **70**, 794.
34. Murnane, M. M., Kapteyn, H. C., Rosen, M. D., and Falcone, R. W. (1991). *Science* **251**, 531.
35. Gray, D. E., ed. (1972). *American Institute of Physics Handbook*, 3rd edition, McGraw-Hill, New York.
36. More, R. M. (1991). "Atomic Physics of Laser-Produced Plasmas," in *Physics of Laser Plasma*, A. Rubenchik and S. Witkowski (eds.), *Handbook of Plasma Physics*, Vol. 3, Chapter 2, pp. 63–110, North-Holland, Amsterdam.
37. Kruer, W. L. (1988). *The Physics of Laser Plasma Interactions*, Addison-Wesley, Reading, MA.
38. Sher, M. H. (1989). "Techniques for Short-Wavelength Photoionization Lasers: A 2-Hz 109-nm Laser," Ph.D. thesis, Applied Physics Department, Stanford University.
39. Ginsburg, V. L. (1964). *Propagation of Electromagnetic Waves in Plasmas*, Pergamon, Oxford.
40. Shay, H. D., Haas, R. A., Kruer, W. L., Boyle, M. J., Phillion, D. W., Rupert, V. C., Kornblum, H. N., Rainer, R., Tirsell, K. G., Slivinsky, V. W., and Koppel, L. N. (1976). *Bull. Am. Phys. Soc., Ser. II* **21**, 1048.
41. Malone, R. C., McCrory, R. L., and Morse, R. L. (1975). *Phys. Rev. Lett.* **34**, 721.
42. Yaakobi, B., Boehly, T., Bourde, P., Conturie, Y., Craxton, R. S., Delettrez, J., Forsyth, J. M., Frankel, R. D., Goldman, L. M., McCrory, R. L., Richardson, M. C., Seka, W., Shvarts, D., and Soures, J. M. (1981). *Opt. Comm.* **39**, 175.
43. Rosen, M., Phillion, D. W., Rupert, V. C., Mead, W. C., Kruer, W. L., Thomson, J. J., Kornblum, H. N., Slivinsky, V. W., Caporaso, G. J., Boyle, M. J., and Tirsell, K. G. (1979). *Phys. Fluids* **22**, 2020.
44. Mead, W. C., Campbell, E. M., Estabrook, K. G., Turner, R. E., Kruer, W. L., Lee, P. H. Y., Pruett, B., Rupert, V. C., Tirsell, K. G., Stradling, G. L., Ze, F., Max, C. E., Rosen, M. D., and Lasinski, B. F. (1983). *Phys. Fluids* **26**, 2316.
45. Ramis, R., and Sanmartin, J. R. (1983). *Nucl. Fusion* **23**, 739.
46. Mead, W. C., Campbell, E. M., Kruer, W. L., Turner, R. E., Hatcher, C. W., Bailey, D. S. Lee, P. H. Y., Foster, J., Tirsell, K. G., Pruett, B., Holmes, N. C., Trainor, J. T., Stradling, G. L., Lasinski, B. F., Max, C. E., and Ze, F. (1984). *Phys. Fluids* **27**, 1301.
47. Carroll, P. K., and Kennedy, E. T. (1981). *Contemp. Phys.* **22**, 61.
48. Nakamo, N., and Kuroda, H. (1983). *Phys. Rev. A* **27**, 2168.
49. Gerritsen, H. C., van Brug, H., Bijkerk, F., and van der Wiel, M. J. (1986). *J. Appl. Phys.* **59**, 2337.
50. Alaterre, P., Pépin, H., Fabbro, R., and Faral, B. (1986). *Phys. Rev. A* **34**, 4184.
51. Mochizuki, T., Yabe, T., Okada, K., Hamada, M., Ikeda, N., Kiyokawa, S., and Yamanaka, C. (1986). *Phys. Rev. A* **33**, 525.
52. Nakano, N., and Kuroda, H. (1987). *Phys. Rev. A* **35**, 4712.

53. Kauffman, R. (1991). "X-Ray Radiation from Laser Plasma," in *Physics of Laser Plasma*, A. Rubenchik and S. Witkowski (eds.), *Handbook of Plasma Physics*, Vol. 3, Chapter 3, pp. 111–162, North-Holland, Amsterdam.
54. Harris, S. E., and Kmetec, J. D. (1988). *Phys. Rev. Lett.* **61**, 62.
55. Charatis, G., Slater, D. C., Mayer, F. J., Tarvin, J. A., Busch, G. E., Sullivan, D., Musinski, D., Matthews, D. L., and Kippel, L. (1981). "Laser-Heated Gas Jet: A Soft X-Ray Source," in *Low Energy X-Ray Diagnostics—1981*, AIP Conference Proceedings, No. 75, D. T. Attwood and B. L. Henke (eds.), pp. 270-274, American Institute of Physics, New York.
56. Fiedorowicz, H., Bartnik, A., and Patron, Z. (1993). *Appl. Phys. Lett.* **62**, 2778.
57. Yamakoshi, H., Chin, C. T., Jaimungal, S., Herman, P. R., Budnik, F. W., Kulcsár, G., Zhao, L., and Majoribanks, R. S. (1993). "Extreme-Ultraviolet Laser Photo-Pumped by a Self-Healing Hg Target," in *Applications of Laser Plasma Radiation*, Vol. 2015, SPIE, Bellingham, WA.
58. Celliers, P., and Eidmann, K. (1990). *Phys. Rev. A* **41**, 3270.
59. Sigel, R. (1991). "Laser-Generated Intense Thermal Radiation," in *Physics of Laser Plasma*, A. Rubenchik and S. Witkowski (eds.), *Handbook of Plasma Physics*, Vol. 3, Chapter 4, pp. 163–198, North-Holland, Amsterdam.
60. Sher, M. H., Macklin, J. J., Young, J. F., and Harris, S. E. (1987). *Opt. Lett.* **12**, 891.
61. Tanaka, K. A., Yamauchi, A., Kodama, R., Mochizuki, T., Yamanaka, T., Nakai, S., and Yamanaka, C. (1988). *J. Appl. Phys.* **63**, 1787.
62. Weber, R., and Balmer, J. E. (1989). *J. Appl. Phys.* **65**, 1880.
63. Sher, M. H., and Benerofe, S. J. (1991). *J. Opt. Soc. Am. B* **8**, 2437.
64. Murnane, M. M., Kapteyn, H. C., Gordon, S. P., Boker, J., and Glytsis, E. N. (1993). *Appl. Phys. Lett.* **62**, 1068.
65. Silfvast, W. T., Richardson, M. C., Bender, H., Hanzo, A., Yanovsky, V., Jin, F., and Thorpe, J. (1992). *J. Vac. Sci. Technol. B* **10**, 3126.
66. Bortz, M. L., and French, R. H. (1989). *Appl. Phys. Lett.* **55**, 1955.
67. Powell, F. R., Vedder, P. W., Lindblom, J. F., and Powell, S. F. (1990). *Opt. Eng.* **29**, 614.
68. Catura, R. C., Joki, E. G., Roethig, D. T., and Brookover, W. J. (1987). *Appl. Opt.* **26**, 1563.
69. Ginter, M. L., and McIlrath, T. J. (1988). *Appl. Opt.* **27**, 885.

2. SYNCHROTRON RADIATION

Peter D. Johnson

Physics Department
Brookhaven National Laboratory
Upton, New York

2.1 Introduction

Over a period of approximately 50 years, synchrotron radiation has moved from being a problem limiting the performance of particle accelerators to a major research tool contributing to most scientific disciplines.

Synchrotron radiation was first described theoretically and observed experimentally immediately following World War II [1–4]. However, exploitation of its experimental capabilities did not take place to any great extent until the late fifties with demonstrations of the possibility of using such radiation for absorption studies in the far-ultraviolet/soft x-ray range [5] and for absorption studies in the gas phase in the VUV range [6]. These initial experiments were followed by pioneering studies throughout the entire spectral range.

Early experimental programs, using synchrotron radiation were carried out in a parasitic mode on accelerators built for various particle physics programs. These sources have subsequently become known as "first-generation" sources. During the seventies, construction began on the first storage rings to be dedicated to the production of synchrotron radiation. Not only have these "second-generation" sources provided access to the radiation to a much larger user community, but they have served as test beds for a new type of radiation source, namely, undulators and wigglers. These "insertion devices," consisting of a linear array of magnets, are placed in the straight sections of the storage ring to produce high-intensity photon beams. With the increase in photon flux, a new class of experiments became possible, and this, in turn, has led to the construction of "third-generation" storage rings, now optimized for photon production from the straight sections via the use of the insertion devices.

There are now more than 30 synchrotron radiation sources in use or under construction worldwide. These tend to be optimized for particular photon energy regimes, from the UV, through soft x-ray, to hard x-ray. The division into these different wavelength ranges was initially a reflection of the different techniques used by the experimentalists to monochromatize the synchrotron radiation. However, with the development of second- and third-generation light sources, the division has carried over into construction of storage rings tailored to the requirements of particular sectors of the user community.

In this chapter we discuss the properties of synchrotron radiation and the different methods used to obtain monochromatized radiation from the continuous spectrum. There are several excellent articles and books that have appeared on these topics, ranging from focused reviews for the specialist [7] to less detailed but complete overviews of the radiation and its application [8]. Here we will present a brief overview and examine those characteristics of synchrotron radiation that are most important to the experimentalist using the radiation.

2.2 Synchrotron Radiation Characteristics

In this section we first review the properties of storage rings and associated synchrotron radiation, including the total emitted power, the photon flux, and the brightness or brilliance of the radiated photons. We also examine the wavelength and polarization characteristics of the photons. In discussing the properties of synchrotron radiation we first consider the radiation emitted at the dipoles or bending magnets and then discuss the characteristics of the radiation emitted by insertion devices.

2.2.1 Dipole or Bending Magnet Radiation

An accelerating electron loses energy by emitting radiation. As shown in Fig. 1, at relativistic velocities, the radiation is emitted into a narrow cone pointing in the direction of the electron's instantaneous motion. The total power $P(\lambda,t)$, in ergs sec^{-1}, radiated into all space at wavelength λ by a single electron moving in a circular orbit of radius ρ with energy γ in rest mass units is given by [4]

$$P(\lambda,t) = \frac{3^{3/2} e^e c \gamma^7}{16\pi^2 \rho^3} \left(\frac{\lambda_c}{\lambda}\right)^3 \int_y^\infty K_{5/3}(\eta) \, d\eta, \qquad (1)$$

where $y = \lambda_c/\lambda$, and $K_{5/3}$ is a modified Bessel function of the second kind. λ_c, the critical wavelength, is a characteristic of the synchrotron radiation source. In terms of the electron beam energy,

$$\lambda_c = \frac{4\pi\rho}{3\gamma^3}. \qquad (2)$$

The radiated power as defined by Eq. (1) has its maximum at a wavelength $\lambda = 0.42\lambda_c$.

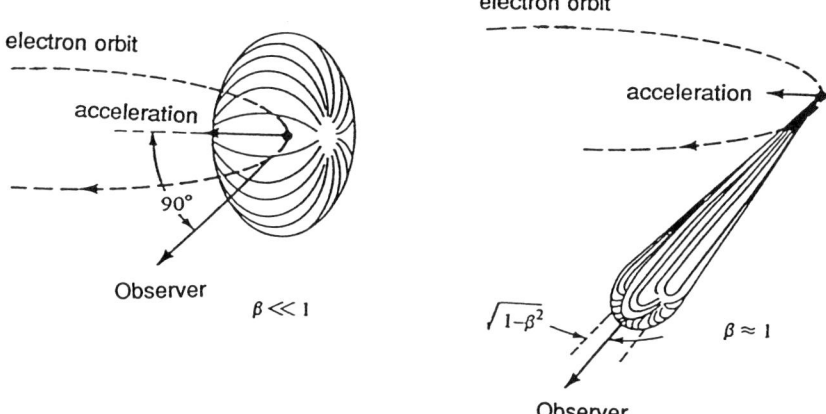

FIG. 1. Radial distribution of the emitted radiation from an accelerating electron as a function of the electron velocity $\beta = v/c$.

The number of photons radiated per second $N(\lambda,t)$ will be related to the total power radiated per unit wavelength simply by the energy of the photon. Thus,

$$N(\lambda,t) = P(\lambda,t)/\hbar\omega. \tag{3}$$

The total power or, equivalently, the total number of photons radiated per second as defined in the above equations are properties of the single electron in the storage ring. Of more interest to the experimental user is the number of photons radiated per mrad in the orbital plane by the total stored electron current. The derivation of the appropriate expressions is reviewed elsewhere [9]. At a given wavelength λ, the total number of photons $N(\lambda,t,\theta)$ radiated per second, per stored current I, and per arc θ are respectively given by

$$N(\lambda,t,\theta) = \frac{3^{3/2} ec\theta I \gamma^4}{40\pi\, h\rho}\, G_2. \tag{4}$$

The functions G_i are defined as [5]

$$G_i = y^i \int_y^\infty K_{5/3}(\eta)\, d\eta, \quad i = 1, 2, 3, \ldots, \tag{5}$$

where again $y = \lambda_c/\lambda$ and $K_{5/3}$ is a modified Bessel function of the second kind.

Rather than showing the spectral distribution as defined in Eq. (4), it is more common to consider the spectral distribution in terms of the photon flux in some finite bandwidth $\Delta\lambda$. For the experimentalist designing a beamline this is more useful because it more clearly maps onto the transmission and resolving power of the monochromators. The photon flux in $\Delta\lambda$ is given by

$$N_k(\lambda,t,\theta) = 0.1732 \frac{ke\theta I\gamma}{h} G_1, \qquad (6)$$

where $k = \Delta\lambda/\lambda$. In real units, this translates into

$N_k = 1.256 \times 10^7 \, \gamma \, G_1$ photons sec^{-1} mrad^{-1} ma^{-1} 0.1%$^{-1}$ bandwidth.

As an example, we show in Fig. 2a [10] the spectral distributions N_k radiated from bending magnets on the UV and x-ray storage rings at the National Synchrotron Light Source (NSLS) and bending magnets for the Advanced Light Source (ALS), and the Advanced Photon Source (APS). This figure clearly shows the maximum in the distributions moving to higher energies as the stored beam energy increases.

It is also quite common to present the brightness or brilliance of the storage ring. Shown in Fig. 2b, the photon flux or spectral distribution is now normalized to the area of the electron beam emitting the photons. This is an important consideration when the beamline is required to focus the monochromatized beam into a small spot in, for example, microscopy applications. Indeed, several of the third-generation sources—including the Advanced Light Source and the Advance Photon Source—have been optimized in terms of their brightness.

Before discussing insertion devices, we note two further properties of the storage ring that are important to experimentalists using the radiation. These properties are the emittance and the lifetime of the stored electron beam.

Equation (6) takes account of the total current in the storage ring but does not take account of the physical properties of the circulating beam. These we describe in terms of emittances that effectively define the orbit in a four-dimensional space. If at each point on the orbit we consider a plane perpendicular to the orbit, we may define the beam in terms of two spatial coordinates within the plane and two angles between the electron trajectory and the ideal orbit. These four coordinates are then a measure of the emittance at each point. The smaller the emittance at any point, the closer the orbit to the ideal.

The other property that we briefly discuss is the lifetime of the stored beam. This lifetime, defined as the time that the beam will take to decay to $1/e$ times its present value, provides an indication of how long the experimenter has to complete an experiment. The lifetime usually bears some form of inverse relationship to the stored current. Thus, the larger the current, the smaller the lifetime.

Scattering of electrons out of the circulating beam may result from two sources. First, the electrons may be scattered by each other. This phenomenon, known as the Touchek effect, is dependent on the electron density in the stored beam. If the electrons are orbiting the ring in a number of bunches, then for the same stored

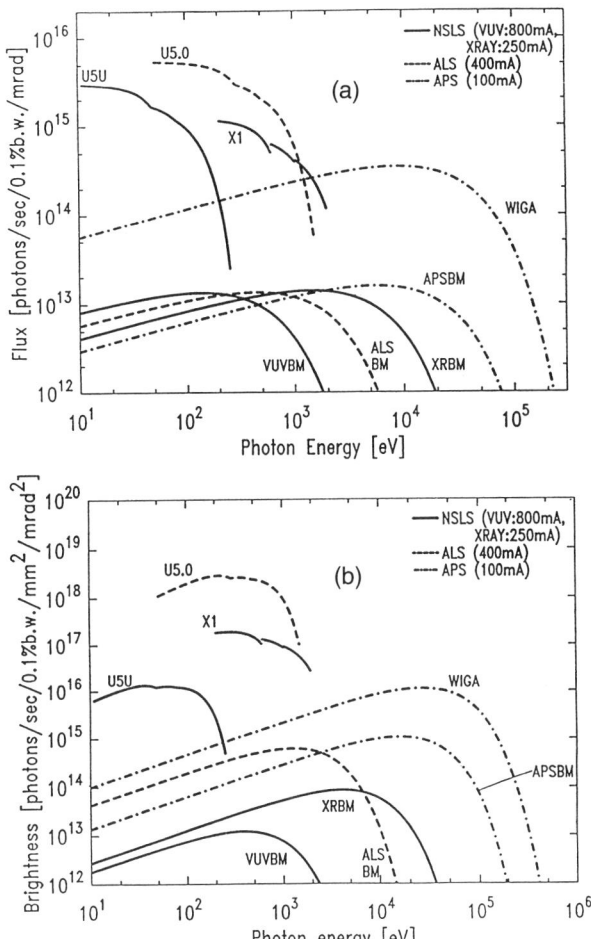

FIG. 2. (a) The photon flux in units of photons sec^{-1} mrad^{-1} ma^{-1} 0.1%$^{-1}$ bandwidth from different sources at the NSLS, ALS, and APS. Solid lines indicate the spectral distribution from the VUV bending magnet, the x-ray bending magnet, the U5 undulator, and the X1 undulator as indicated. Dashed lines indicate the same from the ALS bending magnets and a representative undulator. Dot-dashed lines indicate the same for the APS bending magnet and a representative wiggler. (b) The same as in (a) but now normalized to the area of the emitting electron beam to give the brightness of the different sources.

current the Touchek effect will become worse as the number of bunches is reduced. The second scattering mechanism is collisions with residual gas in the vacuum chamber. This is clearly strongly dependent on the quality of the vacuum in the storage ring.

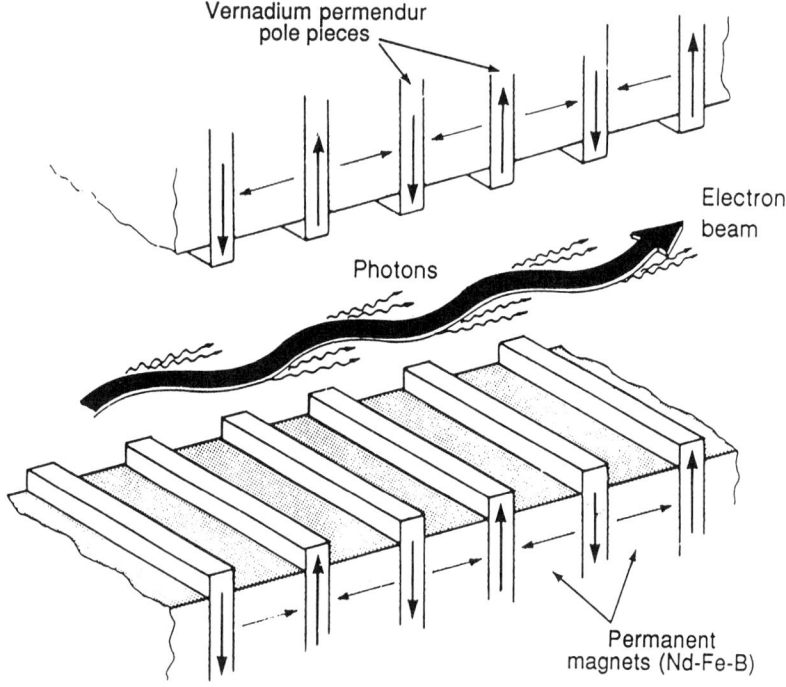

FIG. 3. Electron beam trajectory through the magnetic array of an undulator or wiggler. The photons are emitted on every undulation.

2.2.2 Insertion Devices

At a number of storage rings, insertion devices, including undulators and wigglers, have been installed in the straight sections to produce high-intensity photon beams at particular wavelengths. Undulators and wigglers consist of a linear array of magnets with alternating polarity positioned with their field lines perpendicular to the plane of the storage ring. As illustrated in Fig. 3, this causes the electrons traversing such devices to undulate or wiggle to either side of the straight-through trajectory.

The simplest model for such devices would be to think of them as an array of bending magnets, with the radius of the orbit through each component determined by the strength of the magnet. Indeed, wigglers, which operate with relatively strong magnetic fields, may be thought of as such an array. However, in the limit of weak fields and a large number of component magnets, temporal coherence between wavefronts emitted at different points within the device will lead to a spectral output with characteristic harmonics of the fundamental ($n = 1$) having wavelengths given by [11]

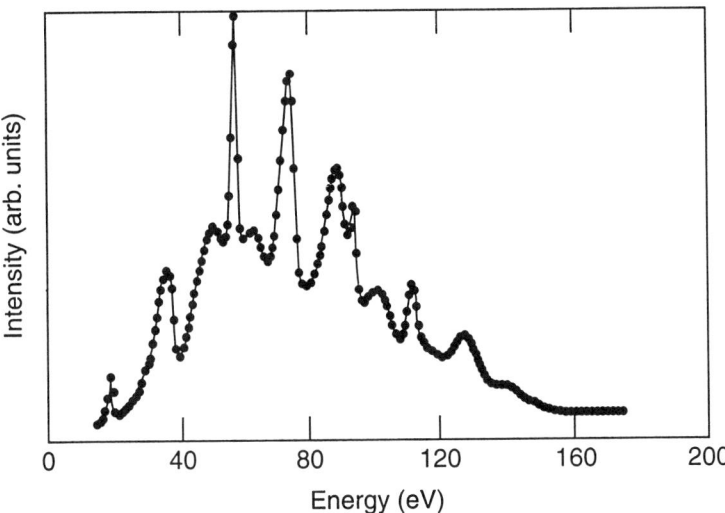

FIG. 4. The spectral output of the 27-period UV undulator installed on the U5 straight section at the NSLS. The fundamental is set at an energy of 19.0 eV.

$$\lambda(K,\vartheta) = \frac{\lambda_0}{2n\gamma^2}\left(1 + \frac{K^2}{2} + \gamma^2\vartheta^2\right), \quad n = 1, 3, 5,\ldots. \quad (7)$$

Here λ_0 is the magnetic period length in the device, γ again is the storage ring energy measured in rest mass units, θ is the observation angle with respect to the undulator axis, and K is a parameter related to the peak magnetic field strength B_0 such that

$$K = 0.934\lambda_0 \text{ (cm) } B_0(T). \quad (8)$$

In an undulator, where $K \sim 1$, the angular divergence of the stored electron beam is comparable to the maximum deflection of the emitted light and the spectral output is dominated by characteristic harmonic peaks. Equation (7) is derived in the zero emittance limit. However, in practice, due to the finite electron beam emittance, even harmonics are observed.

Figure 4 depicts spectra from the undulator installed in the U5 straight section at the NSLS. With the fundamental set at 19 eV, several strong harmonics are clearly observed. The intensity background in the figure reflects the transmission characteristics of the monochromator used to make the measurement.

The total power radiated by an undulator is given by

$$P(W) = 1.9 \times 10^{-6} N\gamma^2 K^2 I \text{ (amps)}/\lambda \text{ (cm)}, \quad (9)$$

where N is the number of magnetic periods in the device. At first sight it appears that the power radiated by an undulator is considerably less than the power radiated by the beam circling the storage ring. However, it should be remembered that for an undulator the flux is highly concentrated in a narrow cone pointing directly along the axis of the device.

Figures 2a and 2b show the relative flux and brightness of representative undulators and wigglers at different storage rings and include the spectral range of the first, third, and fifth undulator harmonics that clearly reflect the stored electron beam energies at the different facilities. The spectral output of the APS wiggler takes the same form as the bending magnets but with a considerable increase in flux.

The term in $\gamma\theta$ in Eq. (7) determines the angular broadening due to observation at angle θ to the undulator axis. On the axis the linewidth of each harmonic n is given by

$$\frac{\Delta\lambda_n}{\lambda_n} = \frac{1}{nN}, \qquad (10)$$

where N is the number of magnetic periods. Thus, even though the undulator output is peaked at the harmonics, it is necessary for most applications to use some form of light monochromatization, as described later in this chapter.

A particular experiment will often require that the output from an undulator be optimized at certain wavelengths. Examination of Eq. (7) shows that it is possible to accomplish this either by tuning the magnetic field parameter K by adjusting the magnetic gap or, alternatively, by changing the energy of the stored electron beam. In everyday operation of a storage ring with several beamlines, the former method is used.

2.2.3 Light Polarization and the Production of Circular Polarization

The synchrotron radiation emitted in the plane of a storage ring is linearly polarized with polarization in the plane. Indeed, for electron velocities v such that $\beta = v/c \sim 1$, the parallel component of the radiation contains 7/8 of the total radiated power emitted by the electron beam [12]. This has proved an important characteristic in the development of a number of experimental techniques, an example being photoemission where the linear polarization of the light has been used to exploit different symmetry selection rules [13]. However, with the development of more powerful sources of synchrotron radiation there have been an increasing number of experimental programs that exploit the possibility of obtaining circularly polarized radiation from either the dipole bending magnets or from more exotic insertion devices. In this section we review the polarization properties of the bending magnet source and examine the new class of insertion devices specifically designed for the production of circularly polarized light.

2.2.3.1 Dipole-Derived Circularly Polarized Radiation
Radiation emitted above and below the plane of the storage ring has polarization components both perpendicular and parallel to the storage plane. The relative concentration of these components is a strong function of both the angle of emission ψ with respect to the orbital plane and the wavelength λ. The parallel and perpendicular components of the emitted power at angle ψ are proportional to [9]

$$F_{\|}(\psi) = (1 + \gamma^2\psi^2)^2 \, K^2_{2/3}\left[\frac{\lambda_c}{2\lambda}(1 + \gamma^2\psi^2)^{3/2}\right],$$

$$F_{\perp}(\psi) = \gamma^2\psi^2(1 + \gamma^2\psi^2) \, K^2_{1/3}\left[\frac{\lambda_c}{2\lambda}(1 + \gamma^2\psi^2)^{3/2}\right]. \quad (11)$$

At any angle, the degree of linear polarization, P_L, is given by

$$P_L = \frac{F_{\|} - F_{\perp}}{F_{\|} + F_{\perp}}. \quad (12)$$

Figure 5 shows the intensity radiated in each polarization component and the sum of these components as a function of $\gamma\psi$. The intensities, plotted for two different wavelengths, are normalized to the intensity of the parallel component along the central axis. By selecting the relevant combination of perpendicular and parallel polarization, the possibility exists of obtaining a tunable source of elliptically or circularly polarized radiation. The degree of circular polarization, P_C, is given by

$$P_C = \frac{2\sqrt{F_{\|}F_{\perp}}}{F_{\|} + F_{\perp}} \quad (13)$$

and is included in Fig. 5. Examination of this figure shows that it is considerably more effective to use a source with a critical wavelength much higher than the wavelength of interest. This allows the experimentalist to take advantage of the fact that the angular distribution of the emitted radiation around the central axis flattens out as the wavelength becomes longer. Operating at wavelengths close to the critical wavelength, on the other hand, the intensity of the radiation will decrease as one moves out of the plane of the storage ring. As an example, we note that a soft x-ray beamline installed on the UV ring at the NSLS produces an 80% polarized beam with 20–30% of the in-plane intensity [14].

A useful way of remembering the sense of circular polarization is that, if when viewed from above the electrons are circulating the storage ring in a clockwise direction, then above the plane of the storage ring the electric field vector of the light will also be rotating in a clockwise direction, that is, the light will be

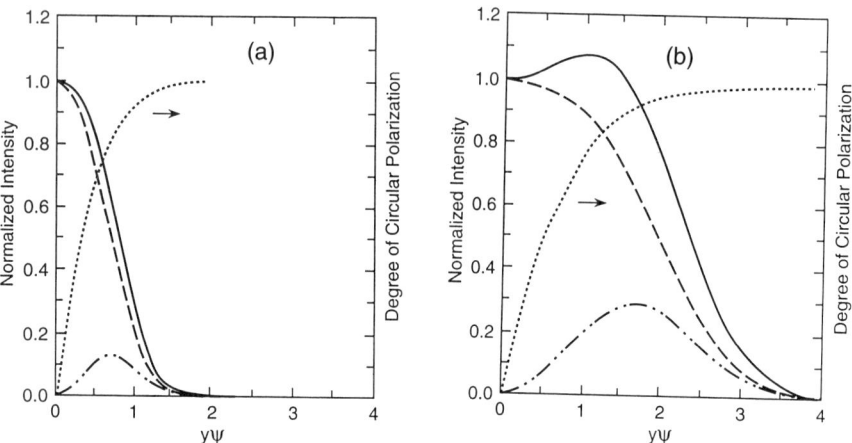

FIG. 5. (a) The perpendicular (dashed line) and parallel (dot-dashed line) polarization components of the emitted radiation and the sum of these components (solid line) plotted as a function of $\gamma\psi$ for $\lambda_c/\lambda = 1$. The degree of circular polarization (short dash line) is also shown. (b) The same as in (a) but now for $\lambda_c/\lambda = 0.1$.

right-hand circularly polarized. Viewed from below, the electric field vector will rotate in the opposite direction and the light will be left-hand circularly polarized.

2.2.3.2 Insertion Devices for Producing Circularly Polarized Light A number of different insertion devices have been designed to produce circularly polarized radiation, particularly in the shorter wavelength range. Here we briefly discuss as examples the Crossed Field Undulator, the Helical Undulator, and the Asymmetric Wiggler.

Shown schematically in Fig. 6, the crossed field undulator [15, 16] consists of two planar undulators with mutually perpendicular magnetic fields placed in series in the straight section of the storage ring. The relative phase difference between the two perpendicular polarization components reflecting emission from the two undulators is controlled by a simple three-pole electromagnet or modulator placed between the undulators. By varying the phase difference it is possible to obtain left or right circularly polarized light from the device. Because of its construction, this type of undulator has also been used as a source that simply switches between two perpendicular components of linear polarization [17].

The two magnetic arrays with vertical and horizontal fields of the crossed field undulator are interlaced in the helical undulator [18]. The phase difference between the two magnetic components now corresponds to a 1/4 magnetic period. The total magnetic field therefore varies helically along the axis, with the result that the electrons emit circularly polarized radiation during their helical motion.

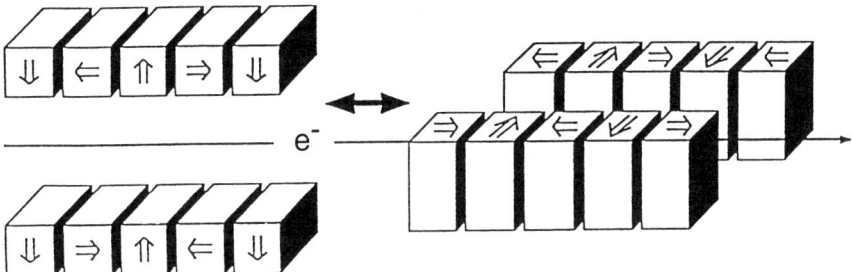

FIG. 6. Arrangement of the crossed field undulator showing the two magnetic arrays with orthogonal fields. Represented by the double-ended arrow, a phase shifter is installed in the region between the two magnetic arrays.

In the previous section, we described the polarization characteristics above and below the plane of the storage ring for dipole or bending magnet radiation. In linear undulators, with the electron beam continuously oscillating from side to side, these polarization characteristics will cancel on successive undulations and it is no longer possible to obtain circularly polarized radiation out of the plane. The asymmetric wiggler, on the other hand, gets around this problem by having a continuous change in the magnetic field strength along the axis of the undulator [19]. In this way, the polarization characteristics of successive undulations no longer cancel each other and the device provides a source of elliptically or circularly polarized light.

2.3 Light Monochromatization

The bending magnet radiation from a synchrotron or storage ring represents a broad continuous spectrum. Thus, in most experimental applications some form of monochromatization is required to select the wavelength of interest. This will still be true when an undulator with well-defined harmonics is used as the source of the radiation, simply because the harmonics will be broadened according to Eq. (10).

In terms of photon optics, a complete beamline will generally include some form of monochromator that contains a wavelength dispersive component plus associated source and image points. The source point may be defined by the stored electron beam itself or a focusing mirror may be used to redefine the source point at the entrance slits of the monochromator. The dispersive component itself or some form of secondary optical element may be used to focus the photon beam directly onto wavelength-selective slits or collimators. An exit mirror may be used to refocus the beam from the exit slits onto the sample under investigation.

In the UV and soft x-ray range, the monochromators used employ diffraction gratings as the dispersive component. In the x-ray range, approximately 1 keV and above, diffraction from appropriate crystals is used. In the following sections we briefly review these different methods of monochromatization.

2.3.1 Grating Monochromators

Diffraction at wavelength λ from a grating with line separation d is governed by the grating equation

$$d(\sin \alpha + \sin \beta) = m\lambda, \tag{14}$$

where α and β represent the incident and exit angles with respect to the surface normal of the grating, and m is the order of diffraction.

A monochromator may be designed around a grating with either a planar or a concave surface. The planar surface has no inherent focusing properties. For concave surfaces on the other hand, the Rowland circle is an important concept [20]. If the grating lies tangentially on the circumference of this circle, which has a diameter R equal to the radius of curvature of the grating, then light from a source point on the circle will be focused back to a point on the same circle. Shown schematically in Fig. 7, the general focusing in the plane of the circle is given by

$$\frac{\cos^2 \alpha}{r} - \frac{\cos \alpha}{R} + \frac{\cos^2 \beta}{r'} - \frac{\cos \beta}{R} = 0, \tag{15}$$

where r and r' are the source-to-grating and grating-to-image distances, respectively. The astigmatism of the optical system is given by the difference between the horizontal focal point defined by r' and the locus of vertical or secondary foci at r'', determined by the related equation

$$\frac{1}{r} - \frac{\cos \alpha}{R} + \frac{1}{r''} - \frac{\cos \beta}{R} = 0. \tag{16}$$

Examination of Eqs. (15) and (16) shows that for a spherical grating r' and r'' will be equal when $\alpha = \beta = 0°$. Moving away from this condition, r' and r'' may only be made equal by using aspheric surfaces.

From Eq. (14) it follows that the wavelength dispersion or resolution, $\Delta\lambda$, of the diffraction grating will be given by

$$\Delta\lambda = \frac{s}{Nmr} \cos \alpha, \tag{17}$$

where N is the line density of the grating and s is the entrance slit width. The magnification, M, of the grating is given by

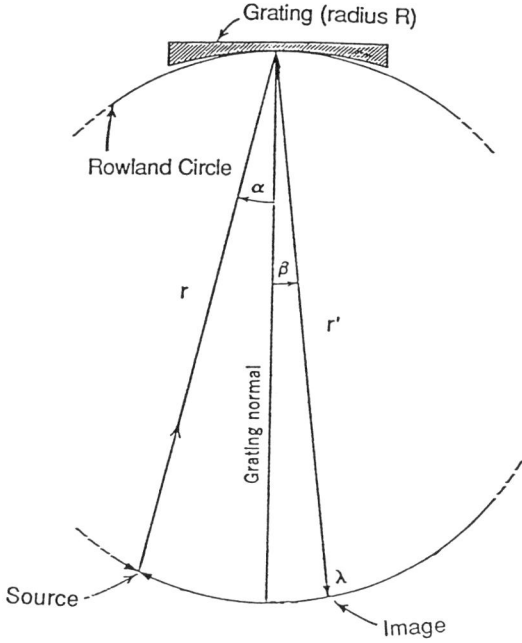

FIG. 7. Schematic diagram showing the relationship of the different parameters describing the properties of a grating of radius R. α and β are the entrance and exit angles, respectively; r and r' represent the source and image distances, respectively.

$$M = \frac{r \cos \beta}{r' \cos \alpha}. \qquad (18)$$

2.3.1.1 Normal Incidence Monochromators A large number of different grating monochromator designs are currently in use in the long-wavelength or UV range. The most commonly used device in this range is the so-called McPhereson or Normal Incidence Monochromator (NIM) [21]. Shown in Fig. 8a, this instrument, which operates in the photon energy range from very low energies to approximately 40 eV, has the source and image points located on the Rowland circle with a total angle of only 15° between them.

Other low-energy monochromators in use include the Seya–Namioka and modified Wadsworth designs. The former instrument represents the same design configuration as the NIM, but the included angle between the entrance and exit arms is now increased to the order of approximately 70° [22]. Monochromators have been used in this configuration to extend the photon energy range up to photon energies of the order of 60 eV. However, the resolution of this instrument is not as good as the McPhereson design.

(a)

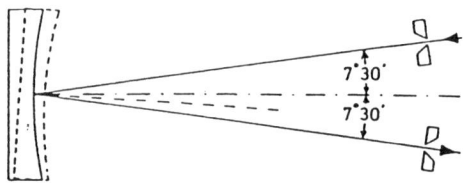

(b)

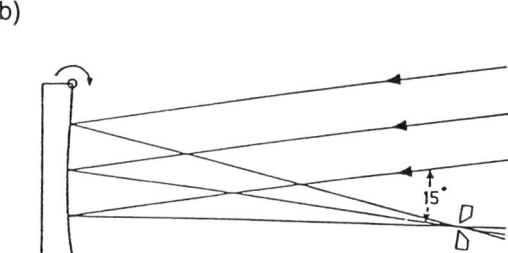

FIG. 8. Normal incidence monochromator geometries. (a) The McPhereson monochromator with 15° between the entrance and exit arms. (b) The modified Wadsworth geometry with the source point at infinity.

In the modified Wadsworth design (Fig. 8b), the source is effectively placed at infinity [23]. The small divergence of synchrotron radiation allows the stored electron beam to serve directly as the source point for the optical system and removes any requirement for focusing optics on the input. Thus, if $r = \infty$ in Eq. (14), then for the Wadsworth mount

$$r' = \frac{R \cos \beta}{\cos \alpha + \cos \beta}. \tag{19}$$

2.3.1.2 Soft X-Ray Monochromators As the wavelength of the light becomes shorter, the requirement for a high reflectivity leads to a larger angle of incidence on the grating. In the shorter-wavelength range, the different monochromator designs are often described by the shape of the grating surface, which may be planar, spherical, or aspherical.

In the Plane Grating Monochromator (PGM), the grating obviously has no inherent focusing properties itself. It is therefore necessary in any design to include entrance and exit focusing optics that collect light from the source and deliver it to the exit slit, respectively. In the most successful design of this type, the SX-700, the stored electron beam itself serves as a virtual entrance slit [24]. The optics consists of a plane premirror delivering the light from the source point to the grating, which is then followed by an elliptical mirror that focuses the dispersed light onto the exit slits.

Moving away from the plane grating, the most commonly employed surface for gratings in the soft x-ray range has been the toroidal shape. In this design, the so-called Toroidal Grating Monochromator (TGM), the intent is that, by using an aspheric surface, it is possible to minimize the aberrations or maximize the resolution in selected wavelength ranges with a minimal number of reflecting surfaces. Intensity losses due to reflectivity are therefore reduced. The standard or most straightforward configuration for a monochromator of this type might consist of an ellipsoidal prefocusing mirror delivering light from the source to the entrance slit, a grating chamber with a selection of gratings, and an ellipsoidal exit mirror refocusing light from the exit slits onto the sample under investigation [25].

A number of changes have been made to the standard TGM configuration to improve and enhance its performance. It has been recognized that it is unnecessary to maintain the horizontal focus at the entrance slit [26]. By moving the latter focus closer to the grating, the exposed surface or illuminated width of the grating is reduced, thereby reducing certain aberrations. The other notable improvement in the design is the introduction of a moving exit slit so that the optical configuration can be redefined for each wavelength. In this way, good resolution may be obtained over a broad wavelength range rather than being restricted to selected wavelengths.

The Spherical Grating Monochromator (SGM) is another important monochromator [27]. This particular design grew out of the recognition that the major factor limiting the performance of monochromators using aspheric optics was an inability to produce aspheric surfaces with good figure error. Thus, in the SGM design each optical component of the beamline represents a spherical or cylindrical surface focusing in only one dimension. Instruments of this type working in the soft x-ray range have been shown to achieve extremely high resolving powers, typically of the order of 8,000–10,000 [28].

The configuration most commonly used for the SGM is shown in Fig. 9. The entrance mirror now consists of two separated focusing elements, one for horizontal focus and the second for vertical focus. To reduce the total number of reflections the horizontal focusing mirror now focuses right through the system to the experimental point. The vertical focusing mirror focuses light from the source point onto the entrance slits, which are then refocused in the vertical by the spherical grating onto the exit slits.

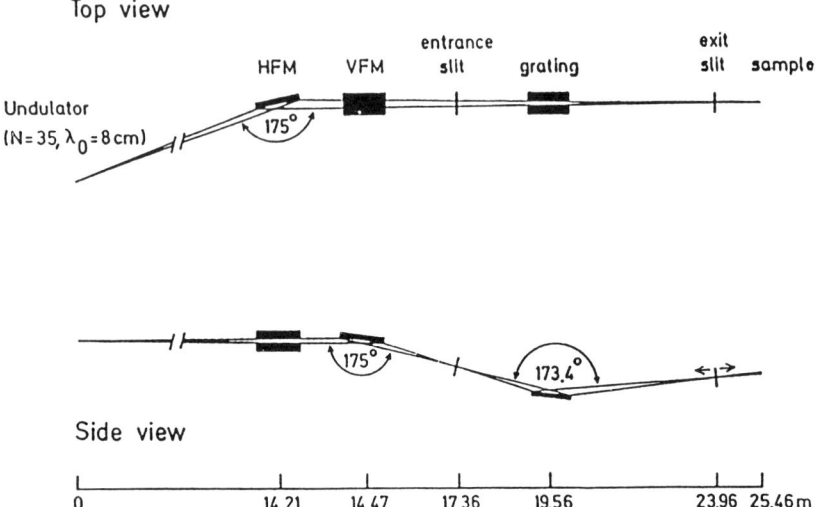

FIG. 9. Top and side views showing the different optical elements of the spherical grating monochromator.

2.3.2 Crystal Monochromators

In the short-wavelength or x-ray range, the diffraction grating is replaced by a crystal as the dispersive element [29]. Now the dispersion is described by Bragg's law. Thus, if the separation between the lattice planes in the crystal is d, then for a given wavelength λ the condition for diffraction is given by

$$2d \sin \theta_B = k\lambda, \quad (20)$$

where θ_B is the angle of incidence with respect to the reflecting planes and k is the order of diffraction. Typical crystals used for the diffraction element include Si and Ge cut in the (111) direction.

The most common configuration used for crystal monochromators consists of two crystals rather than one. Such an arrangement allows the exiting beam to be parallel to the incoming beam. Both crystals are rotated to satisfy the Bragg condition for the wavelength of interest. However, with the first crystal rotating about a fixed axis, the second crystal is often translated as well as rotated to accept the diffracted beam from the first crystal. As shown in Fig. 10, several crystal monochromator designs recognize that focusing can be introduced into the optical design by bending the crystal [30]. The bend is usually in one dimension, giving the crystal similar focusing properties to the optical elements in the spherical grating monochromator described above.

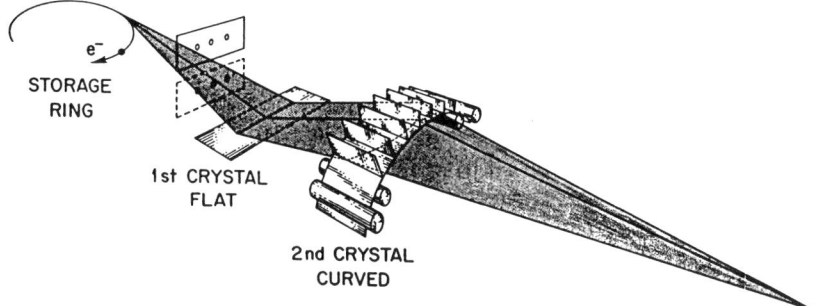

FIG. 10. Example configuration of a two-crystal monochromator for use in the x-ray range from reference [29]. For this particular design the second crystal is bent to provide focusing.

Monochromators may also be designed with the crystal operating in the Laue mode, particularly on beamlines operating in the hard x-ray range [31]. In such monochromators the Bragg condition is still satisfied but in transmission rather than reflection. Typical crystals used in these devices include diamond and beryllium.

The resolving power of a crystal monochromator may be derived from Bragg's law:

$$\frac{\Delta\lambda}{\lambda} = \Delta\theta \cot\theta_B. \qquad (21)$$

Two factors contribute to the angular divergence ($\Delta\theta$): the divergence of the incident beam and the intrinsic width of the Bragg reflection. Less divergence can be achieved in the incident beam by collimation before the first grating. However, with monochromators dispersing vertically, this is often unnecessary as the high electron beam energies in storage rings optimized for the x-ray range naturally produce a photon beam with small divergence in the vertical.

Within a kinematical treatment, perfect reflection occurs in a finite angular width, w, around the angle θ_B, defined by the Bragg condition such that [29]

$$w = \frac{2e^2\lambda^2}{\pi V mc^2 \sin\theta_B} |F_{hkl}| P, \qquad (22)$$

where V is the unit cell volume in the crystal, F_{hkl} is the structure factor of the particular lattice plane scattering the photons, and P is a polarization factor equal to 1 for σ polarization and $\cos 2\theta_B$ for π polarization. Outside of this width, the Darwin width, absorption will attenuate the beam.

2.4 Applications

Experimental techniques using synchrotron radiation as the incident probe are finding widespread application throughout all fields of science, including physics, chemistry, and biology. The incident particle is clearly a photon. However, the emitted particles may be photons, electrons, atoms, ions, or molecules. The techniques may probe either the electronic structure or the atomic structure of the different materials.

We can gain further insight into the possibilities by examining the different experimental programs currently in progress at the NSLS, which has two storage rings spanning the VUV and x-ray ranges and therefore provides an excellent indication of the full range of experiments.

One of the principal techniques used in the UV energy range is photoemission [32]. A photon of energy $\hbar\omega$ is used to excite an electron from some initial state with energy E_i to some final state with energy E_f. The energy conservation in the process is given by a well-known equation:

$$E_f = E_i + \hbar\omega. \tag{23}$$

By measuring the characteristics of the electron in the final state we determine the properties of the electron in the initial state. The technique is used to study the properties of all phases of matter from solids to gases. By measuring the spin of the electrons it is also possible to examine the magnetic properties of materials.

In the case of solids, as illustrated in Fig. 11, photoabsorption spectroscopy can also excite electrons from tightly bound core levels to unoccupied states below the vacuum level, or to final states in the continuum. Such measurements may be used in the near-edge region to determine information about the unoccupied electron states [33] or, by measuring the wavelength dependence of the excitation further from the absorption edge, to obtain information on the local atomic environment. The latter technique is known as extended x-ray absorption fine structure (EXAFS) [34]. The near-edge x-ray absorption fine structure (NEXAFS) has also been used extensively to determine the orientation of molecules adsorbed on surfaces [33].

Atomic or structural information may also be obtained from a number of different scattering experiments. Either the outgoing photoelectron is scattered, giving rise to photoelectron diffraction [35], or, in the shorter wavelength range, the incident photon itself undergoes elastic scattering as in x-ray scattering from surfaces [36]. Such studies may be used to obtain information on adsorption sites or on different phase transition behavior in both the surface and in overlayers deposited on the surface. Magnetic x-ray scattering has also been used to study magnetic properties in the bulk [37]. Here the x-ray scatters the magnetic lattice rather than the atomic lattice. The technique has been particularly successful in studying the magnetic structure of the rare earths [38].

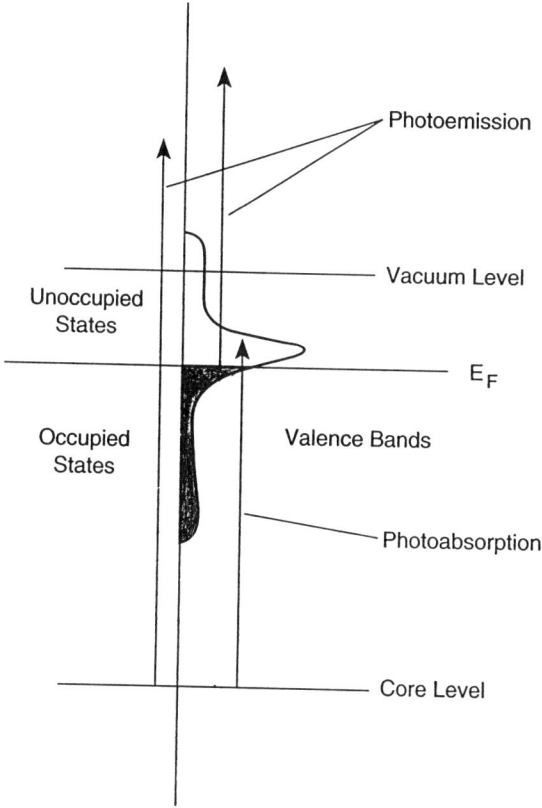

FIG. 11. Diagram showing the photoemission process from both valence and core levels in a solid. Photoabsorption from localized core levels to unoccupied valence states is also indicated. The figure is divided into two halves, representing the two possible spin states of the electron.

There are also a number of experiments in the general area of microscopy. Studies in the soft x-ray range have been applied to examine biological living samples, and by using absorption near-edge chemical contrast it is possible to obtain element-specific micrographs [39].

We have listed just a few of the possible experiments. The interested reader can explore other possibilities by making reference to the numerous annual reports that are now being made available from the different light sources. It is clear that with more facilities being commissioned every year new experimental techniques will evolve and synchrotron radiation will continue to be developed as an important probe of all states of matter.

References

1. Ivancheko, D., and Pomeranchuk, J. (1944). *Phys. Rev.* **65**, 343.
2. Blewett, J. (1946). *Phys. Rev.* **69**, 87.
3. Blewett, J. (1988). *Nucl. Instrum. Methods* **A266**, 1.
4. Schwinger, J. (1946). *Phys. Rev.* **70**, 798; (1949). *Phys. Rev.* **75**, 1912.
5. Tomboulian, D. H., and Hartman, P. L. (1956). *Phys. Rev.* **102**, 1423.
6. Madden, R. P., and Codling, K. (1963). *Phys. Rev. Lett.* **10**, 516.
7. See, e.g., Koch, E.-E. (1983). In *Handbook of Synchrotron Radiation*, D. E. Eastman and Y. Farge (eds.), North-Holland, Amsterdam, and other books in this series.
8. See, e.g., Margaritondo, G. (1988). *Introduction to Synchrotron Radiation*, Oxford Univ. Press, Oxford.
9. Green, G. K., *Spectra and Optics of Synchrotron Radiation*, BNL Internal Report No. 50522, Upton, New York.
10. Hulbert, S. L., private communication.
11. An excellent review of the properties of undulators and wigglers is given by Krinsky, S., Perlman, M. L., and Watson, R. E. (1983). Chapter 2 in *Handbook of Synchrotron Radiation*, D. E. Eastman and Y. Farge (eds.), North-Holland, Amsterdam.
12. Sokolov, A. A., and Ternov, I. M. (1968). *Synchrotron Radiation*, Pergamon, Oxford.
13. Plummer, E. W., and Eberhardt, W. (1982). *Adv. Chem. Phys.* **49**, 533.
14. Chen, C. T., private communication.
15. Moissev, M., Nikitin, M., and Fedorev, N. (1978). *Sov. Phys. J.* **21**, 332.
16. Kim, K. J. (1986). *Nucl. Instrum. Methods* **A219**, 425.
17. Roth, C., Hillebrecht, F. U., Rose H. B., and Kisker, E. (1993). *Phys. Rev. Lett.* **70**, 3479.
18. Madey, J. M. J. (1971). *J. Appl. Phys.* **42**, 1906.
19. Goulon, J., Elleaume, P., and Raoux, D. (1987). *Nucl. Instrum. Methods* **A254**, 192.
20. Samson, J. A. R. (1967). *Techniques of Vacuum Ultraviolet Spectroscopy*, Pied Publications, Lincoln, NE.
21. See, e.g., Saile, V., Gürtler, P., Koch, E.-E., Kozevinkov, A., Skibowski, M., and Steinmann, W. (1976). *Appl. Opt.* **15**, 2559.
22. Seya, M. (1952). *Sci. Light* **2**, 8; Namioka, T. (1954). *Sci. Light* **3**, 15.
23. See, e.g., Howells, M. R. (1982). *Nucl. Instrum. Methods* **195**, 215.
24. Petersen, H. (1982). *Opt. Commun.* **40**, 402; (1986). *Nucl. Instr. Methods* **A246**, 260.
25. Thiry, P., Depautex, C., Pinchaux, R., Petroff, Y., Lepere, D., and Passereau, A. (1980). *Nucl. Instrum. Methods* **172**, 182.
26. Chen, C. T., Plummer, E. W., and Howells, M. R. (1984). *Nucl. Instrum. Methods* **222**, 103.
27. Chen, C. T. (1986). *Nucl. Instrum. Methods* **A256**, 260.
28. Chen C. T., and Sette, F. (1989). *Rev. Sci. Instrum.* **60**, 1608.

29. For a more extensive review of x-ray monochromators, see Matsushita, T., and Hashizume, H. (1983). Chapter 4 in *Handbook of Synchrotron Radiation*, D. E. Eastman and Y. Farge (eds.), North-Holland, Amsterdam.
30. Sparks Jr., C. J., Ice, G. E., Wong, J., and Batterman, B. W. (1982). *Nucl. Instrum. Methods* **195**, 73.
31. Kostroun, V. O., and Materlik, G. (1980). *Nucl. Instrum. Methods* **172**, 215.
32. For reviews of different aspects of photoemission, see Kevan, S. D., ed. (1992). *Angle Resolved Photoemission*, Elsevier, Amsterdam.
33. Stohr, J. (1993). *NEXAFS Spectroscopy*, Springer-Verlag, Heidelberg.
34. See, e.g., Lee, P. A., Citrin, P. H., Eisenberger, P., and Kincaid, B. M. (1981). *Rev. Mod. Phys.* **53**, 769.
35. Woodruff, D. P. (1992). Chapter 7 in *Angle Resolved Photoemission*, S. D. Kevan (ed.), Elsevier, Amsterdam.
36. Robinson, I. K., and Tweet, D. J. (1992). *Rep. Prog. Phys.* **55**, 599–651.
37. Gibbs, D. (1992). *Synchrotron Radiation News* **5**, 18.
38. Gibbs, D., Moncton, D. E., D'Amico, K. L., Bohr, J., and Grier, B. (1985). *Phys. Rev. Lett.* **55**, 234; Gibbs, D., Grubel, G., Harshman, D. R., Isaacs, E. D., McWhan, D. B., Mills, D. L., and Veffier, C. (1991). *Phys. Rev. B* **43**, 5663.
39. Ade, H., Zhang, X., Cameron, S., Costello, C., Kirz, J., and Williams, S. (1992). *Science* **258**, 972.

3. CONTINUOUS WAVE DYE LASERS

Andrew Dienes

Department of Electrical and Computer Engineering
University of California at Davis

Diego R. Yankelevich

Centro de Investigación Científica y de
Educación Superior de Ensenada
Ensenada, México

3.1 Introduction

The continuous wave (CW) dye laser plays an important role as a tunable coherent source from the near-UV to the near-IR. Although it requires an expensive high-power CW ion laser pump, its configuration can be somewhat complicated, and its operation is at times balky, no other optical source provides a comparable combination of tunability, resolution, and power. Today, CW dye lasers cover the spectral range from ~365 to ~1000 nm, with more than 100 mW of power output through this range (and considerably more in the yellow–red region). This is adequate for most spectroscopic and other research applications. The available resolution, while certainly not trivial to obtain, is also very impressive. A true single-frequency CW dye laser, which can be constructed in the laboratory and is also available commercially, has a linewidth of ~0.5–1 MHz. Thus, the tuning range of ~0.5×10^{14} Hz available from a *single* dye contains ~10^8 spectral resolution elements and the multidye CW laser tuning range contains more than 10 times that. Using standard frequency doubling and mixing techniques in nonlinear crystals, the tuning range has been extended down to 260 nm. Additionally, the broad bandwidth gain of a single dye makes possible ultrashort pulse generation by mode-locking, resulting in the shortest optical pulses available today [1].

In what follows, we give a brief description of the basic principles of this device. A simple theory is given to aid the understanding of its design and operation. We discuss the characteristics of practical CW dye lasers and describe the functions of their individual constituent elements. We list the best performing laser dyes with the needed pump sources. Our aim is to advance basic understanding and facilitate both acquisition and use of this important tool. To aid researchers who chose to construct their own laser, we also give a description of an alignment procedure. Due to space limitations, we do not discuss a number of important topics, such as noise characteristics of the dye laser output, extended tuning using nonlinear optical techniques, and ultrashort pulse generation by mode locking.

3.2 Basic Dye Laser Principles

Although dye lasers cover a range of wavelengths from the near-UV to the near-IR, using dyes of widely different chemical compositions, the basic principles are essentially identical for all of them. A description based on a rate equation approach and on stimulated emission and absorption cross-sections is commonly used regardless of the particular material and wavelength. For the dyes used in practical systems, the relevant parameters have rather similar values. The radiative properties of laser dyes depend on the structure of the dye molecule and are also influenced by interaction with the solvent. There is extensive literature (e.g., [2–5]) on the spectroscopic, photophysical, and chemical properties of dyes. No detailed understanding of these is required for the description of the dye laser, and we give here only a brief summary of the properties relevant to lasing behavior sufficient for a basic understanding.

Dye molecules are large and complicated, with many internal degrees of freedom and many possibilities for internal energy conversion. Additionally, there are very frequent collisions with solvent molecules. Figure 1a shows the basic energy level scheme of a generic laser dye in liquid solution form. There are two main manifolds of electronic energy states: singlets and triplets. Corresponding to each electronic state there is a broadband of vibrational–rotational states. Both the process of optical pumping and the laser action take place between the first excited singlet (S_1) and the ground state singlet (S_0) bands. Only the S_1 band is radiative, because a molecule excited to any of the higher energy levels undergoes (mainly by internal conversion) spontaneous nonradiative decay to S_1, which is so rapid as to completely quench any emission. Nonradiative decay from S_1 to S_0 also takes place, but for laser dyes the radiative process is much stronger. Intersystem crossing from singlet to triplet, and vice versa, also occurs nonradiatively, mainly owing to interactions between the dye and the solvent. In the triplet energy manifold, the lowest state is metastable, while the higher energy states are also nonradiative. An important consequence of the rapid nonradiative quenching of all but the S_0, S_1, and T_1 states is that *only these three levels* (*bands*) *have nonnegligible populations* in the dye laser, and only they participate in stimulated emission and absorption. Another important process is the very rapid (for our purposes, essentially instantaneous) intraband thermalization that takes place *within* the

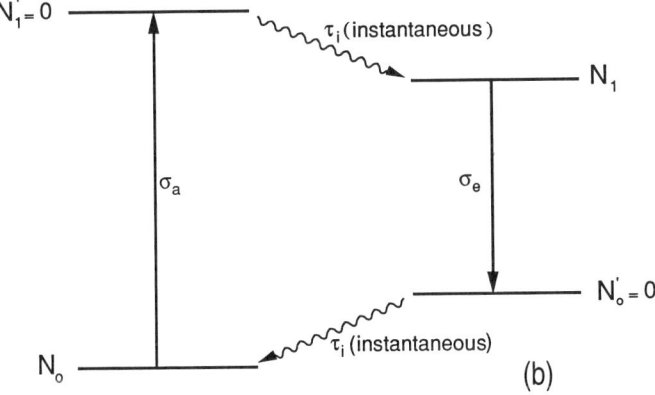

FIG. 1. (a) Schematic of the energy levels of a generic laser dye showing singlet and triplet manifolds, spontaneous rates (lifetimes), and stimulated cross-sections. The nonradiative rates shown as k^*_{nr} are fast enough to quench radiative emission from those levels. (b) Equivalent four-level model corresponding to the singlet states (S_0, S_1). The primed levels have zero population in the steady state.

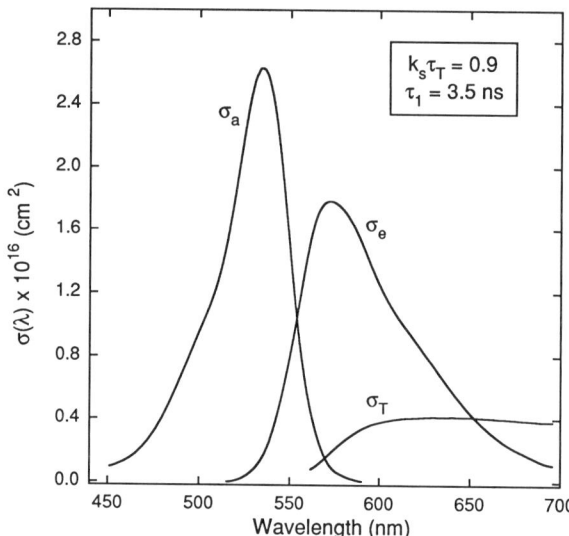

FIG. 2. Singlet absorption and emission plus triplet absorption cross-sections vs. wavelength for the dye R6G. The inset box shows approximate values of the other parameters needed for the theory.

vibrational–rotational continuum band of any of the electronic states. This means that at any time the total population within the band of a given electronic state is in a quasi-thermal equilibrium (Boltzmann distribution). Any population change within a band is rapidly redistributed to the whole band. The strength of the various spontaneous radiative and nonradiative processes are described by lifetimes (τ_{sp}, τ_T, τ_i, etc.) or rates (k_{ST}, k_{nr}, etc.) that are also shown on the diagram. The total lifetime of level S_1 is τ_1, where $1/\tau_1 = 1/\tau_{sp} + k_{nr} + k_{ST}$, which is typically of the order of a few nanoseconds.

The singlet manifold may be modeled as an effective four-level homogeneously broadened system, as shown in Fig. 1b. Homogeneous means that, owing to the very fast intraband thermalization, *all* the molecules in a given band (level) participate in the stimulated transitions from that level. Emission and absorption are spectrally broad, with the emission displaced to longer wavelengths (lower energies). The spectral details of the broad emission and absorption line shapes depend on the photophysics of the vibrational–rotational continuum. In practice, the line shapes are determined experimentally. Figure 2 shows the stimulated absorption and emission cross-sections (σ_e, σ_a) versus wavelength for one of the best known laser dyes, rhodamine 6G (R6G), and the inset gives approximate values of the lifetimes and rates necessary for quantitative calculations. The singlet absorption and emission curves are approximately mirror images of each other.

This can be shown to arise from a superposition of Boltzmann distributions of Lorentzian transitions of various strength [3]. The peak of the emission is energy downshifted from that of the absorption owing to the Frank–Condon principle. In some dyes this shift may be so large that there is only a small overlap between the emission and absorption curves. The emission cross-section is obtained from the spontaneous fluorescence lineshape $E(\lambda)$ by the well-known relationship of the Einstein coefficients A and B. Thus,

$$\sigma_e = \frac{\lambda^4 E(\lambda)}{8\pi\tau_1 cn^2},$$

where

$$\int_0^\infty E(\lambda)\, d\lambda = \varphi. \tag{2}$$

The quantity ϕ is the quantum efficiency of the fluorescence. Its value is $\phi = \tau_1/\tau_{sp}$, and it is also equal to the ratio of the areas under the emission and absorption curves. This important parameter expresses the relative strength of the radiative and nonradiative processes from level S_1 and must obviously be as close to unity as possible for a laser dye. It is important to note that the nonradiative processes, that is, k_{nr} and k_{ST}, are dependent on the dye concentration. For example, in R6G for concentrations larger than ~0.5 × 10^{-3} mol/liter, τ_1 gradually decreases from the low-concentration limit owing to increasing nonradiative decay (concentration quenching). It should also be noted that a value of ϕ close to unity does not guarantee lasing. This is because the excited state (S_1 to S_2) absorption, if it overlaps the emission, may completely prevent lasing. The curve of the excited state absorption cross-section is not shown in Fig. 2 because its location and values are generally not well known. In most laser dyes it can be assumed to lie outside the emission wavelength range. On the other hand, as shown in the figure, the triplet absorption usually at least partially overlaps the emission and thus can be an important loss factor.

It is obvious that, in the longer wavelength region of the fluorescent emission where absorption is very small or zero, even for a small fraction of the total molecules pumped up to the first singlet level dye lasers will have an effective population inversion and therefore optical gain. One of the most important aspects of the dye laser is that the emission spectrum, and consequently the gain, is very broad, resulting in wide tunability. This tunability, however, comes at a price. The relatively high stimulated emission probability (large σ_e) over a wide band means that the lifetime of the lasing level is very short (of the order of a few nanoseconds). This means that high pumping intensity is required in order to maintain a steady-state population in level S_1 for CW gain.

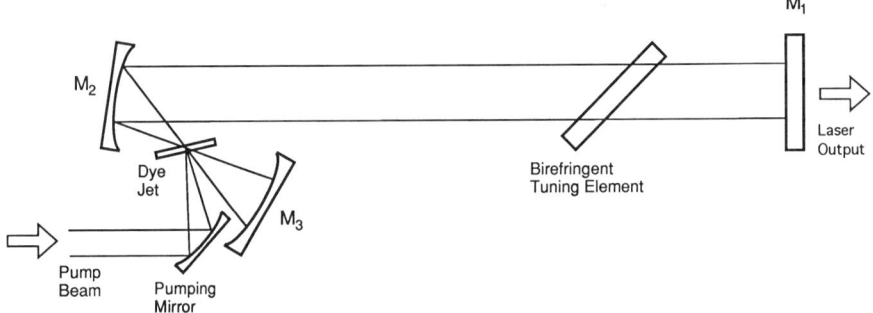

FIG. 3. Schematic of a linear folded three-mirror dye laser cavity. The various elements and their functions are described in the text.

3.3 Simple CW Dye Laser Theory

A unique feature of the CW dye laser is the high pump intensity required to achieve lasing threshold. This necessitates strong focusing of the pump light into a very small active volume, which in turn dictates a special cavity design, as well as a flowing dye cell. Although they may appear rather different, the physical configuration of nearly all the CW dye lasers used today are in fact derived from the same basic folded cavity design [6, 7] with approximately collinear pumping. A commonly used three-mirror standing wave version is shown is Fig. 3; the other important variant is the ring cavity version shown in Fig. 4. Both configurations include various frequency tuning and bandwidth control elements that will be discussed in Section 3.4.6. There are good reasons for the folded cavity design, and a simple general theoretical description shows the necessity for it. While we do not give a detailed and rigorous theory of the CW dye laser here, a brief simplified theoretical description is given that will greatly aid in the understanding of the basic operating principles and of the issues important for design and operation of this device. In what follows, we assume that: (i) the dye is pumped by a CW laser beam collinearly with the lasing radiation, (ii) the excited state absorption cross-section σ^* is zero, and (iii) the pump laser wavelength is in the absorption band where $\sigma_e = 0$. We assume lasing at some arbitrary wavelength within the emission band. Because the transitions are homogeneous, the laser radiation can be considered monochromatic for the purposes of the model. The gain, threshold pump power, as well as the power output can be derived from standard rate equations for the simplified geometry shown in Fig. 5 for a standing wave laser. The probability of any stimulated process is $I\sigma$, where I is laser intensity. Thus, the rate equations for the population densities N_i are

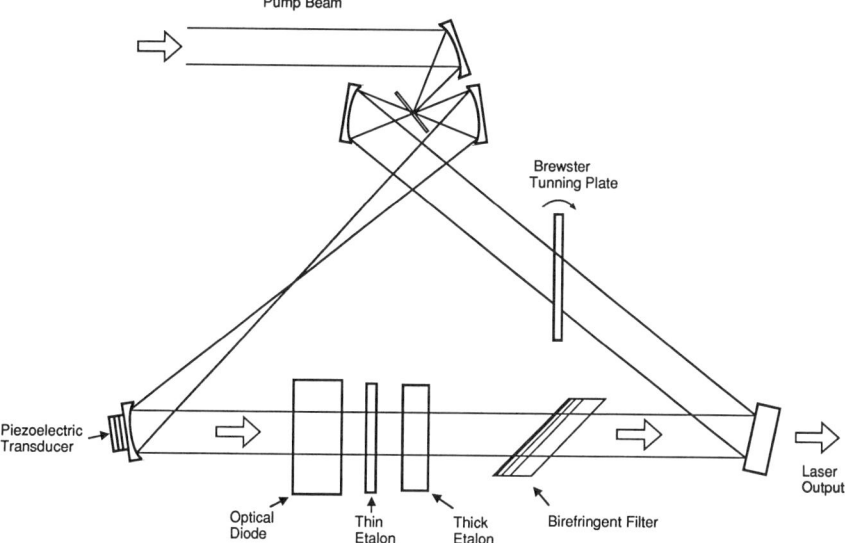

FIG. 4. Schematic of a unidirectional ring dye laser cavity showing frequency control elements that are described in the text.

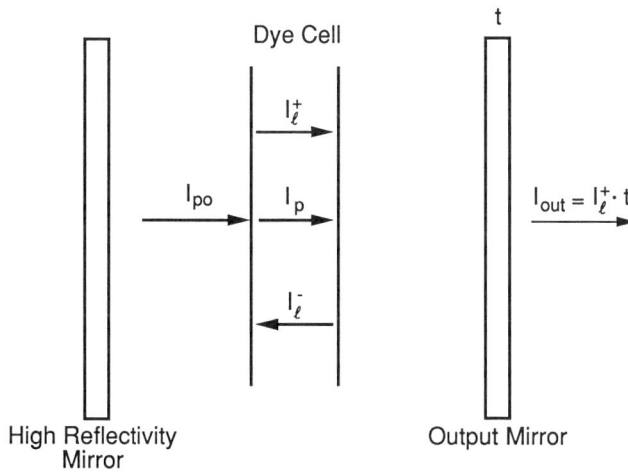

FIG. 5. Simplified schematic of the pumping and lasing geometry. This simple geometry is used in the theory.

$$\frac{dN_1}{dt} = N_0 \sigma_{ap} I_p - \frac{N_1}{\tau_1} - N_1 \sigma_e I_l + N_0 \sigma_{al} I_l, \tag{3}$$

$$\frac{dN_T}{dt} = k_{ST} N_0 - \frac{N_T}{\tau_T}, \tag{4}$$

$$N_0 + N_1 + N_T = N, \tag{5}$$

where the subscripts 0, 1, and T refer to the two lowest singlets and to the triplet states, respectively, I_p is the pump photon intensity, I_l is the *total* laser photon intensity ($I_l = I_l^+ + I_l^-$), and N is the total dye molecular density. We add the equation for the optical gain of the propagating laser emission:

$$\frac{1}{I_l^+}\frac{dI_l^+}{dz} = \sigma_e N_1 - \sigma_T N_T - \sigma_{al} N_0. \tag{6}$$

In the above equations, all the intensities and population densities are functions of time and also of the distance z in the medium. Initially we also make the assumption that the pump and laser beams are plane waves with finite and equal areas A.

3.3.1 Dye Laser Gain and Threshold Pump Power

We now solve the rate equations for the steady state. We find that the steady-state triplet population is $N_T = k_{ST} \tau_T N_2$, and therefore the gain equation becomes

$$g = \frac{1}{I_l^+}\frac{dI_l^+}{dz} = \sigma_e \left[1 - k_{ST} \tau_T \frac{\sigma_T}{\sigma_e} \right] N_1 - \sigma_{al} N_0 = \sigma_e f_T N_1 - \sigma_{al} N_0. \tag{7}$$

This equation demonstrates two important points: (i) for $N_1 < N_0$ gain can exist for the range of wavelengths where absorption is negligible or zero; and (ii) triplet absorption will prevent lasing unless the value of the above defined factor f_T, which represents the reduction in the gain by the triplet loss, is positive. In Section 3.4.3 we discuss how this is achieved in practical CW dye lasers. The parameters given in Fig. 2 yield $f_T = 1.0$ for $\lambda \sim 550$ nm, but it decreases to zero by $\lambda \sim 655$ nm. A typical curve of the gain factor g vs. wavelength is shown in Fig. 6. The sharp cutoff at shorter wavelengths is due to singlet absorption and is typical for most dye laser materials; the long-wavelength cutoff is due to triplet absorption ($f_T < 0$).

Using the steady-state solutions for the population densities, we integrate Eq. (7) over the length of the path (L) in the dye medium to obtain the total single-pass gain $G = I_l^+ (z = L)/I_l^+ (z = 0)$. This eliminates the z dependence of the pump intensity and population densities. First, we calculate the small signal gain by setting $I_l = 0$ in the rate equations. If for simplicity we also assume that the lasing

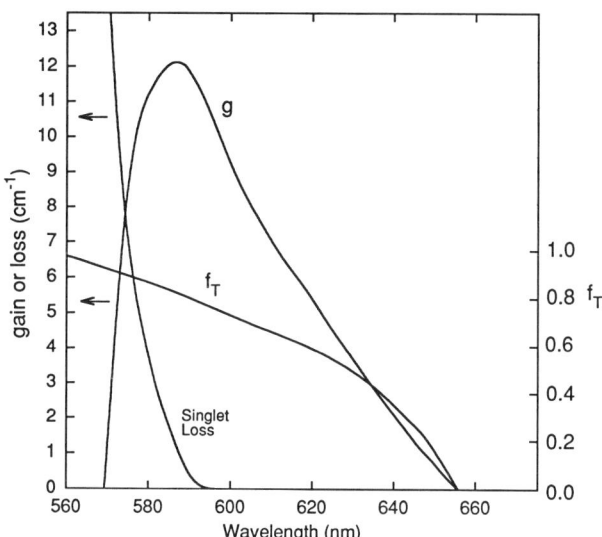

FIG. 6. The differential gain factor (g), the singlet loss factor ($N_0\sigma_a$), and the triplet factor (f_T) shown vs. wavelength for rhodamine 6G. Population densities are $N_1 = 1 \times 10^{11}/\text{cm}^3$, $N_0 = 1 \times 10^{17}/\text{cm}^3$, and $k_{ST}\tau_T = 0.9$.

wavelength is such that $\sigma_{al} = 0$, then straightforward calculations [8, 9] yield a particularly simple equation for the small signal gain:

$$\ln G_0 = \sigma_e f_T \tau_1 I_{po} a = \sigma_e f_T \tau_1 \frac{P_{po}}{h\nu_p A} a, \qquad (8)$$

where P_{po} is the input pump power. The parameter a is the fraction of the pump beam absorbed in the dye cell. For small gain, N_1 is a small fraction of the total population and N_1 can be approximated as constant and equal to N. Then

$$a = 1 - \exp(-\sigma_{ap} NL). \qquad (9)$$

Note that, in reality, N_0 is a function of z, and Eq. (9) is only correct if $N_1 < N_0$. However, it can be shown [9, 10] that Eq. (8) holds true regardless of the form of N_0 as long as the actual value of a is used. Equation (8) shows that, the shorter the lifetime, the more pump intensity is required. Note that the dye concentration (N), the pump absorption coefficient (σ_{ap}), and the cell thickness (L) only enter through the parameter a. When designing a CW dye laser, these quantities are adjusted so as to yield a value of $a \sim 0.9$.

The dye laser threshold is reached when the small signal gain is equal to the cavity loss. If we put the typical parameter values (from Fig. 2) into Eq. (8) and assume 90% absorption of the pump, we find that at $\lambda_l = 590$ nm, for a 15% round-trip gain (i.e., $G_0 = \sqrt{1.15} = 1.072$) we require a pump intensity of 0.87×10^5 W/cm² at $\lambda_p = 514.5$ nm. This requirement dictates tight focusing of the pump beam and restricts possible pump sources to CW ion lasers (argon and krypton), which, in spite of their inefficiency and high cost, have become the workhorses of laser laboratories. (It is likely, however, that diode laser-array-pumped and doubled Nd:YAG laser systems will replace the venerable argon laser for this and many other applications sometime in the near future.) For a round focused beam waist of radius ~15 µm and a total round-trip cavity loss (including output coupling) of 15%, Eq. (8) yields a threshold pump power of 615 mW. Useful output can be obtained at higher pump powers. Since the confocal length of this beam at $\lambda_p = 514.5$ nm is ~2 mm, the dye cell must be less than 1 mm in length for the area to remain constant. Practical cells are flowing dye jets (see Section 3.4.3) with a typical thickness of around 100–200 µm. For the R6G example, at 90% absorption at 514.5 nm, the required dye molecular density N in a 150-µm thick cell is 1×10^{18}/cm³ (assuming $N_0 = N$). This corresponds to a concentration of 1.6×10^{-3} moles/liter. Unfortunately, at this concentration increased nonradiative quenching of the emission begins to happen, reducing both τ_1 and σ_e somewhat. Such additional problems as solubility and aggregation of dye molecules may also appear for this and other dyes, requiring such corrective measures as special additives. Finally, we also check the approximation $N_0 = N$. Equation (3) gives $(N - N_0)/N = (N_1 + N_T)/N$ as 0.18 at $z = 0$ and 0.018 at $z = L$, so that the approximation is reasonable when averaged over the whole length.

3.3.2 Output Power

When pumping is increased above threshold, the dye laser intensity builds up inside the cavity. The steady-state population N_1 is decreased by stimulated emission, so that the gain saturates to the cavity loss (i.e., to the threshold gain). It can be quite accurately approximated as constant over the length of the cell, and the saturated single-pass gain is then found from the steady-state solution for N_1 as

$$\ln G = \sigma_e f_T \tau_1 \frac{1}{1 + I_l/I_s} I_{po} a, \qquad (10)$$

where $I_s = 1/\sigma_e \tau_1$. This equation shows the homogeneous saturation. The output intensity is easily obtained from this by equating the round-trip laser gain to the round-trip cavity loss and using $l_{out} = tI_l^+ = tI_l/2$ (t is the transmission of the output mirror). The result for $\sigma_{al} = 0$ is

$$P_{out} = \frac{\nu_l}{\nu_p} f_T a \frac{t}{t+l} \left[P_{po} - Ah\nu_p \frac{t+l}{af_T\sigma_e\tau_1} \right] = \eta \left[P_{po} - P_{th} \right], \qquad (11)$$

where $1 - t - l/G_0$ is the total round-trip cavity loss (l being the fractional loss other than the coupling out loss). The value of $t + l$ in real systems is less than ~0.1. This equation shows a simple linear dependence of the output power on the pump. The threshold pump power depends linearly on the cavity losses and on the area, and inversely on the fraction of the pump power absorbed by the dye (a), on the emission cross-section, and on the triplet loss factor (f_T). The slope efficiency, on the other hand, only depends on a, f_T, the cavity losses, and, of course, on the energy ratio of the pump and laser photons.

This theory can be made more rigorous by including the effects of the singlet absorption losses. Additionally, we should also take into account that the pump and laser beams are Gaussian beams that can have unequal radii. Following the methods of Pike [8] and of the present authors [9], it can be shown that with pump and laser modes fundamentally Gaussian and $\sigma_{al} \neq 0$, Eq. (11) is still valid, but

$$\eta = \frac{\mu}{\sqrt{1+\mu^2}} \frac{\nu_l}{\nu_p} f_T a \frac{t}{t+l+2s(1-f_T)}, \tag{12a}$$

$$P_{th} = \frac{\mu+1}{2} \pi w_p^2 h\nu_p \frac{t+l+2s}{a f_T \sigma_e \tau_1}, \tag{12b}$$

where s is the singlet absorption loss, given by $s = -\ln(1-a)[(\sigma_{al}/\sigma_{ap})]$, w_p is the pump beam radius, and $\mu = w_l^2/w_p^2$, the ratio of the areas. Singlet absorption adds to the losses in the cavity. For many dyes the singlet and triplet absorption curves have very little overlap, resulting in $s(1 - f_T) = 0$ and no change in slope. The threshold pump power, however, rapidly increases with σ_{al}. For example, for $a = 0.9$ and $t + l = 0.1$, even $\sigma_{al} = \sigma_{ap}/15$ yields $s = 0.15$ and the threshold power more than quadruples. Thus, lasing will cut off in the absorption region. It is also important to point out that *regardless of how hard we pump the dye, the absorption of the pump does not saturate* but remains the same value as at threshold [9, 10]. This is because inside the laser cavity the saturated gain is clamped at the threshold value ($G_{th} < ~1.1$). Thus, the population of the ground state (N_0) is also clamped at its threshold value ($N_0 = N$). As for the effects of unequal Gaussian pump and lasing beams, we see that decreasing the pump beam area relative to the lasing area increases the threshold but also increases the slope efficiency. For strongly pumped lasers a value of μ between 1.5 and 2 is best, that is, a pump beam area slightly smaller than the dye laser beam area. A more accurate description shows departures from an exactly linear slope for $\mu < 1$ [11]. Usually, however, these are overridden by residual thermal effects that will be manifest at high powers. The effects of excited state absorption of the pump light, which also reduces the efficiency, have also been calculated and measured [9].

For the R6G example, pumped by a 514.5-nm argon laser line, with $t = 0.057$, $l = 0.02$, $w_p = 20$ μm, and $\mu = 1.5$ at $\lambda_l = 590$ nm, we calculate $P_{th} = 1.3$ W and $\eta = 0.36$. Experimentally, similar values were achieved early in the history of the CW dye laser [7]. It should be noted that, in practice, a multiple-wavelength pump laser is usually used since this gives more available pump power when more than one line falls within the absorption band of the dye. This alters the theory in obvious ways that need not be elaborated here. The theory given above is also valid for ring dye lasers provided that the factor of two multiplying the singlet loss s is omitted (since the laser beam traverses the dye only once). This theory is simpler, yet more realistic than that given in reference [3] since it does not define a pump-dependent I_s and does properly include the z dependence of the pump and therefore of the gain.

3.4 Actual CW Dye Lasers

The above theory gives a fairly accurate description of most CW dye lasers. The major exception is a standing-wave type laser that has been forced into single-mode operation. The reasons for this will be discussed in Section 3.4.5. Since practical single-mode CW dye lasers are always of the unidirectional ring type and for them the theory *is* valid, this limitation can be ignored. In practical lasers there are some additional effects that result in somewhat lower efficiencies than those predicted by this theory. They will be discussed briefly in Section 3.4.6. While the theory is good enough to be used to design and optimize actual systems, in practice the parameters of the dye are not well known and most of the design procedure is carried out in a trial-and-error fashion [12, 13]. Here we use it for "reverse engineering," through which we aim to understand the principles underlying the design and operation of currently used systems. The first and most important step is the understanding of the CW dye laser resonator.

3.4.1 The Laser Resonator

The necessary very small mode area can be achieved by using a small concentric cavity, and that is how the first CW dye laser was made [14]. However, for practical lasers we need a longer resonator [6, 7] with a well-collimated beam in one part, allowing placement of elements for frequency tuning and control, loss modulation [15], or an additional focal spot for passive mode-locking [16], etc. The only configurations capable of fulfilling these requirements are the folded cavities of the type shown in Figs. 3 and 4. The basic design and the advantages of these configurations can be easily understood by simple ray tracing in a three-mirror cavity, as shown in Fig. 7a. The parallel rays from mirror M_1 are focused to the focal point of mirror M_2. This is where the dye cell is placed (at

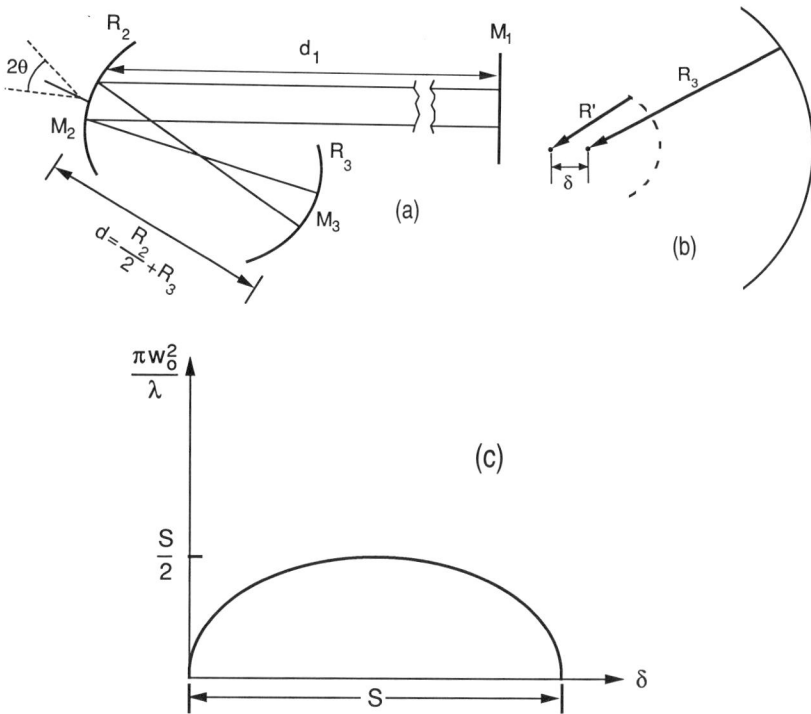

FIG. 7. (a) Schematic of the three-mirror folded cavity showing mirror distances, and geometrical optics ray tracing of the laser beam in the focusing fold and in the long collimation arm. (b) Equivalent two-mirror cavity obtained by imaging the flat mirror (M_1) through the folding mirror (M_2). The image mirror has its center of curvature at the focal point of M_2, which is at a distance δ from the center of curvature of M_3. (c) Minimum beam radius w_0 (in units of confocal distance) vs. adjustment distance δ for the three-mirror laser cavity shown in (a) and its two-mirror equivalent shown in (b).

Brewster's angle to minimize the internal losses). If this point is also the center of curvature of mirror M_3, then the diverging rays are returned upon themselves. Thus, for $R_3 = R_2/2 = f$, the folded arm has total length $d = R_2/2 + R_2/2 = R_2$, and the cell is located in the middle. The length of the long arm (d_1) is not critical but should be much longer than d, and $R_1 = \infty$ (a flat mirror) or large. If we need an additional long arm or want to have a ring cavity, then R_2 and R_3 must be made equal for a cell located in the middle (see Fig. 4). Additional folded focusing arms can also be easily added [16]. To find the actual mode parameters and the stability range of the resonator, standard Gaussian beam analysis is used. The basic three-mirror cavity and its more complicated extensions have all been analyzed in detail

[17, 18]. We give here the main results for the basic three-mirror version only, which is shown simplified in Fig. 7a.

It is logically easy to comprehend (and has been rigorously shown [19]) that an equivalent two-mirror cavity is found by *imaging both mirror M_1 and its center of curvature through folding mirror M_2*. We now assume that $d = R_2/2 + R_3 + \delta$. Using the imaging rule and ignoring the tilt of mirror M_2, it follows that our cavity is equivalent to a short two-mirror resonator formed by mirror M_3 and a nearly concentrically located image mirror of radius $R' = f^2/d_1$, as shown in Fig. 7b. Standard-mode analysis [20] then gives the Gaussian "waist" size and the stability range δ for the position of mirror M_3. These are shown graphed in Fig. 7c. The stability range $S = R' \sim f^2/d_1$, and in the middle of this range the beam waist area is given by $\pi w_0^2/\lambda = S/2 = f^2/2d_1$. For $R_1 = \infty$, $R_2 = 10$ cm ($f = 5$ cm), $R_3 = 5$ cm, and $d_1 = 100$ cm, this gives $w_0 = 15$ μm, a typical value needed for lasing. By contrast, the beam in the long arm has a minimum beam radius of $w_0 \sim 0.6$ mm (on the flat mirror), a well-collimated beam. The above calculated beam waist radii are in free space. Inside the dye cell, after refraction into the Brewster-angle cell, the beam becomes elliptical and the area πw_p^2 in Eq. (12b) must be corrected by a multiplying factor n (the index of refraction of the cell). Clearly, due to threshold requirements, R_2 cannot be much larger than 10 cm. Practical geometrical considerations, as well as beam aberrations (coma and astigmatism) restrict it to values larger than 5 cm. *All practical CW dye lasers use R_2 (and R_3) between 5 and 10 cm* in the fold around the dye cell. The pump laser beam must be focused to a spot size approximately equal to that of the laser mode. Exact colinearity is obviously ideal and can be accomplished using the cavity folding mirror itself and a prism [7]. However, this is not essential, since the very thin dye cell allows excellent overlap even with a fairly substantial angle between the two beams. Independent focusing with a lens or a mirror is most commonly used, with a focal length about the same or slightly smaller than that of the fold mirror, as shown in Fig. 3. There are many variations on this basic configuration, but the above analysis is sufficient for a basic understanding. The resonant frequencies of the cavity, of course, are still determined by its overall dimensions, that is, the longitudinal (fundamental Gaussian) mode spacing is $\Delta f = c/2\mathcal{L}$, where $\mathcal{L}$ is the total cavity length.

3.4.2 Astigmatism and Compensation

An additional feature of a folded tightly focusing cavity is that the folding mirror introduces an astigmatism between the tangential (in the plane of the diagram) and sagittal (perpendicular) planes, owing to the different effective focal lengths of folding mirror M_3 in these two planes. Astigmatism not only results in elliptical-shaped beams but can greatly reduce or even kill the stability of the resonator by displacing the stability regions of the two planes with respect to each other. Interestingly, a Brewster-angle cell in the focal region also causes an astigmatism, owing to the refraction of the beam in the cell, which results in different effective cell lengths in the two planes. It turns out that these two astigmatisms are of different signs, allowing compensation by adjusting the cell thickness or the folding angle. Detailed analysis is given in reference [17]. The main result is that, using a folding focusing mirror with radius of curvature R, a folding angle 2θ, and a cell with index of refraction n and thickness L, the condition for zero net astigmatism is

$$\sin\theta \tan\theta = \frac{2L}{R}\left[\frac{1}{n^2} - \frac{1}{n^4}\right]\sqrt{n^2+1} \; . \tag{13}$$

All current systems use thin free-flowing dye cells. For $L = 150$ μm and $R = 10$ cm, we find $2\theta = 4.5°$ is needed. Such a small angle is not practically feasible, so current systems either live with some residual astigmatism (which increases the lasing area and reduces the stability range of δ) or use added compensation. The simplest of these is an extra Brewster plate near the dye cell or inside an additional fold [12]. Astigmatism is also present in the pump beam if a tilted focusing mirror is used instead of a lens, and this increases the pump beam focal area somewhat. For the above reasons, in a practical dye laser the beam areas and the parameter μ are usually only approximately known.

3.4.3 The Flowing Dye Cell and Triplet Quenching

Although the conversion efficiency of this laser is quite good, CW operation means continuous power dissipation. Much of the power loss takes place in the active dye volume through internal nonradiative processes. This means that several watts of heat are continuously input to the active volume of $\sim 1 \times 10^{-7}$ cm^3. To carry this heat away a flowing dye cell is needed, together with a heat exchanger in the flow system. Although early designs used cells with glass windows [6, 7], all modern systems use free-flowing dye jets [21, 22] whose advantages of simplicity, high-speed flow, and long lifetime outweigh the disadvantage of the additional low-frequency noise introduced by the physical motions and local ripples of the jet. A jet nozzle consists simply of flattened stainless steel metal tubing. Since the nozzle opening is very narrow, filtration of the flowing dye is also needed in order to avoid clogging. For high enough viscosity to produce a smooth surfaced ribbon, the solvent is typically ethylene glycol, sometimes mixed with water or other organic solvents and certain additives (to improve solubility and reduce aggregation). Using a centrifugal pump to produce a pressure of ~35 psi, an approximately rectangular ribbon ~3 mm in width and ~100–200 μm in thickness is typically produced with a flow speed of up to 10 m/s. Even at this high flow speed, for the highest pump powers some residual thermal "blooming" [12,

23] will remain and limit both beam quality and efficiency. Having established a stable good-optical-quality jet, the dye concentration is adjusted to give around 90% absorption of the pump. For multiline pumping, some pump wavelengths may be weakly absorbed. If the concentration cannot be increased enough owing to solubility limitations, concentration quenching, and to increased singlet losses, then the net pump absorption may be somewhat lower.

The high-speed flow and circulation of the dye automatically solves one of the other important requirements necessary for CW lasing of all dyes: triplet quenching [13, 22], that is, shortening the lifetime of the metastable T_1 state. This is accomplished mostly by natural aeration of the solution in the flow-circulation system (oxygen is a well-known triplet quencher), and also by actually moving the molecules accumulated in T_1 out of the active volume. The combination of these two effects shortens the effective triplet lifetime in the dye solutions to give a triplet loss factor f_T sufficiently close to unity for CW operation over most of the emission spectrum. In spite of this, triplet absorption remains a limiting effect in most dyes. It has been found that the additive cyclooctatraene (COT) further decreases triplet lifetime and thus improves laser efficiency for most dyes [12, 13]. However, this chemical is not only rather unstable (both in the solution and on the shelf) but also smells very bad. Its use is generally avoided when possible.

3.4.4 Mirror Transmission

It is clear from the theory and from practical considerations of the cavity that there is only small flexibility for most of the design parameters of a CW dye laser. An exception is the output mirror transmission (t), which must be carefully designed for optimum lasing efficiency for each laser dye. As seen from Eqs. (12a,b), the slope efficiency increases monotonically with t, but so does the threshold. For a given dye and a given pump power, there is an optimum value of transmission (t_{opt}) at *each wavelength*. From Eqs. (11) and (12) we find by differentiation with respect to the mirror transmission t that

$$t_{opt} = \sqrt{[l + s(l - f_T)] [g_0 - sf_T]} - [l + s(1 - f_T)], \qquad (14)$$

where $g_0 = \ln G_0$ is the small signal gain at the given pump power (Eq. (8), with a $2/(1 + \mu)$ correction factor). For $\sigma_{al}, \sigma_T = 0$, this reduces to

$$t_{opt} = \sqrt{lg_0} - l,$$

in agreement with the well-known general theory of homogeneously broadened lasers [24] and with experiments [22]. A smaller effective cross-section ($f_T \sigma_e$) obviously means less gain and therefore a lower value of t_{opt}. In practice, while a certain amount of tailoring of mirror transmission is possible, compromises have

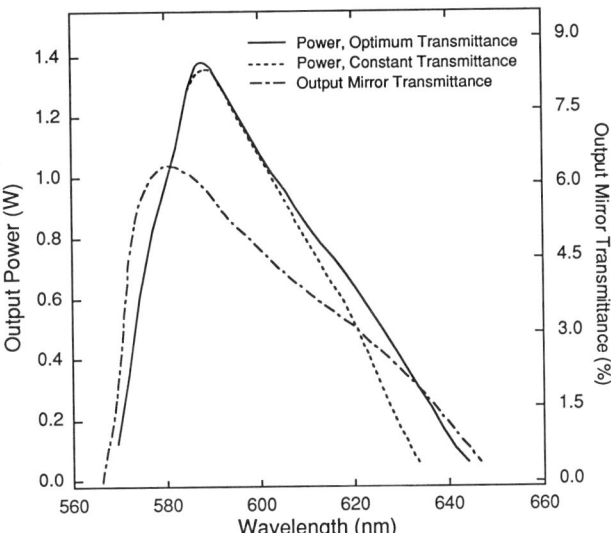

FIG. 8. Calculated output power vs. wavelength for an R6G CW dye laser in a linear cavity, using the data shown in Fig. 2. Pump beam area = 1.25×10^{-5} cm^2 (w_p = 20 µm), beam area ratio µ = 1.5, residual loss l = 2%. The output mirror transmission is optimized at each wavelength for one curve and is constant for the other. The curve of the optimized mirror transmittance is also shown.

to be made. In Fig. 8 we show the calculated power output of an R6G dye laser vs. wavelength with a mirror transmission that is constant 5.7% over the entire gain region. The output with mirror transmission optimized at each wavelength is also shown, together with the t_{opt} used. (It is equal to 5.7% at 590 nm). With the fully optimized mirror, improved performance is clearly seen on the long-wavelength side. The transmission is largest at 580 nm, where the linear gain is the largest and decreases at both shorter and longer wavelengths. Note, however, that on the short-wavelength side rapidly increasing absorption dominates the total loss and optimizing the mirror transmission does not result in any improvement. The important point for output mirror design is to optimize the transmission as well as possible on the long-*wavelength* side of the peak gain. At any wavelength, the calculated output is a linear function of the input power. At 590 nm, the values of the threshold pump power and of the slope efficiency are 1.3 W and 36%, respectively.

3.4.5 Wavelength Tuning and Lasing Bandwidth Control

In order to make use of the large gain bandwidth of the dye laser, we must be able to tune its wavelength. Additionally, different applications have various requirements as to the bandwidth of the output laser radiation. For many applications, a bandwidth of ~0.5 nm (~500 GHz), or even larger, is quite sufficient. On the other extreme, some spectroscopic experiments require a bandwidth as low as ~1 MHz. This means that oscillation must be restricted to a single cavity mode (since cavity mode spacing is typically ~150 MHz). Frequency stabilization may also be required, because if the single oscillating mode "hops" between modes the effective bandwidth is obviously greatly increased. We discuss only tuning and bandwidth narrowing here. Frequency stabilization is described elsewhere in this volume.

Without any wavelength control, a standing-wave-cavity CW dye laser operates at the peak of the gain with a bandwidth typically as broad as ~1–2 nm (1–2 THz). The reasons for this broad oscillation bandwidth need to be understood. We have already stated and used in the theory the fact that dyes are spectrally homogeneous. To put it in other terms, this means that all the lasing frequencies compete for the same excited singlet population, that is, for the same gain. This would imply, according to well-known fundamental laser theories, that the strongest lasing mode suppresses the others and single-mode operation results. However, in linear cavities this is not the case, owing to an effect commonly referred to as spatial hole burning. The two oppositely traveling laser beams form a standing wave pattern inside the cavity. For each of the cavity resonance frequencies there is a different standing wave pattern, the adjacent ones having a number of nodes differing by one. On the other hand, the gain has no such pattern, since the pump is a traveling wave. The obvious outcome is that the different laser modes do not interact with the same molecules and saturate (or "burn down") different spatial locations, and mode competition is thus diminished. This effect is strongest if the thin dye laser medium is at the center of the cavity (where standing wave patterns of adjacent modes are orthogonal) and weakest if it is at the end [25]. But regardless of the location of the dye, mode competition is weakened enough that many laser modes oscillate. Another way to describe this effect is that oscillation at one mode is able to extract energy only at the antinodes. The laser will oscillate in many modes, so that an overall uniform intensity is established and maximum energy is extracted. This is a built-in assumption in our theory, since standing wave patterns of the laser intensity were ignored.

It is possible (although by no means easy) to force a linear-cavity CW dye laser into single-mode operation by placing a hierarchy of ever narrower bandwidth filters inside the cavity. In such a case, however, the output power of the laser is considerably diminished because only about half of the actual inverted medium is utilized. It has been found that the single-mode output of such a laser is in practice only ~40% of that obtained in broadband operations. Fortunately, this problem can be overcome by using a ring configuration [12, 26] in which the laser radiation is forced to travel unidirectionally. In such a laser the frequencies of the resonant modes are still separated by $\Delta f = c/2\mathcal{L}$, but are pure traveling waves. Thus, all modes compete for the same populations and single-mode operation results. Owing to the very broad gain and to mechanical instabilities, however, the output will still be broadband because the frequency of lasing will rapidly change among many modes. However, it is now possible to narrow this bandwidth by the addition of tunable frequency-selective filters to a single, much more stable, mode without any loss of output (except, of course, that due to the insertion loss of the filters). It is appropriate for us to discuss and describe the elements used to achieve single-mode operation. It should be noted that the selectivity of these elements inside the dye laser cavity is much stronger than would be in normal extracavity use. This owes to the fact that the spectrally homogeneous dye laser is very sensitive to small variations of the net gain [27]. Typically, a net gain reduction of 0.5–1% on a propagation direction or on a group of modes is sufficient to suppress the lasing. This is fortunate, for it allows the use of relatively simple elements with much lower actual selectivity than what needs to be achieved.

An element commonly called an optical diode is used to force a ring dye laser into unidirectional operation. This element is essentially lossless for a laser beam propagating in one direction but provides a loss in the other direction. This is achieved by rotating the polarization of the laser beam a few degrees for one direction of travel only. For this beam, small loss is then experienced at the number of Brewster-angle surfaces of the other cavity elements that follow the "diode." A differential loss of about 0.5% is sufficient to suppress lasing in this direction. Polarization rotation for only one direction of propagation is accomplished by a series combination of two well-known elements. One is a half-wave plate rotator that is a birefringent quartz plate with an optic axis in the plane of the plate and of such thickness as to provide a 180° phase delay between the ordinary and extraordinary waves. The other is a Faraday effect rotator that utilizes the circular polarization birefringence resulting from application of a strong axial magnetic field to a crystal. The resulting rotation of linear polarization is independent of the direction of propagation. The rotations due to the above two elements cancel for one direction of propagation and add for the other. Details of the design of an optical diode can be found in the literature [28].

There exist many different types of tunable-frequency (wavelength) selective filters, but only a few are suitable for use inside the CW dye laser cavity. The most popular by far are the birefringent (Lyot) filter and simple Fabry–Perot etalons. While other devices have been successfully used, these have proven to be the simplest, cheapest, and most reliable. They are the only ones we describe here. A birefringent filter consists of several quartz plates, with thicknesses related by integral multiples, cut with the optic axis parallel to the surface, and tilted at Brewster's angle relative to the direction of propagation. Details of the design can be found in the literature [29, 30]. For our purposes it is sufficient to know that the frequency selectivity results from the birefringence of the plates combined with the polarization-dependent loss from the Brewster-angle surfaces. Each of the plates acts as a wavelength-dependent wave plate that changes the polarization of the light passing through it. The filter is tuned by rotating it about the normal (while maintaining the Brewster-angle orientation) and thereby changing the amount of birefringence. In general, the polarization exiting each plate is elliptical and loss is experienced at the Brewster surfaces. At one wavelength, however, the light exiting each plate is linearly (p) polarized and thus experiences no loss. This is the center of the main passband. Typical birefringent elements for a dye laser have one to three plates. For a three-plate filter the main passband is about 1.5 THz wide. There are other passbands, but those have higher losses and no lasing occurs in them.

A Fabry–Perot etalon for the CW dye laser may simply be a single glass plate placed at near normal incidence and coated on each side for a desired reflectivity. The transmittance of such a parallel plate interferometer is

$$T(\delta) = \left[1 + \frac{4R}{(1-R)^2} \sin^2(\delta/2) \right]^{-1}, \tag{15}$$

where $\delta = (4\pi nL/\lambda) \cos \theta$, R is the power reflectance of the etalon surfaces, L is its thickness, n is its index of refraction, and θ is the angle between the etalon normal and the inside direction of propagation. This angle should be small to minimize "walkoff" losses. The finesse F, which is the ratio of the free spectral range (mode spacing) to the half-power bandwidth, is given by $F = \pi\sqrt{R}\,[1-R]^{-1}$. Obviously, the center frequency of an etalon can be tuned by changing either the length L or the angle θ. Typically, a hierarchy of three frequency-selective elements are used to achieve single-mode operation. Since each element has repeating passbands, the resolving powers must be chosen so that one (coarser) element is able to select a single resolution element of the next (finer) component. The successive stages of frequency selection are shown in the diagrams of Fig. 9. The element with the coarsest resolution is usually a birefringent filter with a half-power bandwidth of about 1.5 THz. This is usually followed by two etalons. The first of these is a single-plate one with an FSR of around 200 GHz, tunable by rotation. The second is a thicker air-spaced etalon made of two parallel glass surfaces between Brewster cut glass prisms with an FSR of about 10 GHz and tunable by changing the spacing. The etalons are coated for a finesse of about 3. This element thus has a half-power bandwidth of about 3 GHz, but due to the sensitivity of the laser to small variations of the net gain it is able to select a single one of the cavity modes that are typically spaced about 150 MHz apart. The bandwidth of this single-mode laser is now determined by the mechanical stability of both the optical elements and the gain. Unless specially engineered and stabilized for true single-mode operation, some mode hopping will still take place, resulting in a bandwidth that may be as small as ~400 MHz or as large as ~5 GHz. Feedback-stabilized true single-mode CW dye lasers have a bandwidth (deter-

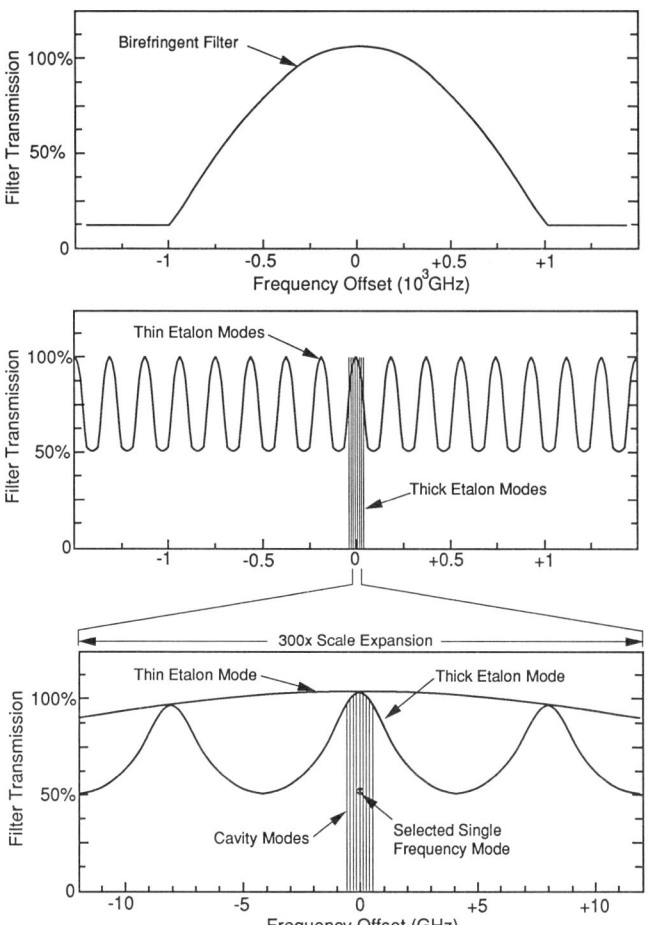

FIG. 9. A three-element frequency selection and tuning scheme shown schematically for a ring dye laser. Each finer selective element has a passband tuned to the center of the next coarser element. Lasing in only one passband of each element results from the sensitivity of the dye laser to small gain variations. The final thick etalon selects a single-cavity mode, but the lasing mode can hop around between several ones.

mined by the jitter of the mode frequency) typically of the order of ~1 MHz, but record stability and bandwidth in the kHz range has also been achieved.

From the above discussion it is easy to understand why the following two tunable CW dye laser designs are used most often. One is a relatively simple compact three-mirror standing-wave cavity laser with a single wide-band frequency tuning element that is mechanically controlled. The output is broadband, with its value depending on the bandwidth of the tuning element. A single-plate birefringent filter typically results in a ~400 GHz bandwidth, but ~40 GHz can be achieved with a three-plate filter. The second common design is a unidirectional ring laser with a hierarchy of frequency-selective elements, depending on bandwidth requirement, but capable of single-frequency operation. For either of these designs we must be able to tune and reproducibly track the operating frequency. Since the reproducibility needs to be only as accurate as the bandwidth, this is a relatively easy task for the broadband linear geometry laser. Tuning is accomplished by mechanically rotating the birefringent filter. The wavelength of lasing is approximately calibrated and can be externally measured. For the much narrower band ring laser, however, sophisticated electromechanical design is required. The hierarchic multi-tuning-element design described above allows for continuous tuning. Only a brief description of the tuning and scanning controls can be given below.

Any Fabry–Perot interferometer (main cavity, etalon) is tuned by changing its effective length. A length change of one half a wavelength tunes the frequency through one free spectral range. Thus, for a red laser a typical piezoelectric displacement of a mirror of the main cavity by ~6 μm will result in a tuning range of ~20 mode spacing or ~3 GHz. The finest tuning adjustment, within a portion of the thick etalon bandwidth, is made by changing the effective cavity length. Practically, this is accomplished in two stages: by piezoelectric translation of one cavity mirror and by slight rotation of a thick Brewster-angle plate (or double Brewster plates) using a galvo-motor. The next coarser adjustment is made by piezoelectric tuning of the length of the thick etalon. The next stage of tuning is by rotation of the thin etalon (also using a galvo-motor). The coarsest adjustment is made with the birefringent filter, usually purely mechanically. Tracking the oscillating frequency requires simultaneous and precise control of several elements. Description of a tracking system design is beyond the scope of this chapter. Suffice it to say that the continuous scanning range is typically limited to about 30 GHz. This interval can then be positioned anywhere within the laser tuning curve by adjusting the birefringent filter and the thin etalon, but the center wavelength must be independently measured. Computer-controlled tuning and tracking over the entire tuning range of a given dye has been developed [31], and such systems are commercially available.

3.4.6 Laser Dyes and Commercial Systems

In the previous sections we used the R6G dye laser as our example, since the parameters of this dye are reasonably well known. This was the first CW dye laser, and it is still one of the best. Pumped with a powerful 24 W multiline argon ion laser, and with a special solvent mix to alleviate thermal problems, a record high multimode power output of 6.5 W and single-mode power of 5.6 W has been obtained for this dye [12]. Many CW laser dyes are now available from the near-UV to the near-IR, and the performance of the best ones approach, and in some respect surpass, that of the R6G. Before we describe various other dyes and commercial systems, we should briefly discuss some additional factors that limit the performance of CW dye lasers but are not included in the simple theory. (i) At the concentrations needed to absorb a high fraction of the pump in the thin dye jet, aggregation of the dye molecules may occur for many dyes. This problem may be partially alleviated by the addition of a small amount of surfactant such as Ammonyx LO [12, 14], but this can also bring in such other problems as foaming. (ii) At high pumping powers, a thermal lens is created in the pumped region. As discussed in the literature [32], this lens is also astigmatic due to the flow, and cannot be compensated. For relatively low-power systems this problem is not very serious and results in only minor degradation of performance. The record powers reported in the literature, however, can only be obtained with special solvent mixtures and additives [12]. (iii) Residual aberrations (astigmatism and coma) in both the pump and laser beams increase the effective areas and reduce beam overlap, thus increasing threshold and decreasing slope efficiency. These effects also prevent the use of focusing mirrors (for both pump and cavity) with $R < 5$ cm. (iv) Transverse modes may be partially present in the pump and may also appear in the dye laser beam. In this regard, it should be noted that the nearly confocal effective resonator does not discriminate well against higher-order transverse modes, which can appear in case of misalignment or in case of the pump beam having transverse variation other than fundamental Gaussian mode. Owing to the above effects, actual threshold powers can be higher and slope efficiencies can be somewhat lower than that calculated by the theory given in Section 3.3.

We do not give a comprehensive listing of dyes here but show the tuning range of some of the best performers with their common abbreviations in Fig. 10. These dyes are available from the large chemical companies and are made especially for laser applications. It is not necessary to know their exact chemical composition. The overlapping lasing ranges cover the wavelength range from ~360 to ~1000 nm. Quite well-optimized systems using these dyes have been reported [12, 13, 26, 33]. Since the output power depends on the power of the pump source, as well as on many details of the specific design, we do not give output power values. We do, however, indicate which pump source and which line (or group of lines) is needed for each dye. The commonly available pump lasers and their useful strong lines are listed in Table I. The obvious requirement is that the absorption band must overlap with one of these lines. A regular argon laser has the two familiar strong blue-green lines. For blue- and ultraviolet-emitting dye lasers, a UV-capable argon laser is needed (which can also provide the blue-green lines with a change of mirrors). The longest-wavelength dyes can only be pumped with the red lines of the krypton laser. This laser also has a group of violet lines useful for pumping certain blue-green laser dyes and yellow-green lines of reasonable power, which

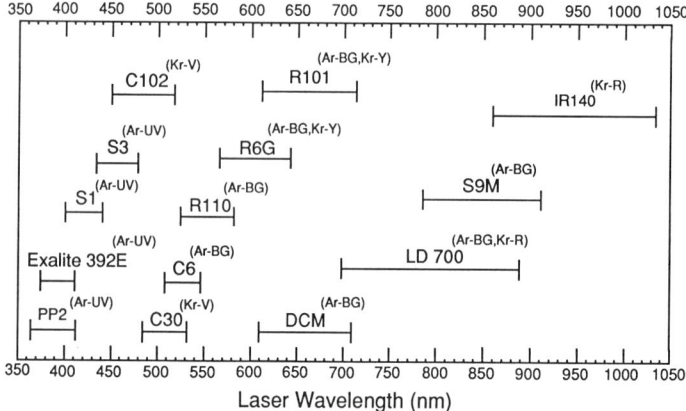

FIG. 10 Schematic diagram of the tuning ranges of the best laser dyes. For each dye the possible pump ion laser and group of lines are also shown with the codes given in Table I.

can be the excitation source for some red lasing dyes. Thus, while a single ion laser obviously is not capable of pumping the entire range of laser dyes, a large fraction can be covered. It should be noted that dye lasers between 530 and 930 nm are the easiest to design, set up, and operate, and they provide the highest power outputs as well. This is due partly to the use of the two blue-green (480 and 514.5 nm) high-power argon laser lines, and partly to the high emission cross-section and low triplet absorption of these dyes (high σ_e and f_T close to unity). By contrast, the blue- and UV-emitting dyes have lower values of the effective emission cross-section ($f_T \sigma_e$), and, unfortunately, the available pump power in the UV is also much less. As an example, we briefly examine the results reported for the shortest-wavelength dye, PP2 [32]. This dye was used in a ring cavity with both the folding mirror and the focusing mirror having $R = 7.5$ cm to minimize spot size. The pump was a large argon laser (Coherent Innova 100 UVE) operated at a very high discharge current (65 A). A total of 3.4 W were obtained on six UV lines 300.3–335.8 nm and focused to $w_p \sim 20$ μm; 84% of this pump power was absorbed. With a mirror transmission of 1.9%, the lasing range was 364–408 nm with a maximum output of 255 mW at 383 nm, where the threshold pump power was 1.15 W. The slope efficiency of only 11% can be semiquantitatively understood as follows. The given values of $a = 0.84$ and $t = 0.019$, and the estimated values of $\mu = 1.5$, $v_l/v_p = 0.84$, $l = 0.025$, and $s = 0$ would give a slope efficiency of 0.25 if the triplet factor f_T was unity. Thus, a value of $f_T \sim 0.5$ is indicated. This is within the known range for blue-emitting dyes. From the threshold pump power we also calculate $f_T \sigma_e \tau_1 = 4 \times 10^{-16}$ cm^2 sec, about the same as for R6G. The main problem, obviously, is the low slope efficiency.

TABLE I. Most Important Ion Laser Lines for Pumping CW Dye Lasers

Ion laser type	Wavelength of line	Notes	Code for Figure 10
Argon or UV-capable argon	476.5 488.0 496.5 501.7 514.5	a a	Ar-BG
UV-capable argon	300.3–335.8 333.6–363.8 363.8–385.8	b b	Ar-UV
Krypton	647.1–799.3 520.8–568.2 406.7–468.0	c c c	Kr-R Kr-Y Kr-V

[a]These two lines are much stronger than the others, and contribute most of the pump power.
[b]These groups of lines can be obtained only at very high tube currents (e.g., 65 A in a Coherent Innova 100 UVE). Change of mirrors is also needed.
[c]Change of mirrors is needed for each of these groups of lines.

Another example is given in Fig. 11. This example shows the output power and approximately optimized mirror transmission vs. wavelength as well as output vs. input powers at peak wavelength. The dye is DCM, one of the more efficient ones, lasing in a single-mode ring configuration, similar to the one shown schematically in Fig. 4. Since the parameters of the dye are not well known, no quantitative comparison with the theory is made. However, the behavior is clearly similar to what is described by the theory. The threshold pump power of 400 mW at the peak lasing wavelength is smaller than that calculated for the R6G example (Section 3.4.4). This indicates a higher effective emission cross-section, or a longer singlet lifetime, or stronger focusing, or a combination of all of these. The peak wavelength slope efficiency is 0.78 times smaller than we calculated for R6G. Part of this difference is due to the 1.1-fold longer wavelength of DCM emission and part to the smaller output mirror transmission. We note that output mirror transmission increases toward the peak on the long-wavelength side, as required by the theory. Unlike the calculated optimum mirror, however, mirror transmission here continues to increase on the short-wavelength side. This is because of mirror design practicalities. This departure from the theoretical optimum means no practical difference, because, as noted in Section 3.4.4, on the short-wavelength side of the peak the output is nearly independent of mirror transmission. We also note that for this dye the short-wavelength cutoff is not as rapid as that for R6G. This owes to

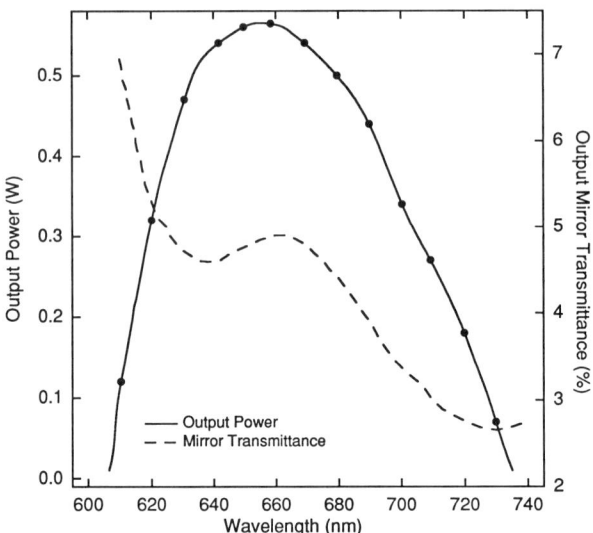

FIG. 11. Experimentally measured output power and mirror transmittance vs. wavelength for the dye DCM in a single-mode ring cavity. The mirror transmission is approximately optimized. Data provided by T. Johnston, Coherent Inc.

a displacement of the emission and absorption spectra (a Frank–Condon shift) that is somewhat larger in this dye than in R6G.

Given the availability of a suitable pump laser, one can build a CW dye laser to match a given set of requirements of tuning range, power, and linewidth. This is especially advantageous if only a broadband output and a restricted tuning range, which can be covered by one dye, are needed. Then a simple linear cavity configuration can be optimized for the given dye. Commercial systems at higher expense, and somewhat lower (but guaranteed!) output and overall tuning range are available. Some have a single fixed geometry, that is, fixed mirror radii and spacing. Different mirror coatings and dyes are supplied to cover specified wavelength ranges. The obvious advantage of such a system is that, after a simple mirror and dye change, only minor realignment is needed to restart operation in a new frequency range. The disadvantage is that significant compromises have to be made in the design, and the laser is by necessity not optimized at any wavelength. A well-designed commercial system with adjustable geometry (e.g., Coherent model 599) is well optimized for all the dyes and not at all difficult to realign. Frequency and bandwidth control and the associated tracking is likely to be the main issue in the decision of whether to build a system at lower cost or to buy a commercial one. As discussed in Section 3.4.5, elements for tuning and bandwidth restriction are added to the cavity according to the specific requirements. A linear-cavity broad-bandwidth (~50–500 GHz) laser with external measurement of wavelength is quite easy to build from components, but the construction of a ring-cavity narrow-band (~1–4 GHz) system is considerably more difficult. Building a true single-mode (~1 MHz bandwidth) system with continuous tracking of the frequency is a very complicated task. Frequency control and tracking systems of various degrees of sophistication (and proportional cost) are available with the commercial systems. Given the complexity of the required electromechanical controls, it usually pays to take advantage of the guaranteed performance of a commercial system if the requirements on the bandwidth and frequency tracking are stringent.

3.5 Alignment of a CW Dye Laser

Building a CW dye laser from components requires knowledge of its alignment, which is a somewhat different procedure from that of other lasers. It is in part easier, owing to the strong fluorescence, wide gain bandwidth, and relatively high gain, and at the same time more difficult, owing to the small lasing area and the large number of elements. We describe an alignment procedure based on the experience accumulated by the authors for the three-mirror folded cavity shown in Fig. 3 and R6G as the dye. It is important to remember that the pump laser is a powerful source whose output should be handled with extreme care. Blue-green block goggles should be used at all times during the alignment procedure. The fluorescence from the pumped dye stream is also very bright. Looking at the fluorescence for prolonged periods of time is harmful. A piece of smoked glass should be used when verifying the position of the pumping beam in the jet. Also recommended is a smoked Plexiglas enclosure surrounding the dye jet (with the proper apertures for the laser and pump beams) during normal operation of the laser.

We refer to Figs. 3 and 7a. The mirrors are assumed to have radii of $R_1 = \infty$, $R_2 = 10$ cm, and $R_3 = 5$ cm. The pump is focused by either a lens or a mirror with high reflectivity at the Ar pump wavelengths and $f = 5$ cm ($R = 10$ cm). It will be referred to as the pump mirror or lens. M_1 and M_2 should be mounted in kinematic holders fixed to linear translation stages for precise adjustment of the mirror–jet distance. M_3 should be mounted on a kinematic stage only. The pump mirror should be mounted on a x–y–z linear translation stage. The first part of the procedure has the objective of adjusting the dye jet and focusing the pump beam centered in the dye jet (approximately 2 cm from the tip of the jet nozzle). The nozzle should have an approximately 150 μm thick opening and should be mounted in a rotation stage. The operating pressure in the fluid line should be between 30 and 40 psi. The viscous solvent–dye solution tends to hang in a film on the upper wall of the catcher tube, and turbulence and bubbles are created if the high-velocity jet stream impacts into this area. Adjust the position of the strike point for minimum bubbles and turbulence. If the strike point is properly set, the return dye flow will look like a static orange dye film. A pump beam power between 50 and 100 mW is enough to perform the following alignment steps. Place the pump mirror in an approximately correct position. Rotate the nozzle so that the angle between the pump beam and the normal of the dye jet surface is close to Brewster's angle (~56°). Now insert a glass slide in the path of the pump beam a few centimeters away from the pump lens toward the pump laser, and look with an index card for a round yellowish fluorescence spot reflected from the surface of the slide. If the lens is at the correct distance, the reflected fluorescence spot is approximately collimated, that is, it should keep its size and shape as we move the card away from the glass slide. If this is not the case, adjust the pump lens to the dye jet distance until a collimated spot is observed. The pump beam should not be focused in ripples or turbulence of any kind in the surface of the jet. Verify that the reflected pump beam from the surface of the dye jet is a nice circular green spot.

The next part is the correct setting of mirrors M_2 and M_3. Continuing with low pump power, remove the glass slide and place M_2 as close as possible to the correct (~5 cm) distance from the dye jet and look with an index card for the yellowish fluorescence spot reflected from mirror M_2. Steer M_2 so that this spot travels close to the dye jet towards the location where M_1 will be positioned, but make certain that it does not traverse the dye jet. If the distance between M_2 and the dye jet is correct, this fluorescence spot should be collimated, and its size and shape (as we move the index card away from M_2 and toward M_1) should be slightly smaller than that of a one-inch diameter folding mirror (~2 cm in diameter is about right). If mirror M_2 is too far from the jet, the fluorescence spot remains collimated but becomes somewhat larger. If it is too close, then the spot will be refocused between M_2 and M_1 and grow again. Such positionings should be avoided because they yield unstable cavities that will not lase. Now we can place M_3 at a distance of ~5 cm from the dye jet and look for the reflected fluorescence spot from *this* mirror. The spot from M_3 should travel through the dye jet and reflect from M_2 toward M_1. Since this spot is going through the dye jet, it will have a brownish-orange color (not yellow as the spot from M_2). Steer M_3 until the brown-orange spot overlaps the fluorescence spot from M_2 and fine adjust the M_3 mount until the two spots are concentric. The location of M_3 is correct if, as in the case of the fluorescence spot from M_2, the spot from M_3 is collimated but its diameter is smaller (~1.5 cm) than that of the M_2 spot.

The final steps involve placing and aligning mirror M_1 and starting laser action. If the previous steps were correctly completed, the two collimated fluorescence spots described above travel colinearly toward where M_1 will be located. Place M_1 in the path of the overlapping spots approximately 1 m away. (Unlike other distances, this one is not critical.) Now we must reflect the spots on M_1 that come from M_2 and M_3 back towards M_2. Adjust M_1 so that the back-reflected spots are steered into the surface of M_2. To perform fine adjustments on M_1, punch a 2- to 3-mm diameter hole in an index card and center the hole in the spots traveling towards M_1. Adjust M_1 until the back-reflected spots travel through the hole towards M_2. At this point the pump laser power should be increased to ~2 W. The cavity should be close to lasing when we observe a bright reddish specular pattern on the surface of M_1 that changes its shape when we perform adjustments on M_1. When a lasing flash occurs, adjust the lasing spot approximately to the center of M_3 using the controls of M_1 and adjusting the pump mirror only as necessary to bring the power up. Also, now set the jet stream accurately at Brewster's angle to the dye laser beam, proceeding as follows. Hold an index card beside the dye jet to view the dye laser spots reflected off the jet. Then rotate the dye jet holder very gradually to observe a minimum in the intensity of the spots seen on the card. Set the rotation stage at this minimum, which is Brewster's angle.

These are the basic introductory steps to obtain laser action. As experience is accumulated, the user will be able to optimize power and mode shape. To build and align more complicated cavities, such as the ring cavity shown in Fig. 4, it is recommended that a three-mirror cavity is first lased. Then additional sections are gradually added. To start this procedure, mirror M_3 is replaced by a 10-cm radius one and a new long arm is added to the cavity. One proceeds in an obvious and logical manner to build up the ring cavity.

Acknowledgments

The authors gratefully acknowledge numerous useful discussions with Timothy Johnston of Coherent Inc., who also provided the data for Fig. 11.

References

1. Fork, R. L., Brito-Cruz, C. H., Becker, P. C., and Shank, C. V. (1987). "Compression of Optical Pulses to 6 fs Using Cubic Phase Compensation," *Opt. Lett.* **12**, 483–485.
2. Schafer, F. P., ed. (1990). *Dye Lasers*, 3rd enlarged and revised edition, Springer-Verlag, Berlin, New York.
3. Shank, C. V. (1975). "Physics of Dye Lasers," *Rev. Mod. Phys.* **47**, 649–657.
4. Hillman, L. W. (1990). In *Dye Laser Principles, with Applications*, F. J. Duarte (ed.), Academic Press, Boston.
5. Maeda, M. (1984). *Laser Dyes, Properties of Organic Compounds for Dye Lasers*, Academic Press, Tokyo.
6. Kohn, R. L., Shank, C. V., Ippen, E. P., and Dienes, A. (1971). "An Intracavity Pumped CW Dye Laser," *Opt. Comm.* **3**, 177–178.

7. Dienes, A., Ippen, E. P., and Shank, C. V. (1972). "High Efficiency Tunable CW Dye Laser," *IEEE J. Quantum Electronics* **QE-8**, 388.
8. Pike, H. A. (1971). Ph.D. thesis, University of Rochester, New York.
9. Teschke, O., Dienes, A., and Whinnery, J. R. (1976). "Theory and Operation of High Power CW and Long-Pulse Dye Lasers," *IEEE J. Quantum Electronics* **QE-12**, 383–395.
10. Chan, I. M. (1979). "A Theory of CW Dye Laser Characteristics," M.Sc. Thesis, University of California at Davis.
11. Dienes, A., Couillaud, B., and Ducasse, A. (1978). "Effects of the Gaussian Dependence of the Fields on the Output Power of the CW Dye Laser," *IEEE J. Quantum Electronics* **QE-14**, 702–704.
12. Johnston Jr., T. F., Brady, R. H., and Proffitt, W. (1982). "Powerful Single-Frequency Ring Dye Laser Spanning the Visible Spectrum," *Appl. Opt.* **21**, 2307–2316.
13. Yarborough, J. M. (1974). "CW Dye Laser Emission Spanning the Visible Spectrum," *Appl. Phys. Lett.* **24**, 629–630.
14. Peterson, O. G., Tuccio, S. A., and Snavely, B. B. (1970). "CW Operation of an Organic Dye Solution Laser," *Appl. Phys. Lett.* **17**, 245–247.
15. Dienes, A., Ippen, E. P., and Shank, C. V. (1971). "A Mode-Locked CW Dye Laser," *Appl. Phys. Lett.* **19**, 258–260.
16. Ippen, E. P., Shank, C. V., and Dienes, A. (1972). "Passive Modelocking of the CW Dye Laser," *Appl. Phys. Lett.* **21**, 348–350.
17. Kogelnik, H., Ippen, E. P., Dienes, A., and Shank, C. V. (1972). "Astigmatically Compensated Cavities for CW Dye Lasers," *IEEE J. Quantum Electronics* **QE-8**, 702–704.
18. Li, K. K., Dienes, A., and Whinnery, J. R. (1981). "Stability and Astigmatic Compensation of Five-Mirror Cavity for Mode-Locked Dye Lasers," *Appl. Opt.* **20**, 407–411.
19. Kogelnik, H., and Li, T. (1966). "Imaging of Optical Modes—Resonators with Internal Lenses," *Appl. Opt.* **5**, 1550–1567.
20. Kogelnik, H. (1966). "Modes in Optical Resonators," in *Lasers: A Series of Advances*, A. K. Levine (ed.), Vol. 1, Dekker, New York.
21. Runge, P. K., and Rosenberg, R. (1972). "Unconfined Flowing-Dye Films for CW Dye Lasers," *IEEE J. Quantum Electronics* **QE-8**, 910–911.
22. Shank, C. V., Edighoffer, J., Dienes, A., and Ippen, E. P. (1973). "Evidence for Diffusion-Independent Triplet Quenching in the Rhodamine 6G Ethylene Glycol CW Dye Laser System," *Opt. Comm.* **7**, 176–177.
23. Teschke, O., Whinnery, J. R., and Dienes, A. (1976). "Thermal Effects in Jet-Stream Dye Lasers," *IEEE J. Quantum Electronics* **QE-12**, 513–515.
24. Rigrod, W. W. (1963). "Gain Saturation and Output Power of Optical Masers," *J. Appl. Phys.* **34**, 2602–2609.
25. Pike, C. T. (1974). "Spatial Hole Burning in CW Dye Lasers," *Opt. Comm.* **10**, 14–17.
26. Jarret, S. M., and Young, J. F. (1979). "High-Efficiency Single-Frequency CW Ring Dye Laser," *Opt. Lett.* **4**, 176–178.

27. Klein, M. B., Shank, C. V., and Dienes, A. (1973). "Detection of Small Laser Gains in a Helium–Selenium Discharge Using a Dye Laser Assisted Oscillation," *Opt. Comm.* **7**, 178–180.
28. Johnston, T. F., and Proffit, W. (1980). "Design and Performance of a Broad-Band Optical Diode to Enforce One-Direction Traveling-Wave Operation of a Ring Laser," *IEEE J. Quantum Electronics* **QE-16**, 483–488.
29. Bloom, A. L. (1974). "Modes of a Laser Resonator Containing Tilted Birefringent Plates," *J. Opt. Soc. Am. B* **64**, 447–452.
30. Teschke, O., and Holtom, G. (1974). "Design of a Birefringent Filter for High-Power Dye Lasers," *IEEE J. Quantum Electronics* **QE-10**, 577–579.
31. Williams, G. H., Hobart, J. L., and Johnston, T. F. (1983). "A 10-THz Scan Range Dye Laser with 0.5 MHz Resolution and Integral Wavelength Readout," Paper No. 12 Presented at the SICOLS Conference, Interlaken, Switzerland; also CLE, Baltimore, MD, paper WS-1; also U.S. Patent 4,864,578.
32. Johnston, T. F., and Sasnett, M. W. (1993). "The Effect of Pump Mode Quality on the Mode Quality of the CW Dye Laser," *Proc. SPIE* **1834**, 218–226.
33. Johnston, T. (1988). "High Power Single Frequency Operation of Dyes over the Spectrum from 364 nm to 524 nm Pumped by an Ultraviolet Argon Ion Laser," *Opt. Comm.* **69**, 147–152.
34. Johnston, T., private communications.

4. SEMICONDUCTOR DIODE LASERS*

R. W. Fox and L. Hollberg

National Institute of Standards and Technology
Boulder, Colorado

A. S. Zibrov

Lebedev Institute of Physics
Moscow, Russia

4.1 Introduction

This chapter deals with the technology of applying semiconductor lasers to scientific and technical fields. Diode lasers are prevalent in many applications because of their well-known attributes: high reliability, miniature size, relative simplicity of use, and relatively low cost. These factors are important, but the use of diode lasers in technical fields is also increasing because of their unique capabilities, such as: tunability, high efficiency, useful power levels, reasonable coherence, and excellent modulation capabilities. Volumes of information are already available on the technology of semiconductor lasers [1–8]. Their application to scientific fields has been discussed in a few review articles [9–13], and at least one special issue of a journal focused on the spectroscopic detection of atoms and molecules using diode lasers [14]. As in any rapidly advancing field, this information becomes outdated quickly and will need to be supplemented with current journal publications. We will concentrate here on the basic laser characteristics and general principles of operating single-frequency, tunable, diode laser systems.

4.2 General Characteristics of Diode Lasers

A wide variety of diode lasers are now commercially available. These range from low-power/high-speed communications lasers to high-power/wide-stripe devices that run multimode (both spatial and temporal) and are used mainly for pumping solid-state lasers. For most scientific applications it is much easier to use

---

*Contribution of NIST, not subject to copyright.

single spatial-mode lasers, and we will concentrate on these devices here. In addition to the usual Fabry–Perot type lasers, distributed feedback (DFB) and distributed Bragg reflector (DBR) lasers are now commercially available at some wavelengths. These lasers have a more complex resonator structure that incorporates an optical grating fabricated within the semiconductor chip. Bragg reflection from the internal grating provides wavelength-selective optical feedback that forces single-longitudinal-mode operation at a wavelength within the reflection bandpass of the grating.

Figure 1 gives a rough sketch of the current (1995) distribution of commercially available continuous wave (CW) single spatial-mode semiconductor lasers. There are gaps in the wavelength coverage (and actual availability) and significant differences in performance across this distribution. Some typical characteristics of semiconductor lasers are outlined in Table I. More details can be found in manufacturer's catalogs and an excellent recent handbook [15].

Another family of semiconductor lasers (outside the scope of the present chapter) is based on group IV–VI elements (most common are Pb-salt lasers). These lasers span the wavelength range from 3.3 to 30 µm and operate at cryogenic temperatures. The lead-salt lasers have not experienced the same degree of commercialization as the room-temperature devices, primarily because they are much more expensive and they have poor mode quality and relatively low output powers (≤5 mW). However, high-quality devices using Sb-based semiconductors that operate at wavelengths that range from ~2 to 4.5 µm have been developed [16]. These promising new devices, unfortunately, are not yet commercially available.

4.2.1 Tuning Characteristics

Often of paramount importance for scientific applications are the tuning characteristics of a laser. A diode laser's wavelength is determined by the semiconductor material and structure, and is a function of both temperature and carrier density. For a typical single-mode laser the wavelength increases monotonically with increasing temperature and then suddenly jumps to another mode at a longer wavelength. Although this jump is most often to the next cavity mode (~0.3 nm cavity mode spacing), it is not at all unusual for one or several modes to be skipped. The tuning range of each mode is typically on the order of 25% of the mode spacing, but might vary by a factor of 5 or more. Subsequently decreasing the temperature causes the operating wavelength to shift downward with similar behavior, but hysteresis will be observed at the mode transitions. Details of the tuning characteristics are available in the general references [1–8] and [17].

Operating a simple diode laser at the wavelength of an atomic or molecular transition usually requires iterative selection of injection current and temperature settings. However, for any given laser the optimum tuning to a specific wavelength cannot always be achieved. Possible alternatives include trying several lasers or

GENERAL CHARACTERISTICS OF DIODE LASERS 79

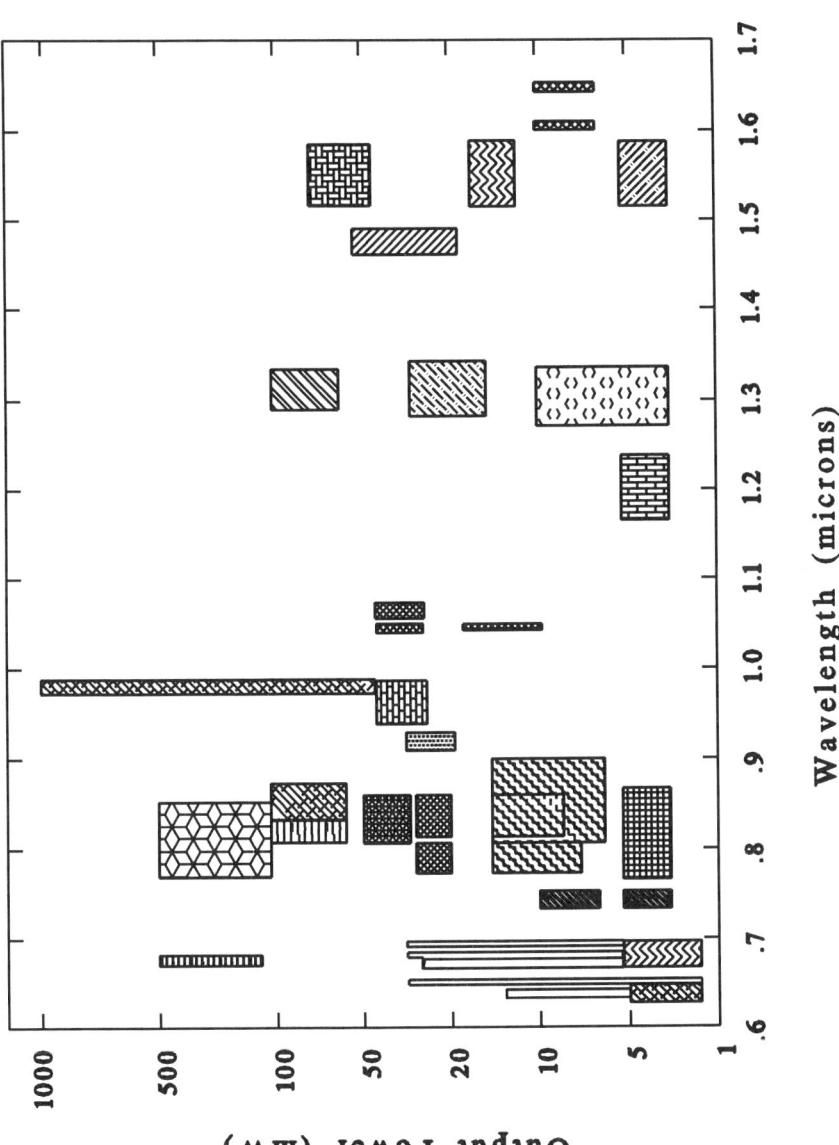

FIG. 1. Power vs. wavelength of commercial single spatial-mode CW room-temperature semiconductor diode lasers. Much of this information has been taken from various literature sources and manufacturer's catalogs and has not necessarily been verified.

TABLE 1. Typical Semiconductor Laser Parameters[a]

Semiconductor	$(Al_xGa_{1-x})_yIn_{1-y}P$	$Al_xGa_{1-x}As$	$Ga_xIn_{1-x}P_yAs_{1-y}$	MOPA
Wavelength	635–670 nm	750–850 nm	1.3–1.5 μm	670, 780–850, 980 nm
Output power (mW)	3–30	5–200	3–100	500–1000
Far-field divergence FWMI (degrees)	8 × 40	11 × 33	30 × 35	0.3 × 35
Waveguide mode dimensions (μm)	4 × 1	3 × 1	1.25 × 1	1 × 100
Astigmatism (μm)	~10	1–5	1–5	~500
Threshold current	30–90 mA	20–60 mA	20–50 mA	~0.5 A
Operating current	50–120 mA	50–200 mA	40–120 mA	~2.5 A
$I_2 = I_1 e^{(T2-T1)/T0}$	$T_0 \approx 100$ K	$T_0 \approx 150$ K	$T_0 \approx 60$ K	
Slope efficiency (mW/mA)	0.5–0.7	0.7	0.2	1
Refractive index	3.1–3.5	3.3–3.6	3.2–3.5	3.3–3.6
Frequency vs. injection current (GHz/mA)	~5	~3	1	
Frequency vs. temperature				
Large-scale (nm/K)	~0.2	~0.25	~0.3	
Small-scale (GHz/K)	~30	~30	~10	
Gain bandwidth (nm)	20	30	50	20
Typical linewidth (MHz)	200	5–20	100	
Alpha factor (α)		3–6	4–8	

[a] These values are representative examples and can vary significantly with structure and composition. The data applies to index-guided single spatial- and spectral-mode Fabry–Perot lasers. Much of this information has been taken from various literature sources and manufacturer's catalogs and has not necessarily been verified.

using optical feedback techniques to control the wavelength. Extended-cavity lasers (Section 4.3) allow tuning to any wavelength within the gain curve of the laser. Another option is to place a small mirror near (~100 μm away) the laser's facet, forming an etalon that acts as a mode selector [18]. An attractive alternative (but improbable for other than a few standard wavelengths) is to find a monolithic DBR laser that operates at the desired wavelength.

4.2.2 Output Power

A typical output power vs. injection current (P–I) curve for a CW diode laser is shown in Fig. 2. The slope of the P–I curve above threshold gives the laser's slope-efficiency (in mW/mA). Semiconductor lasers operate as forward-biased diodes with the voltage drop fixed by the bandgap and an additional series resistance of about 2 to 50 Ω. As expected for a semiconductor diode, the laser's

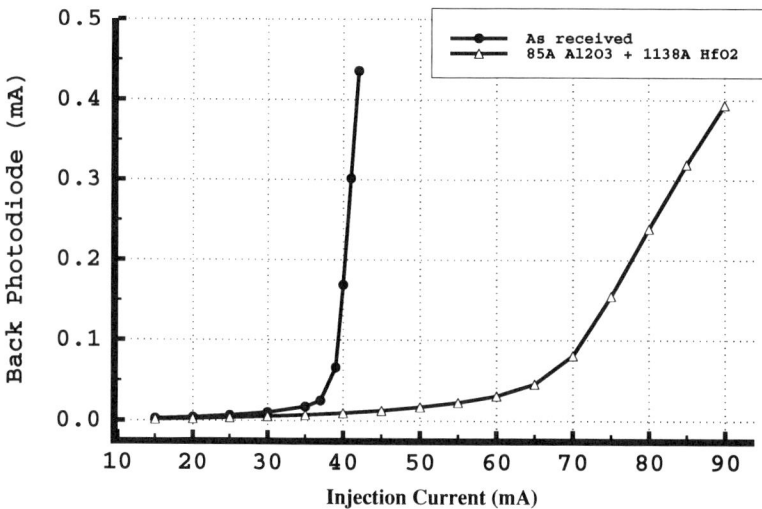

FIG. 2. Output power vs. injection current for a typical AlGaAs diode laser operating near 780 nm. The vertical axis is photocurrent from the photodiode that monitors power out of the laser's back facet. The solid dots show the P–I curve of the standard commercial laser as received ($I_{th} \approx 38$ mA), while the open triangles show the P–I curve of the laser after we coated the front facet with an antireflectance (AR) coating ($I_{th} \approx 68$ mA). As fractionally more power is coupled out the front facet, the slope efficiency of the back facet power is decreased.

operating characteristics are strongly temperature-dependent. Using a simple model for a diode laser [3, 5] gives a threshold gain of

$$g_{th} = a_L + \frac{1}{L} \ln\left[\frac{1}{R_1 R_2}\right], \quad (1)$$

and a total output power above threshold of

$$P_{total} = \frac{\eta_i h\nu(I - I_{th})}{e\, g_{th}\, L} \ln\left[\frac{1}{R_1 R_2}\right], \quad (2)$$

where R_1 and R_2 are the facet reflectances, a_L is the loss internal to the laser, $h\nu$ is the photon energy, I is the injection current, I_{th} is the threshold injection current, η_i is the internal quantum efficiency, e is the electron charge, and L is the laser length. The threshold will vary with temperature as $I_2 = I_1 \exp[(T_2 - T_1)/T_0]$, where

T_0 is a characteristic temperature for the laser. A more precise theoretical treatment allows parameters in these equations to depend on wavelength, carrier density, semiconductor material, and laser structure. If the injection current is increased to excessively high values, the *P–I* curve becomes sublinear, with the output power eventually decreasing due to gain saturation, heating, or damage resulting from high optical powers on the facets. Operation at overly high currents shortens the lifetime and can seriously damage or destroy the laser. An important measure of a laser's operating characteristics is the pumping rate, $R = (I - I_{th})/I_{th}$, which is the injection current above threshold normalized to the threshold value. Generally in the safe operating region of temperature and current, the laser's characteristics improve with pumping rate: the power increases, the linewidth decreases, and the relative amplitude noise decreases.

Recent advances in diode laser technology have produced impressive results in high-power devices based on tapered amplifier designs. By starting with a narrow-stripe single-mode waveguide and then expanding the width of the waveguide out to larger dimensions, the output power can be increased while retaining single spatial-mode operation. Thus, monolithic master-oscillator/power-amplifier (MOPA) systems can be integrated into a single semiconductor chip [19]. The increase in power obtained with tapered amplifiers scales roughly as the ratio of the widths of the output to input waveguide dimensions. Single spatial-mode devices with watt-level output powers are now appearing on the commercial market. Obviously, high-power devices will open up many new possibilities. Of particular interest will be their use with nonlinear optical materials to access other spectral regions.

4.2.3 Beam Quality

The spatial mode produced by diode lasers depends on the laser structure and the collimating optics. Since the output beam diverges rapidly from the small semiconductor waveguide, a high-numerical-aperture lens is needed to collect the light into a collimated beam. Beam quality is not always what we might hope for, but at least in the case of single-mode lasers it can be corrected to be nearly cylindrically symmetric Gaussian. Output beams are typically asymmetric, with a larger divergence in the direction perpendicular to the junction (see Table I). Well above threshold, most are strongly polarized with the electric field parallel to the junction. The modes can also have astigmatism, which manifests itself as the apparent axial separation of the source of rays in the planes parallel and perpendicular to the junction. Astigmatic distances for Fabry–Perot lasers are generally in the range of 1 to 10 µm, while for tapered amplifiers the astigmatism can be ~1 mm, about half the length of the chip. Left uncompensated, this aberration may degrade the beam quality and result in an aberrated far-field intensity distribution. A cylindrical lens can be used to correct the astigmatism, while the asymmetry in divergence is often corrected with a pair of anamorphic prisms that are oriented to angularly magnify the smaller divergence and hence produce a nearly circular beam [20]. Alternatively, a cylindrical telescope can be used to correct the angular asymmetry. In addition, there are sometimes small changes in the spatial mode parameters with large changes in the pumping rate R.

4.2.4 Laser Amplitude Noise

An important attribute of semiconductor lasers is that they typically have much lower amplitude noise then other tunable laser systems. For example, a single-mode laser operating well above threshold might typically exhibit amplitude fluctuations that approach the shot-noise level for frequencies above ~1 MHz. The noise does increases toward lower frequencies with an approximately $1/f$ dependence. A common measure of the amplitude noise is the relative intensity noise: RIN = $\langle \Delta P^2 \rangle / \langle P \rangle^2 \approx [2 \cdot \Delta f S_p(f)]^2 / \langle P \rangle^2$, where Δf is the detection bandwidth, $S_p(f)$ is the spectral density of the power fluctuations, and $\langle P \rangle$ is the mean power. In general, RIN will decrease with increasing pumping rate. If the laser is operated in a multimode regime, the amplitude noise may be much more pronounced, especially when there are only a small number of modes oscillating. Optical feedback can also cause increased amplitude noise.

4.2.5 Optical Feedback

The effects of optical feedback on diode lasers can be profound and depend on a number of factors: the amount of feedback, the facet reflectance, the pumping rate R, and the distance from the laser to the source of feedback (feedback time delay). In some regions of this multidimensional space, the feedback can narrow the linewidth and stabilize the laser's wavelength, while in other regions optical feedback causes such instabilities as erratic mode jumping and coherence collapse [5, 21–24]. In general, optical feedback (even small amounts $P_{\text{feedback}}/P_{\text{out}} \leq 10^{-5}$) from uncontrolled sources should be avoided. For many applications this simply means tilting optical elements to avoid direct reflections. In other cases we are forced to use Faraday isolators, which unfortunately tend to be bulky and expensive for wavelengths less than 1 µm. In some of the next sections we discuss ways to use optical feedback to advantage.

Even though the terminology used to describe diode lasers is inconsistent and imprecise, it may be useful to review some of the nomenclature. For some reason the description "solitary laser" has come to mean a semiconductor diode laser that has not been modified by external effects such as optical or electronic feedback. "External cavity laser," on the other, hand describes a system that lases only because of optical feedback from external optical elements [25]. The name "extended-cavity diode laser" (ECDL) has come to mean something between these two cases; that is, when the laser is operated in a regime where strong feedback from one side of the chip (from an external element such as a grating) is used to control the laser's mode. Sometimes the nomenclature and jargon are more confusing than useful.

4.2.6 Laser Degradation

When properly protected from electrical damage, semiconductor diode lasers are reliable and long-lived. However, there are variations from laser to laser and some variations of laser characteristics with time. Quite a lot of research has been done on the degradation of laser diodes (there is even a book on the subject [26]). However, we still lack detailed understanding of some aspects of laser aging, such as the change in laser wavelength with time. Some characteristic modes of laser degradation are now recognized, including "dark line defect" and facet degradation (which can be caused by gradual photochemical effects or catastrophic optical damage, COD). There seems little that users can do about these changes other than providing good electrical protection and a clean environment for lasers whose facets are open to the air. High currents and high temperatures will increase the rate of degradation. When treated properly, modern lasers have very long natural lifetimes (~10,000 h), but unnatural lifetimes can be quite short!

4.3 Extended-Cavity Lasers

Tuning the wavelength of diode lasers and narrowing their linewidth can be accomplished using some form of dispersive optical feedback [27]. Designs that employ relatively high optical feedback power, $P_{\text{feedback}}/P_{\text{out}} \gtrsim 0.05$ (which is the power fed back into the laser's waveguide mode relative to the output power without feedback), can have a large tuning range and be very stable. In addition to broad tuning, extended-cavity lasers typically have much narrower linewidths than solitary lasers (fast linewidths ~50 kHz as opposed to ~20 MHz). Diffraction gratings are the most commonly used dispersive feedback element, although intracavity etalons, prisms, and birefringent filters have also been used successfully.

Good-quality commercial extended-cavity diode laser systems are now available from a few manufacturers, but distribution in terms of power and wavelength are still somewhat limited. Because of special requirements or cost constraints, users may find it desirable to build their own extended-cavity systems. This can be done relatively simply, as described in a number of papers [7, 12, 28, 29] and as we outline below.

4.3.1 ECDL Construction

There are many good designs for extended-cavity lasers, but no particular design that is optimum for all applications. In Fig. 3 we see five different configu-

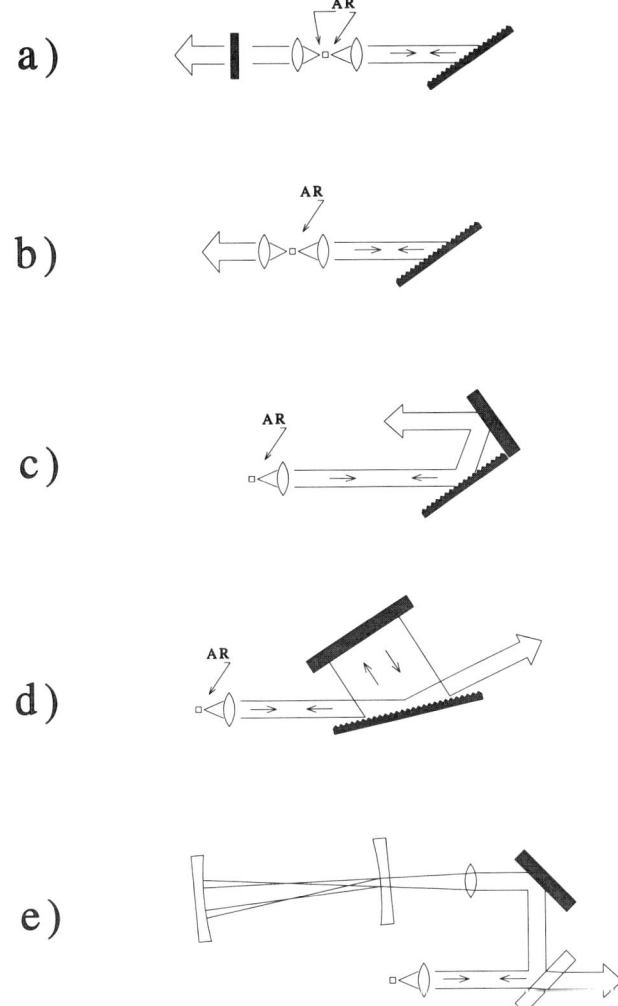

FIG. 3. Various optical feedback configurations. The first design (a) is a two-sided external-cavity design, (b,c) are single-sided extended-cavity designs in Littrow configuration, (d) is an extended-cavity laser that uses a grating in a grazing-incidence configuration, and (e) is a resonant optical-locking configuration using a confocal Fabry–Perot cavity. Design (b) is appealing because there is no movement of the output beam when the grating is rotated to tune the wavelength, and it can also be optimized to deliver high output power. On the other hand, it has the disadvantage that it requires a special laser package (or a rather severe modification of standard laser packages) to gain unobstructed access to both facets. In present usage, the two most common tunable laser designs are (c) and (d) because they are relatively easy to implement using commercial lasers.

rations of optical feedback that can be used to control diode lasers. For optimum performance many of these designs require good antireflection coatings on the output facet of the chip (see Section 4.5).

When designing a laser cavity, it is important to remember that the diffraction efficiency of typical gratings has a strong polarization dependence. For beams polarized perpendicular to the rulings we expect high diffraction efficiency, while for beams polarized parallel the efficiency can be significantly reduced. In the Littrow configuration (Fig. 3b,c), the number of grating lines covered by the laser mode is constrained by the focal length and numerical aperture of the collimating lens, the orientation of the laser spatial mode, and the line spacing and corresponding cutoff wavelength of the grating. Since the beam from a laser waveguide is normally polarized in the direction parallel to the junction, in the far field the laser beam is polarized along the narrow dimension of the spatial mode. When the laser's asymmetric mode is incident on a diffraction grating, there is a compromise between high resolution and high diffraction efficiency. The user has very little freedom of design, which means that the resolution is sometimes not as high as we would like. In some cases it can be advantageous to use a half-wave plate between the laser and the grating to decouple the polarization (hence diffraction efficiency) from the spatial mode orientation. An alternative approach is to use prisms to expand the small direction of the spatial mode [20, 24, 30]. With the one-sided Littrow configuration (Fig. 3c) there is also the disadvantage that the beam moves as the laser's wavelength is tuned. A remedy to the problem of beam movement with tuning (present in the Fig. 3c configuration) is to use a mirror mounted with its surface perpendicular to the grating surface so as to form a retroreflector. Rotating both the grating and mirror together leaves the output beam direction unchanged, although there will be some beam displacement if the axis of rotation is not defined by the intersection of the grating and the mirror. This arrangement is not easily compatible with long continuous scans.

Even with these limitations we often use the standard Littrow configuration (Fig. 3c) because of its simplicity, and most of the time it works reasonably well. For many applications (atomic physics in particular) large tuning ranges are not very important. In practice one can achieve continuous scans of ~2 to 10 GHz by simply arranging the grating pivot point so that the grating angle tunes synchronously with the laser's frequency as controlled by the extended-cavity length. Coarse tuning can be accomplished manually with a fine pitched screw. A typical 800-nm system might consist of an 8-mm focal length collimating lens (N.A. = 0.5), an 1800-line/mm grating in Littrow configuration with a cavity length of about 5 cm. In a typical Littrow system we normally orient the laser mode so that the larger dimension is orthogonal to the direction of the grating rulings. This orientation gives better resolution and better output power than the opposite orientation.

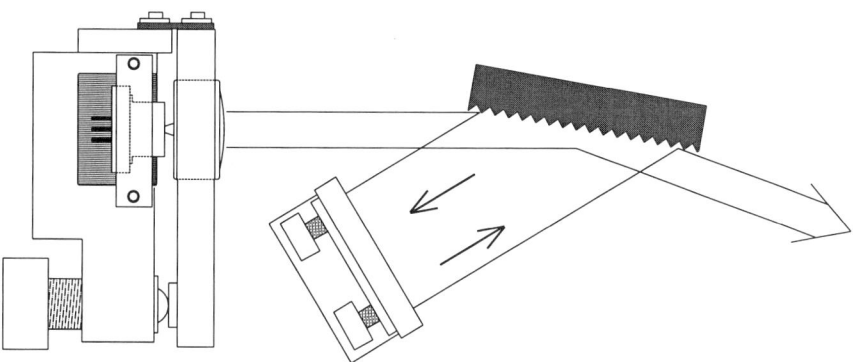

FIG. 4. Diagram of a simple grazing-incidence grating-tuned extended-cavity diode laser. This extended-cavity system is constructed of three basic units: the laser mount, the grating mount, and the feedback mirror. The diode laser, Peltier cooler, and collimating objective are all integrated together in a stable laser mount that uses a flexure and fine-pitched screw for precise focus adjustment. The feedback mirror and the angle of the grating are adjusted relative to laser beam direction to provide just enough optical feedback power for stable single-mode operation.

The grazing incidence arrangement (Fig. 3d) as proposed by Harvey and Myatt [31] was shown to have the important advantages of significantly higher spectral resolving power and no movement of the output beam when the laser is tuned. This design, implemented with diode lasers by Harvey, is often called the Littman (or Littman–Metcalf) configuration because they introduced the use of grazing-incidence gratings to control dye lasers [32]. Good discussions of the grazing incidence design have also been given by Day [33]. A simple mechanical layout for a grazing-incidence grating-tuned ECDL is shown in Fig. 4.

In designing all of these systems we need to address a few basic requirements. We need to precisely align the laser cavity, to tune the wavelength, and to precisely control the laser chip temperature and extract excess heat. Care must be taken to ensure a stable thermal and mechanical structure by using good materials and kinematic design principles. Ironically, ECDLs provide narrow spectral linewidths, but they are also much more susceptible to external perturbations than were the solitary lasers that we started with! For stable single-frequency operation, ECDLs need to be isolated from vibrations and pressure fluctuations. A useful principle to keep in mind in this design is: *if something can move, it will*. Allowing a large dynamic range of motion competes directly with stability.

The simple optomechanical design sketched in Fig. 4 is an example of a versatile system that satisfies most (but never all) of our basic requirements. It consists of a copper baseplate (not shown) that acts as a rigid backbone for the laser resonator and at the same time serves as the heat sink for the temperature control system. A commercial laser is mounted in a small copper fixture that bridges the Peltier cooler and attaches to the laser base mount (also copper). The base mount is then rigidly attached to one end of the baseplate. The collimating lens is connected to the laser base mount by a stiff spring-steel flexure that is clamped in place using locating pins and multiple screws. The lens is mounted in an eccentric ring that is clamped in place after initial coarse alignment. The transverse alignment can be set accurately enough with careful hand alignment and visual inspection of the beam direction and shape. In the example of Fig. 4, the eccentric on the lens mount is used to adjust the vertical height of the lens, while the horizontal position is set by the transverse displacement of the laser on the base mount. Critical adjustment of the focus is done with a high-quality fine-pitched screw that translates the lens against the restoring force provided by the flexure. We typically use 6.3-mm diameter screws with 3.1 threads/mm (1/4-80 screw). The lever arm from the flexure pivot point (Fig. 4) gives further drive reduction. This simple design tilts the lens slightly as it is translated for focusing; however, this is not a serious limitation because the lens can be set very close to the optimum position before it is clamped in place. In constructing this system we tolerate the thermal expansion of copper in order to have the good thermal conductivity and low mechanical Q. For very-high-frequency stability the cavity length can be controlled with a PZT. For some high-power lasers we use water cooling in the laser base mount. The water can be circulated without pump vibrations by using a simple thermal syphon.

One of the requirements for stable operation of ECDLs is to return sufficient optical feedback power to the diode laser waveguide mode. The optimal feedback power depends on the characteristics of the laser and, in particular, the reflectance of the output facet. Typically feedback power ratios ($p_{\text{feedback}}/p_{\text{out}}$) of from 5 to 50% are appropriate. In grazing-incidence systems the feedback mirror and the angle of the grating (relative to the laser beam direction) are adjusted to provide just enough feedback power for stable single-mode operation (typical incidence angles for the grating might be ~75–85°). Good-quality high-numerical-aperture objectives (N.A. ~ 0.5) are recommended for the collimating lens. In most cases it is not necessary to correct for the laser's astigmatism for the laser to function properly in extended-cavity mode. However, the output beam will have some aberration if only spherical optics are used. Objectives lenses specifically designed for diode laser collimation often assume that there is a thin window between the laser chip and the lens. For best mode quality, the lens design needs to take into account any window on the laser mount.

4.3.2 Alignment

The optical alignment of extended-cavity lasers requires precision, but with practice alignment can be accomplished in a simple common-sense manner. Infrared-sensitive cards that fluoresce in the visible are extremely useful when working with near-IR lasers. Electronic IR viewers or CCD cameras can also be used. For the initial alignment, the objective lens is carefully centered on the laser's output so that the beam propagates in the same direction as the unperturbed laser emission. The beam should appear roughly collimated over a distance of several meters. The grating is then positioned so that the first-order diffracted beam returns to the laser (taking care that the grating blaze is toward the lens). A card with a small aperture held in front of the lens is useful to roughly position the returned beam on the outgoing beam. Fine tuning of the focus and alignment will then be necessary to obtain stable single-mode operation.

The method that we use most often to align ECDLs was derived from work describing the output power-vs.-injection current characteristics of ECDLs [21, 34, 35]. The method is simple and requires only monitoring the output power as a function of the swept injection current. Following the coarse alignment procedure described above and with the laser operated near threshold, a triangular ramp is applied to the injection current. Monitoring the output power of the ECDL with a large-area photodiode and an oscilloscope (synchronized to the triangular wave) yields a classic diode laser P–I curve with no optical feedback (see Fig. 2). When there is feedback from the extended cavity, there will be abrupt discontinuous changes in threshold behavior. Iterative adjustment of the focus and extended-cavity alignment will result in oscilloscope traces similar to those in Fig. 5.

4.3.3 Tuning

The maximum tuning range of an ECDL depends on the design of the cavity and on the detailed characteristics of the diode laser chip. Factors that affect the tuning include: the facet reflectances, the degree to which the spectral broadening is homogeneous, the side-mode suppression ratio, the pumping rate R, the spatial mode control (how well the mode is index guided), and the propensity for nonlinear coupling between modes (for example, four-wave mixing). As users, the facet reflectance is almost the only parameter that we have any control over. If the reflectance of the laser facet that faces the grating is very low ($<10^{-4}$), the maximum tuning range can be as large as the laser's gain curve (approximately ±20 nm for AlGaAs lasers and even larger for InGaAsP lasers). Continuous single-mode scans over many nanometers require very low facet reflectance and synchronization of the tuning of the grating passband with change in cavity length.

4.4 Electronics

The electronics required to operate semiconductor lasers are relatively straightforward and can be very simple if we do not require precise tuning. Since the

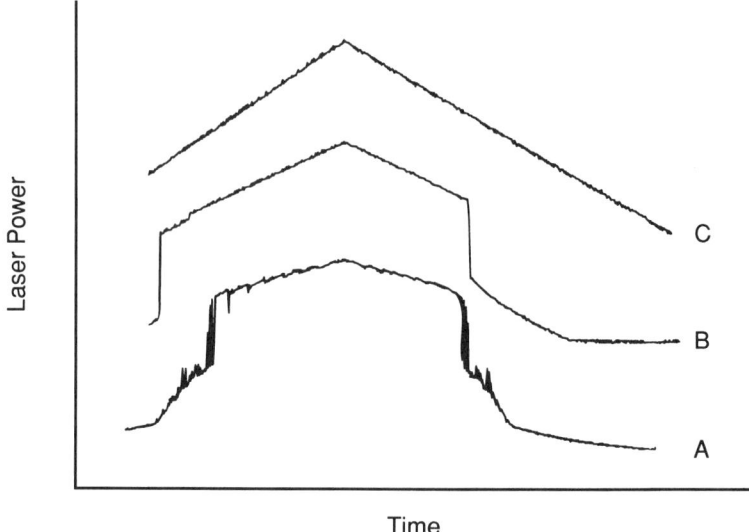

FIG. 5. Output power of ECDL vs. injection current, which is swept near threshold with a triangle wave. Three different P–I curves are shown for different cavity alignments; the curves have been offset vertically for clarity. The horizontal axis (labeled here as time) is synched with a triangle wave ramp, with maximum injection current corresponding to maximum output power. The characteristic features of the P–I curve are a good indicator of the quality of the extended-cavity alignment. As the focus of the collimating lens and feedback from the grating are aligned, the output power increases in abrupt steps. The rounded corners and noise on curve A corresponds to poor cavity alignment and are indicative of multilongitudinal mode operation. Curve B shows the type of abrupt jumps that are indicative of single-mode operation. Curve C is near-optimum alignment, where the injection current has very little effect on the laser's frequency and no mode jumps are apparent.

laser's output power and frequency depend sensitively on temperature and injection current, we need to control both with some precision. The typical laser parameters in Table I can be used to estimate the stability and noise performance that will be required of the current source and temperature controller for a given application.

4.4.1 Current Sources

The most important requirement of current sources used for diode lasers is that they **MUST BE FREE** of electrical transients that can seriously damage the laser. CW narrow-stripe lasers require an injection current of about 20 to 200 mA and have a forward voltage drop of about 2 V. In contrast, some of the high-power devices—such as wide-stripe lasers, MOPAs, and diode laser arrays—might require a few amperes of injection current. The simplest diode laser current supply is just a battery and a current-limiting resistor (or potentiometer). In practice we usually want to do more than just turn the laser on, and so the current source becomes somewhat more complicated. A number of good diode laser current sources are now commercially available. These range from simple integrated circuit chips that are designed for low-cost consumer electronics to very elaborate microprocessor-controlled systems. If a laboratory with a limited budget is planning on operating a significant number of precision diode laser systems, it can be cost-effective to produce some simple circuit boards for current sources and temperature controllers. Following this approach, we have found a few general principles to be useful.

- When possible do not put the AC-to-DC power converter in the same box with the current source electronics; very good filtering and regulation of the DC supplies are also important.
- It is important to keep the diode laser current path, both to and from the laser, independent of other signal and return paths.
- Pay careful attention to avoiding spikes on startup, shutdown, and even power failures. Before endangering a valuable laser, time is well invested in testing for transients (under all conceivable operating conditions) with an inexpensive imitation laser load (such as 10 LEDs in parallel and these in series with a 1 Ω current sensing resistor to ground; you can also use one or more rectifier diodes in place of the LEDs).
- To protect against damage caused by transients and operator errors, it is important to attach some passive protection near the laser chip (a typical protection circuit is shown in Fig. 6). Protection circuits can also have adverse effects, such as slowing down modulation response and increasing thermal instability due to diode leakage current. This last effect can be eliminated when necessary by controlling the temperature of the protection diodes by placing them on the laser heat sink.
- Watch for potential ground loop problems, particularly when connecting multiple cables between the laser, current source, sweep source, temperature controller, optical table, detectors, oscilloscope, etc. Our approach has been to keep the laser heat sink at ground potential but to isolate it from all other grounds (including that of the box that we use to enclose the laser), except at the laser current source where there is a common ground.
- Look for other useful hints and examples in catalogs supplied by the laser manufacturers.

An example of a good-quality diode laser current source is in Fig. 6. Other examples can be found in the literature [36, 37]. This circuit is designed to have good stability, low noise, and modulation capability.

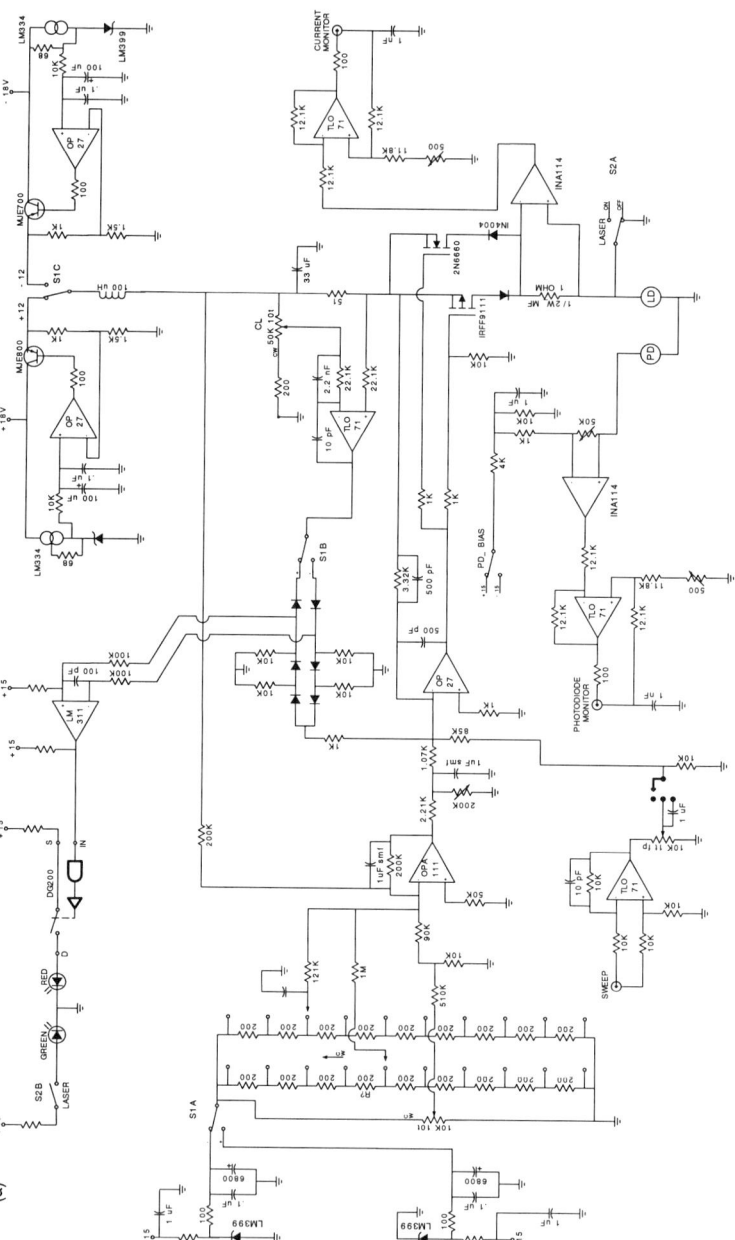

FIG. 6. Diode laser current source (a) and laser protection circuit (b). The operation of the current source is based on a summing amplifier that balances the voltage drop across a sense resistor (R_s) with a voltage supplied by a precision variable-voltage reference. To keep the diode laser mount at ground potential, the sense resistor is placed between the summing amplifier output and the supply rail. A typical modulation (sweep) input is also shown as a signal that adds to the summing amplifier input. A variable current limit (CL) sets the maximum current that can be supplied to the laser (a red LED indicates that the current limit is reached). A three-pole switch (S1A–C) changes the source polarity. Photodiode bias polarity is set by PD-BIAS. When the laser is off, a short (S2A) is connected across the laser. The laser protection circuit is located at the laser mount and consists of a simple low-pass filter and clamping diodes.

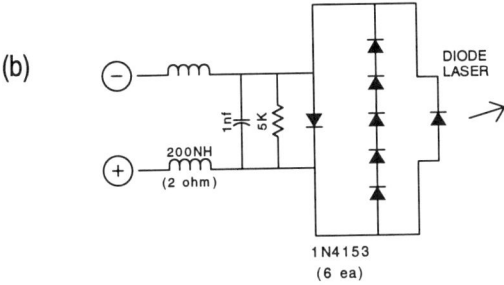

Fig. 6b

4.4.2 Temperature Controllers

The high sensitivity of diode laser wavelengths to temperature (~30 MHz/mK) places stringent requirements on their temperature control. For fixed-wavelength operation, solitary lasers require better temperature stability than ECDLs. With ECDLs the temperature stability is more of a concern because of thermal expansion of materials than it is because of the temperature sensitivity of the laser's gain. In practice the most difficult task in maintaining temperature stability is not the electronics, but rather it is the design of the thermal environment. A stable thermomechanical environment will reduce problems associated with such important, but often less obvious, perturbations as: beam-pointing stability (remember the angular magnifying properties of the short-focal-length collimating objective) and fluctuations of even the small optical feedback that is invariably present. Putting ECDLs in hermetically sealed containers greatly improves both thermal and optomechanical stability.

Important tradeoffs have to be balanced in designing temperature control systems for diode lasers. For instance, we would like to have the thermistor very close to the laser to accurately measure the laser's temperature, but we also want the thermistor close to the thermoelectric transducer so that the time delay in the servo is small and the loop can have fast response and high gain. These considerations indicate that a very small thermal mass should be used to support the laser. But the mount should also be mechanically stable and have a large thermal mass so that rapid temperature fluctuations (outside the loop bandwidth, e.g., from moving air) do not perturb the laser's temperature. For practical reasons we would also like the entire system to be relatively light and compact. Obviously, compromises are in order.

Figure 7 depicts an example of a temperature controller that uses a Peltier cooler to control the temperature of a diode laser; others can be found in the literature [12, 29]. An excellent analysis of the use of Peltiers in temperature control systems has been provided by Van Baak [38]. As with current sources, there are some basic design principles that are useful in constructing thermal control systems:

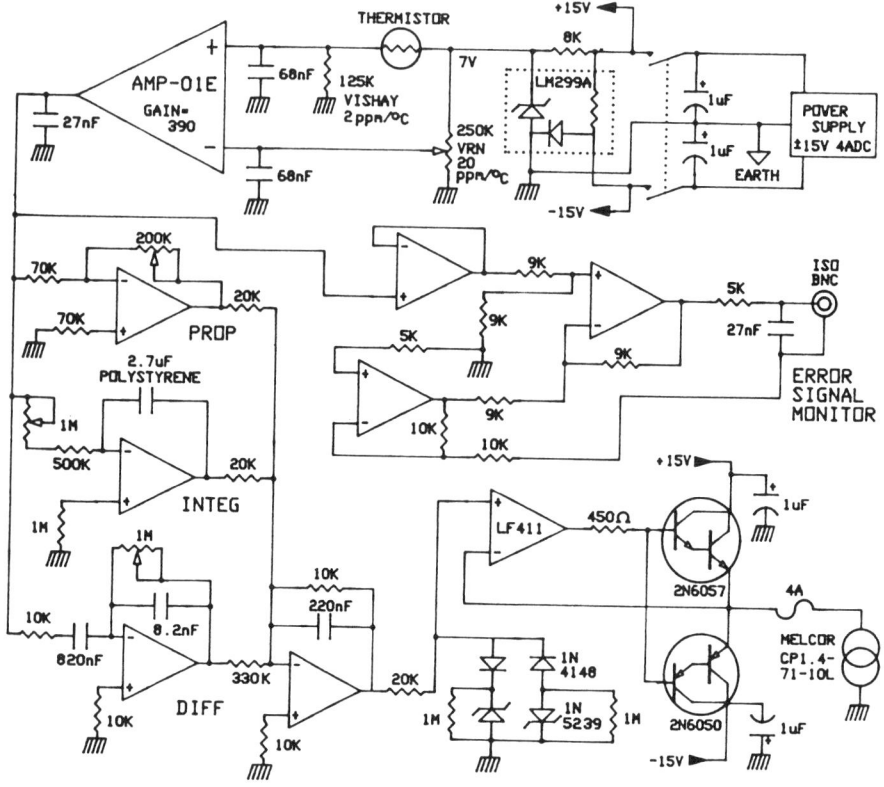

FIG. 7. Diode laser temperature controller. A precision thermistor (nominal value 125 kΩ) is used in a resistance bridge to sense the laser's temperature, and a Peltier (thermoelectric) cooler is used for cooling or heating. The servo-loop is a PID (proportional integral differential) control that processes the error signal and then sends it to a driver amplifier which provides sufficient power to run the Peltier cooler (typically a few volts and a few amperes). This circuit is from reference [36].

- Two-stage thermal control can be very effective in improving the stability of ECDLs. One controller is used for the extended cavity and the other for the laser chip.

- Thermistors should be buried in holes below the surface (& 5 mm) to avoid perturbations from temperature differences with the surrounding air.

- High-thermal-expansion materials (e.g., plastics, aluminum, copper) require better temperature control to maintain mechanical stability and alignment. Invar and fused silica can be useful for some components.

- The combined thermomechanical properties of the materials and construction must be considered. This includes the dynamic response of the material used within the thermal control loop (aluminum, copper, and silver have large coefficients of thermal diffusivity, which is good for control systems, but they also have large coefficients of thermal expansion, which is bad for mechanical stability).
- Peltier coolers are not very efficient, so a relatively large thermal reservoir may be required to dissipate excess heat.
- Watch out for thermal drifts and emfs in the electronics. Even good-quality resistors and capacitors can have significant temperature coefficients. Potentiometers can be particularly troublesome.

4.5 Optical Coatings on Laser Facets

Operating a diode laser in an extended-cavity configuration works best if the reflectance of the laser's output facet is very low. ECDLs are inherently coupled cavity systems and are known to have operating regions of stability and instability [21–23]. By reducing the reflectance of the output facet the performance of the ECDL is improved: the size of the stable operating region (in temperature, current, and feedback power) is increased, the usable output power is increased, and, most importantly, the laser's tuning range is increased.

In Eq. (1) we see the strong dependence of the laser threshold current on the reflectance of the facets. This is illustrated in Fig. 2, which shows the P–I curves of a laser before and after the output facet is antireflection (AR) coated. Almost all commercial lasers come from manufacturers with coatings already on the laser's facets. These coatings serve two purposes: first, they protect the facets from degradation, and, second, they adjust the facet reflectance to optimize the output power. Since the early days of diode lasers, Al_2O_3 coatings have been used to reduce facet degradation and increase the lifetime of diode lasers. We typically find that commercial low-power (P . 10 mW) AlGaAs lasers have single-layer coatings on both facets that are approximately $\lambda/2$ in optical thickness. In most cases we do not know with certainty what coating material has been used. Empirically we often (but not always) find that the laser coatings have an index of refraction of about 1.60, which is consistent with e-beam vapor-deposited Al_2O_3. Higher-power lasers (& 15 mW) typically have a similar coating on the output facet, except that the thickness is approximately $\lambda/4$ (which results in a reflectance of a few percent). The back facet of the high-power lasers typically has a multilayer coating with a reflectance of 90% or more.

Most commercial lasers can be improved for use in extended-cavity configurations by reducing the reflectance of the output facet. The majority of laser manufacturers are not interested in putting special coatings on diode lasers for those users who need a few inexpensive lasers. In order to transform general-purpose lasers into ECDLs, we have found it advantageous to develop some simple coating capabilities in our own lab. Fortunately, the equipment needed to coat diode lasers is readily available from the optical coating/semiconductor industry. For example, inexpensive (can opener–like) tools are available for removing the caps from the standard hermetically sealed semiconductor packages. Otherwise, all that is really needed is a simple vacuum coating system with electrical feedthroughs.

Our coating techniques are based on traditional dielectric coating methods but are directed toward the specific problem of modifying the facet reflectance of semiconductor lasers. We generally use common coating materials (e.g., Al_2O_3, HfO_2, SiO) and well-established techniques (thermal evaporation, electron beam deposition, and RF sputtering). If one does not want to spend time characterizing the unknown coatings on the lasers and carefully calibrating thickness monitors, it is easiest to do the coatings by monitoring the change in threshold while the laser is being coated. During the coating process the laser injection current is ramped through threshold with a repetitive triangular sweep and the power out of the back facet is monitored. Many lasers already have a monitor photodiode behind them that can be used for this purpose.

Diode laser coatings have been studied in detail, and good-quality AR and mirror coatings have been developed. Some very useful information can even be found in the published literature, but often details are lacking because they are proprietary in nature. The lowest-reflectance coatings that have been reported on diode lasers range from 10^{-4} to 10^{-5}. These results have often been achieved by laboratories working on optical amplifiers, in which case it is necessary to suppress lasing altogether while maintaining very high optical gain.

Two relatively simple AR coatings that work reasonably well on commercial lasers are: (i) thermally evaporated silicon monoxide and (ii) e-beam-deposited HfO_2 and/or Al_2O_3. Either of these coatings can achieve reflectances below 10^{-3}, even on commercial lasers that already have some coating on their facets. Silicon monoxide is convenient because the equipment required for thermal evaporation is relatively simple. It also has the useful (but challenging) property that changing the oxygen pressure in the coating chamber changes the oxygen composition (x) in the film (SiO_x), and hence the index of refraction from about 1.6 to 2.0. With all coatings, but with SiO_x in particular, the apparent index of refraction of the coating can change over time as the laser is exposed to and operated in air.

An AR coating that we use regularly for semiconductor lasers is a dual coating of Al_2O_3 and HfO_2. We apply this coating using a standard electron beam evaporation source. Both of these materials produce good-quality optical coatings that are compatible with most of the commercial lasers that we have tested. To reduce the facet reflectance of commercial lasers we first add Al_2O_3 (when needed) to bring the laser's base coating thickness to approximately $\lambda/2$, and then we put down a λ/4 layer of HfO$_2$ (deposited index approximately 1.89). The optical thickness of coatings is monitored by watching the change in laser threshold while the laser is being coated. This particular two-layer coating is relatively easy to monitor because the threshold reaches an extremum at or near the proper thickness for both layers.

4.6 Diode Laser Frequency Noise and Stabilization

We draw a clear distinction between two aspects of laser frequency stabilization. The first concerns the reduction in the drift and low-frequency fluctuations (jitter) of the laser's center frequency, and the second concerns the faster fluctuations that are responsible for the "fast linewidth" of the laser. The general aspects of frequency stabilization of lasers are addressed in other chapters (Zhu and Hall, Chapter 5 in this volume) and elsewhere. We will concentrate here on those aspects particularly relevant to diode lasers [5–7]. An excellent discussion of the subject is given by Telle [39].

First, the optical cavity of a solitary diode lasers is very small and the facet reflectance is not very high, which means that the Q of the resonator is low and the resulting linewidth of the laser is relatively large (typically 10–200 MHz). In addition, a number of physical processes affect the spectral width of diode lasers. These include carrier density and temperature fluctuations, pump fluctuations, and spontaneous emission. These effects can be amplified by the strong coupling between the phase and amplitude of the lasing field. This coupling causes the linewidth of diode lasers to be increased above the fundamental Schawlow–Townes limit by a factor of $(1 + \alpha^2)$, where α (typically 3–10) is the ratio of the change in the real to the imaginary part of the susceptibility of the gain medium.

4.6.1 Electronic Control

An inexact but useful rule of thumb for frequency control servo systems is that the bandwidth of the servo needs to be somewhat greater than the linewidth of the laser before the servo can actually reduce the laser's linewidth. Solitary lasers have linewidths that are typically tens of MHz wide, which means that it is difficult, but not impossible, to achieve servo-loop bandwidths that are high enough to narrow the linewidth. Even though the frequency of the laser light responds on nanosecond time scales to changes in the injection current, other factors limit the loop response time. Very fast electronics must be used, and the physical size of the system must be small to minimize propagation delays. It is necessary to compensate phase shifts in the transfer function of injection current to laser frequency and also the time delays inherent in the frequency discriminator (for example, a Fabry–Perot cavity).

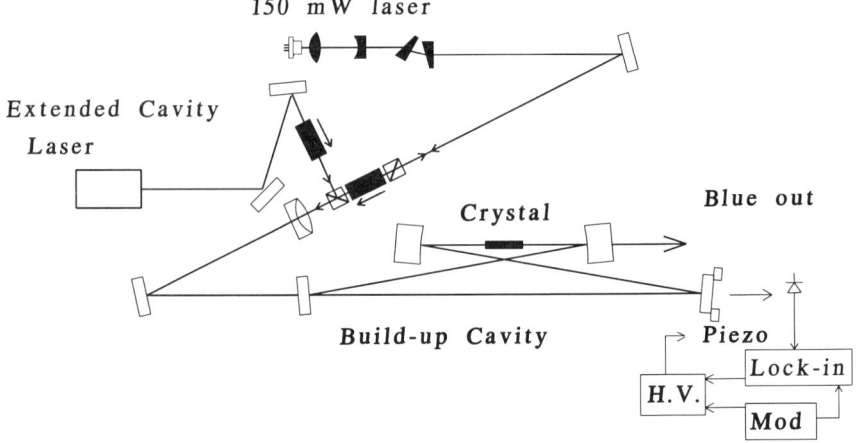

FIG. 8. Frequency-doubled diode laser. In this system a grating-tuned ECDL provides tuning and high spectral purity, which is then used to control the frequency of a higher-power diode laser by injection locking. Injection about 5 mW from the ECDL into the slave laser provides about 150 mW of tunable output power. The spatial mode of the slave laser is asymmetric ($\theta_\perp/\theta_\parallel \sim 3$) and has a small amount of astigmatism (about 3 μm); these are corrected by using an anamorphic prism pair and a cylindrical lens. After spatial mode correction and passing through an optical isolator, there are about 105 mW remaining in the beam that is incident upon the ring buildup cavity that contains the $KNbO_3$ crystal. The combined mode matching and coupling efficiency of this beam into the ring is about 75%. Usable blue output powers as high as 45 mW are obtained in a good stable spatial and temporal mode near 425 nm.

4.6.2 Optical Control

Resonant optical feedback from a confocal cavity (Fig. 3e) can be used to narrow the linewidth and simultaneously stabilize the center frequency of diode lasers [2]. With this method optical feedback occurs only in the narrow spectral windows defined by the confocal Fabry–Perot resonances. When the solitary laser's frequency is tuned within the optical capture range of a Fabry–Perot resonance, the resulting resonant feedback causes the laser to lock its frequency to the nearby cavity resonance. The laser's fast linewidth is thus reduced by a factor of about 1000 (to <10 kHz), and the center frequency is then automatically stabilized to the cavity resonance. This system does not require modification of the solitary diode laser and typically operates in the weak feedback regime, $P_{\text{feedback}}/P_{\text{out}} \cdot 10^{-3}$. A limitation of this optical locking technique is that the tuning range of the laser is still restricted to those wavelengths accessible by the solitary laser alone. Although no active control is needed to optically lock a laser to a cavity, electronic servos will be necessary for stable long-term operation at a specific frequency.

4.6.3 Combined Optical and Electronic Control

As with most tunable lasers, the lasing frequency of ECDLs is controlled by a mechanical system (the grating angle and cavity length) that is susceptible to perturbation and drift. Active frequency stabilization can be done using some sort of frequency reference (such as a Fabry–Perot cavity or an atomic resonance) and an electronic control system. The error signal from the frequency reference is processed by a loop filter and fed back to control the laser frequency. The main feedback path is usually to a PZT that controls the laser cavity length. The frequency stability that can be achieved using only a PZT loop is usually adequate for most applications, but when linewidths less than about 50 kHz are needed it is relatively easy to use an additional feedback loop to the injection current (or to an intracavity modulator) to further narrow the laser linewidth. These additional feedback paths have higher servo bandwidths because they are not limited in speed by having to move the mass of a mirror or grating. With servo bandwidths of ≥1 MHz it is even possible to phase-lock a diode laser to a reference laser source.

4.7 Extending Wavelength Coverage

Present technology is now reasonably good at controlling the spectral characteristics of diode lasers, but these lasers may not be available at the wavelengths that we require (Fig. 1). Fortunately, new wavelengths are on the horizon. There are promising research results with blue lasers using ZnSe and GaN, and the Sb-based IR lasers also have great potential. In the meantime, we can use the existing diodes and nonlinear optical techniques to extend wavelength coverage.

Good results have been reported by using diode lasers and difference-frequency generation (DFG) to produce single-frequency tunable infrared light [40]. The chalcopyrite crystals $AgGaS_2$ and $AgGaSe_2$ and periodically poled lithium niobate are well suited to this application because they have large nonlinear coefficients and they phase match for diode laser wavelengths. In fact, it appears that almost the entire spectral region between 2 and 18 μm could be covered by using these crystals and diode lasers as the input.

In the opposite wavelength direction there is a need for good tunable lasers in the blue and UV based on solid-state sources. Fortunately, nonlinear crystals such as $KNbO_3$, KTP, $LiNbO_3$, BBO, and $LiIO_3$ can be used for second-harmonic generation (SHG) and sum frequency mixing of diode laser light [41]. With these techniques and materials the region between 200 and 500 nm is accessible. For example, the SHG system developed in our lab by C. Weimer is diagrammed in Fig. 8. This system uses $KNbO_3$, which is a particularly good example because it has a large nonlinear coefficient, and it noncritically phase matches for AlGaAs diode laser wavelengths. Starting with an injection-locked 150 mW diode laser, this system produces more than 40 mW of tunable light near 425 nm. An analogous system that uses angle phase matching in $LiIO_3$ produces about 200 µW of 405 nm light for an input power of 40 mW at 810 nm.

From recent publications and conference proceedings we can anticipate that higher-power lasers will soon be available in many wavelength regions. As the technology improves we will be able to incorporate these lasers and MOPAs with various nonlinear optical materials to greatly extend useful wavelength coverage.

Acknowledgments

Our thanks to V. L. Velichansky, C. W. Weimer, N. Mackie, T. Zibrova, H. G. Robinson, J. Marquardt, S. Waltman, M. Stephens, and L. Mor for contributions to this work.

References

Given the length and general nature of this chapter, we provide only a very basic list of references that might serve as a starting point for further study.

1. Kressel, H., and Butler, J. K. (1977). *Semiconductor Lasers and Heterojunction LEDs*, Academic Press, New York.
2. Casey Jr., H. C., and Panish, M. B. (1978). *Heterostructure Lasers, Parts A and B*, Academic Press, New York.
3. Thompson, G. H. B. (1980). *Physics of Semiconductor Laser Devices*, Wiley, New York.
4. Yariv, A. (1989). *Quantum Electronics*, 3rd edition, Wiley, New York.
5. Petermann, K. (1988). *Laser Diode Modulation and Noise*, Kluwer Academic, Dordrecht, The Netherlands.
6. Yamamoto, Y., ed. (1991). *Coherence, Amplification, and Quantum Effects in Semiconductor Lasers*, Wiley, New York.
7. Ohtsu, M. (1992). *Highly Coherent Semiconductor Lasers*, Artech House, Boston.
8. Zory, P. S., ed. (1993). *Quantum Well Lasers*, Academic Press, Boston.
9. Camparo, J. (1985). *Contemp. Phys.* **26**, 443.
10. Ohtsu, M., and Tako, T. (1988). In *Progress in Optics*, Volume 25, E. Wolf (ed.), Elsevier, Amsterdam.
11. Lawrenz, J., and Niemax, K. (1989). *Spectrochim. Acta* **44B**, 155.
12. Wieman, C., and Hollberg, L. (1991). *Rev. Sci. Instrum.* **62**, 1, and the references therein.
13. Tino, G. (1994). *Phys. Scripta.* **T51**, 58–66.

14. Niemax, K., ed. (1993). *Spectrochim. Acta Rev.* **15**(5).
15. Suematsu, Y., and Adams, A. R., eds. (1994). *Handbook of Semiconductor Lasers and Photonic Integrated Circuits*, Chapman and Hall, London.
16. Choi, H. K., Walpole, J. N., Turner, G. W., Eglash, S. J., Missaggia, L. J., and Connors, M. K. (1994). *IEEE Photon. Tech. Lett.* **6**, 7, and the references therein.
17. Franzke, J., Schnell, A., and Niemax, K. (1993). *Spectrochim. Acta Rev.* **15**(5), 379–395; see also general diode laser references [1–7, 14].
18. Bonnell, L. J., and Cassidy, D. T. (1989). *Appl. Opt.* **28**, 4622–4628; Ruprecht, P. A., and Brandenberger, J. R. (1992). *Opt. Comm.* **93**, 82–86.
19. Welch, D. F., Parke, R., Mehuys, D., Hardy, A., Lang, R., O'Brien, S., and Scifres, D. S. (1992). *Electron. Lett.* **28**, 2011–2013.
20. Kasuya, T., Suzuki, T., and Shimoda, K. (1978). *Appl. Phys.* **17**, 131–136.
21. Lang, R., and Kobayashi, K. (1980). *IEEE J. Quantum Electronics* **QE-16**, 347–355.
22. Tkach, R., and Chraplyvy, A. (1986). *J. Lightwave Tech.* **LT-4**, 1655–1661.
23. Tromborg, B., and Mørk, J. (1990). *IEEE J. Quantum Electronics* **26**, 642–654; Mørk, J., Tromborg, B., Mark, J., and Velichansky, V. L. (1992). *Proc. SPIE* **1837**, and the references therein.
24. Zorabedian, P. (1994). *IEEE J. Quantum Electronics* **30**, 1542, and the references therein.
25. Fleming, M. W., and Mooradian, A. (1981). *IEEE J. Quantum Electronics* **QE-17**, 44–59.
26. Fukuda, M. (1991). *Reliability and Degradation of Semiconductor Laser and LEDs*, Artech House, Boston.
27. Belenov, E. M., Velichansky, V. L., Zibrov, A. S., Nikitin, V. V., Sautenkov, V. A., and Uskov, A. V. (1983). *Sov. J. Quantum Electronics* **13**, 792; Akul'shin, A., Bazhenov, V., Velichansky, V., Zverkov, M., Zibrov, A., Nikitin, V., Okhotnikov, O., Sautenkov, V., Senkov, N., and Yurkin, E. (1986). *Sov. J. Quantum Electronics* **16**, 912.
28. Boshier, M. G., Berkeland, D. J., Hinds, E. A., and Sandoghadar, V. (1991). *Opt. Comm.* **85**, 355–359.
29. MacAdam, K. B., Steinbach, A., and Wieman, C. (1992). *Am. J. Phys.* **60**, 1098–1111.
30. Duarte, F. J. (1993). *Laser Focus World*, pp. 103–109, February
31. Harvey, K., and Myatt, C. (1991). *Opt. Lett.* **16**, 910.
32. Liu, K., and Littman, M. G. (1981). *Opt. Lett.* **6**, 117; McNicholl, P., and Metcalf, H. J. (1985). *Appl. Opt.* **24**, 2757.
33. Day, T., Luecke, F., and Brownell, M. (1993). *Lasers Optronics*, pp. 15–17, June; Day, T., Brownell, M., and Wu, I.-Fan (1995). *Proc. SPIE* **2378**, 37–41.
34. Velichansky, V. L., Zibrov, A. S., Molochev, V. I., Nikitin, V. V., Sautenkov, V. A., Tyurikov, D. A., and Kharisov, G. G., (1981). *Sov. J. Quantum Electronics* **11**, 1165–1171.
35. Bazhenov, V. Yu., Bogatov, A. P., Eliseev, P. G., Okhotnikov, O. G., Pak, G. T., Rakhvalsky, M. P., Soskin, M. S., Taranenko, V. B., and Khairetdinov, K. A. (1982). *IEEE Proc.* **129**(1), 77–82.

36. Bradley, C. C., Chen, J., and Hulet, R. G. (1990). *Rev. Sci. Instrum.* **61**, 2097.
37. Libbrecht, K. G., and Hall, J. L. (1993). *Rev. Sci. Instrum.* **64**, 2133–2135.
38. Van Baak, D. A. (1992). *Am. J. Phys.* **60**, 803–815.
39. Telle, H. (1993). *Spectrochim. Acta Rev.* **15**, 301–327.
40. Simon, U., Waltman, S., Loa, I., Tittel, F. K., and Hollberg, L. (1994). *J. Opt. Soc. Am. B* **12**, 323–327; Petrov, K. P., Waltman, S., Dlugokencky, E. J., Arbore, M., Fejer, M. M., Tittel, F. K., and Hollberg, L. (1997). *Appl. Phys. B*, and the references therein, to be published.
41. Risk, W. P., and Kozlovsky, W. J. (1992). *Opt. Lett.* **17**, 707–709, and the references therein.

5. FREQUENCY STABILIZATION OF TUNABLE LASERS

Miao Zhu

Hewlett-Packard Laboratories
Palo Alto, California

John L. Hall

JILA, National Institute of Standards and Technology
and University of Colorado
Boulder, Colorado

5.1 Introduction

Many applications require a laser system whose frequency stability is better than that of a free-running laser or a commercial laser system. These applications include the development of optical atomic frequency standards, tests of fundamental physical principles, quantum optics, coherent transient phenomena, and various precision measurements, to name just a few. Here we try to prepare the reader to understand and overcome the laser problems so as to get on with the "good stuff."

Although it would appear to be the most direct and most effective method to just eradicate the laser frequency noise sources, for practical reasons, either due to technical complexities or fundamental physical limitations, these frequency noise sources usually cannot be completely eliminated. Also, with the diminishing cost and increasing capability of optoelectronic technology, it may well be a better choice to use a simpler laser design—thereby ideally winning lower cost, lower weight, and perhaps robustness—and use a high-performance servo system to force the laser output to conform to our expectations. Therefore, after a laser has been chosen for a particular application, based on a large number of considerations (e.g., the desired wavelength and optical power), usually a servo system is exploited to stabilize this laser's frequency. In this chapter we will concentrate on the subject of using a servo system to stabilize the laser frequency.

In beginning our discussion of laser frequency stabilization, it is useful to observe that different laser systems have their dominant frequency noises arising from different physical origins. In the case when the laser linewidth is limited by so-called "technical noise," which is true for most lasers, it is relatively easier to reduce the frequency noise by using a servo system. Basically these technical noises are slow in their time frame, even though they may be of large amplitude.

In this case, the required servo corrections will need to be accurate, but might not need to extend to a frequency equal to the original linewidth. The resulting linewidth decreases with increasing servo gain until reaching a plateau; the ultimate linewidth limit is then the measurement noise (ideally only shot noise) introduced in the process of realizing the servo system. On the other hand, if the laser linewidth is limited by quantum fluctuations (the Schawlow–Townes limit [1] is easily visible in solitary diode lasers), a servo system capable of reducing the laser's linewidth will need to be very fast. The servo bandwidth needs to exceed the original linewidth by a significant factor. Electronic time delays begin to become a problem, and it accordingly becomes attractive to take optical measures first to overcome this limit. In the diode laser case one typically would use external optical feedback to extend the effective laser cavity [2]. These measures may actually introduce additional technical noise, but this could be suppressed by using a servo system with a narrower bandwidth.

In the discussion of laser frequency stabilization, one is interested in the optical power spectral density of the laser's output electric field. We can view this electric field as an ideal monochromatic electric field with a constant amplitude E_o and a constant carrier frequency Ω_0, which is then modulated in both amplitude and phase [3–9], that is,

$$\mathbf{E}(t) = \mathbf{e}E_0 \left[1 + V_N(t)\right] \exp\left[-i\Omega_0 t - i\varphi_N(t)\right], \quad (1)$$

where e is a unit vector indicating the polarization, $V_N(t)$ and $\varphi_N(t)$ are noise amplitude modulation and phase modulation, respectively. The case of large amplitude modulation (over-modulation) is ignored in Eq. (1). The (noise) modulation frequency is $f_N(t) \equiv (1/2\pi)(d/dt)\,\varphi_N(t)$. In the case of $|V_N(t)| \ll 1$, the spectral characteristics of $f_N(t)$ or $\varphi_N(t)$ determines the laser's optical power spectral density, that is, the power spectral density of $\mathbf{E}(t)$. For the simplicity of the discussion, $\varphi_N(t)$ can be seen as a sum of two terms. The first term may be taken to be sinusoidal with one or more Fourier frequency components, which generate sidebands around the laser carrier frequency. If these sidebands are far enough away from the carrier, say more than the width of the spectral feature of the physical quantity to be measured, they are not so important in the spectroscopic application except for the loss of the optical power in the carrier due to their existence. However, if these Fourier frequencies are lower than the width of the spectral feature of interest, they show up as a broadened linewidth in the measurement. The second term in $\varphi_N(t)$ may be characterized as being due to stochastic processes. This part broadens the linewidth of the laser output carrier (and the sidebands generated by the first noise term). The task of the exploited servo system is to suppress this undesired noise frequency/phase modulation, ideally to the level limited only by shot noise associated with the error signal derivation process. From Eq. (1) one can readily calculate the optical power spectral density, or various

INTRODUCTION

integral quantities such as the fractional power in the optical carrier, or within a narrow bandwidth around the carrier.

In many physical cases the power spectral density of the frequency modulation noise $f_N(t) \equiv (1/2\pi)(d/dt)\,\varphi_N(t)$ can be approximated as

$$P_{\text{FM noise}}(\omega) = \begin{cases} \dfrac{\pi \langle f_N^2 \rangle}{B}, & |\omega| \leq 2\pi B \\ 0, & |\omega| > 2\pi B \end{cases}, \qquad (2)$$

where $\langle f_N^2 \rangle / B$ is the noise spectral density (in units of Hz^2/Hz). In one special case where

$$B \ll \sqrt{\langle f_N^2 \rangle},$$

B is the cutoff frequency and the laser exhibits a Gaussian line shape with a full linewidth of $[8 \ln(2) \langle f_N^2 \rangle]^{1/2}$:

$$P_E(\Omega) = \frac{E_0^2}{\sqrt{8\pi^3 \langle f_N^2 \rangle}} \exp\left[-\frac{(\Omega - \Omega_0)^2}{8\pi^2 \langle f_N^2 \rangle}\right]. \qquad (3)$$

In the other special case, where

$$B \gg \sqrt{\langle f_N^2 \rangle},$$

the laser power spectral density is Lorentzian with a full linewidth of $\pi \langle f_N^2 \rangle / B$:

$$P_E(\Omega) = \frac{E_0^2}{\pi} \frac{\dfrac{\pi^2 \langle f_N^2 \rangle}{B}}{(\Omega - \Omega_0)^2 + \left(\dfrac{\pi^2 \langle f_N^2 \rangle}{B}\right)^2}. \qquad (4)$$

These two special cases [8] are often useful limits for comparison with experimental results.

Of course, for processing and careful study of these optical frequencies, we have in mind using radio frequency (rf) analysis methods. Operationally this is accomplished by heterodyne mixing, so that the optical frequency is mapped down to dc. Noise from symmetric detuning around the carrier Ω_0 will map to the same difference frequency, ω. The resulting power spectral density will be denoted $P_{f_N}(\omega)$. Figure 1 shows a simplified block diagram in which the laser frequency noise, $P_{f_N}(\omega)$, is reduced by the gain $G(\omega)$ of the servo loop to

$$P_{f_{Ncl}}(\omega) = \frac{P_{f_N}(\omega)}{|1 + G(\omega)|^2} + \frac{|G(\omega)|^2}{|1 + G(\omega)|^2}\left\{P_{f_{Nref}}(\omega) + P_{f_{Nmeas}}(\omega)\right\}, \qquad (5)$$

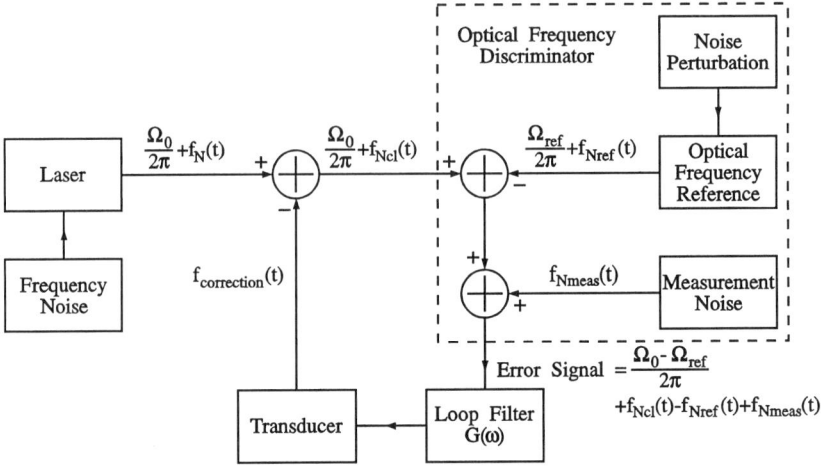

FIG. 1. Symbolic block diagram for laser frequency stabilization, indicating signals and information flow rather than real connections and the laboratory layout.

where $P_{fNcl}(\omega)$ is the closed-loop noise spectral density, $P_{fNref}(\omega)$ and $P_{fNmeas}(\omega)$ are the frequency noise density of the optical frequency reference and the measurement noise, respectively. As a general rule for any servo system, one would like to have as large a gain $|G(\omega)|$ as possible to suppress $P_{fN}(\omega)$. This means that the largest possible bandwidth is needed, which in turn requires the smallest time delay in the servo system. On the other hand, neither $P_{fNref}(\omega)$ nor $P_{fNmeas}(\omega)$ can be reduced by increasing $|G(\omega)|$. The contribution of these two terms will determine the limit we can achieve by using a servo system to stabilize laser frequency, fixing the noise plateau referred to in the introduction.

As shown in Fig. 1, there are three necessary parts in the system: an optical frequency reference for deriving the error signal, a transducer to reduce or correct the optical frequency error, and a so-called loop filter to supply the necessary gain and phase compensation. These three parts will now be discussed in some detail in the following sections.

5.2 Optical Frequency References

Measurement of either an optical frequency or an optical phase can be used to derive the error signal for laser frequency stabilization. For the discussion here we do not distinguish the interesting difference between these two cases [9]. An ideal optical frequency reference (including the method to derive the error signal) should have the following properties: the reference frequency has low intrinsic noise; the reference frequency is insensitive to environmental disturbances; and the error signal is proportional to the detuning and has a known dependence upon the input optical power. It is also required that the reference discriminator have a high signal-to-noise ratio (S/N) and a broad bandwidth, and that its reference frequency can be precisely tuned. Usually for any given optical frequency reference not all of the requirements can be satisfied simultaneously. Thus, some approximations and compromises have to be applied. For example, it is reasonable to accept an optical reference system that gives an error signal proportional to detuning when it is small and that gives an arbitrary value with correct sign when the detuning becomes large. For some applications a multiple-reference system has to be used. The most commonly used optical frequency references are Fabry–Perot cavities, atomic (molecular, ionic) transitions, and other stable lasers.

5.2.1 Fabry–Perot Cavity

The Fabry–Perot cavity, which falls into the category of multibeam interferometers [10], is probably the most commonly used optical frequency reference. For detailed mathematical formulation of a Fabry–Perot cavity, the reader is referred to the literature [11, 12]. Here we limit our discussion to the two-(spherical)-mirror stable cavity (Fig. 2) serving as the optical reference cavity. For a more complicated geometry, the optical configuration can be mapped to an equivalent two-mirror cavity, although sometimes two orthogonal planes, a tangential plane and a sagittal plane, may be needed for the discussion.

5.2.1.1 Optical Properties
The criterion that a two-mirror cavity has a stable optical mode can be written as

$$0 \leq \left(1 - \frac{L}{R_1}\right)\left(1 - \frac{L}{R_2}\right) \leq 1, \tag{6}$$

where R_1 and R_2 are the radii of the curvature of the mirrors, and L is the distance between the two mirrors on the cavity axis, which is defined by a straight line connecting the centers of the curvature of the two mirrors. The resonance frequency of the TEM_{qmn} mode is given by

$$f_{qmn} = \frac{c}{2L}\left(q + 1 + \frac{1}{\pi}(m + n + 1)\cos^{-1}\sqrt{\left(1 - \frac{L}{R_1}\right)\left(1 - \frac{L}{R_2}\right)}\right), \tag{7}$$

where c is the speed of light. The other important parameters of a Fabry–Perot cavity are the free spectral range, finesse, and input coupling efficiency. The free spectral range (FSR) of the TEM_{00} mode, defined as the frequency separation of two adjacent longitudinal modes, is given by

$$\text{FSR} = \frac{c}{2L}. \tag{8}$$

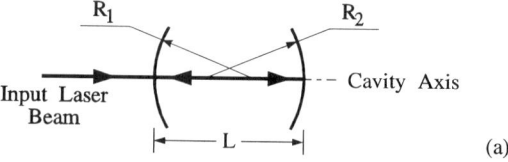

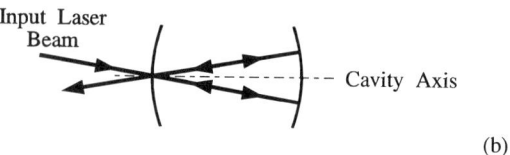

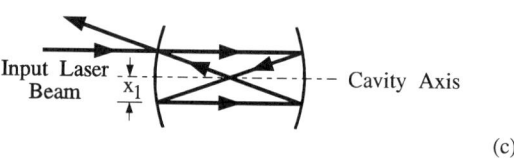

FIG. 2. Fabry–Perot cavities. (a) General form: R_1 and R_2 are radii of the curvature for the mirrors, L is the distance between the mirrors. (b,c) Confocal cavity with $L = R_1 = R_2$.

The finesse, F, and input matching efficiency, η_{in}, are given as, respectively,

$$F = \frac{\text{FSR}}{\Gamma} = \frac{\pi (R_r)^{1/4}}{1 - \sqrt{R_r}}, \tag{9}$$

and

$$\eta_{in} = 1 - \frac{P_{\text{reflected}}}{P_{\text{input}}} = 1 - \left[\frac{(1 - A_{in}) \sqrt{R_r / R_{in}} - \sqrt{R_{in}}}{1 - \sqrt{R_r}} \right], \tag{10}$$

where $\Gamma = 2\gamma$ is the linewidth (FWHM) of the cavity, R_r is the round-trip power reflectance, R_{in} and A_{in} are the power reflectance and the power loss of the input coupler mirror, and P_{input} and $P_{\text{reflected}}$ are the input optical power and reflected optical power, respectively.

In some applications, one wants to know the FSR of a cavity with a precision better than the one obtained by measuring the separation of the mirrors. This can be achieved by locking two lasers to the adjacent TEM_{00} modes and measuring the heterodyne frequency. An alternative method is to lock the laser on the cavity and scan a frequency-shifted beam, which could be generated either by an acoustooptic modulator (AOM) or an electrooptic modulator (EOM), across the adjacent mode. DeVoe et al. [13] introduced another method that locks the laser on the cavity and uses a second set of frequency-modulation (FM) sidebands to determine the FSR of the cavity. Precision at the level of a few Hz is readily obtainable.

The cavity linewidth can be measured by comparing the observed linewidth with the interval between orders (the FSR) when either the laser frequency or cavity length is scanned. For high finesse a smaller scan range is needed, and it is convenient to supply sidebands with known frequency to provide the local calibration intervals. Using this method, the observed lineshape is actually the convolution of the laser linewidth and the cavity linewidth. If the dominant Fourier frequency of the laser frequency noise is comparable to or higher than the cavity linewidth, this method will not reveal the true cavity linewidth. As an alternative method, one can measure the decay time of the stored field in the cavity, especially for determining the loss of very-high-reflectivity mirrors [14, 15]. The coupling efficiency of the cavity could be obtained by measuring the resonant dip of the reflected beam when the laser frequency scans across the resonance frequency of the cavity. Attention should be paid to the bandwidth of the photodetector and mode matching condition, and to using a sufficiently slow scan so that one samples the equilibrium response.

The mode stability of an optical cavity [16] is very sensitive to the mirror alignment near both the concentric cavity ($L = R_1 + R_2$) and the parallel plane cavity ($R_1 = R_2 = \infty$) limits. Accordingly, these two mirror configurations are not commonly used as optical frequency references for laser frequency stabilization. The symmetric confocal cavity ($L = R_1 = R_2$) is least sensitive to the mirror alignment, while an unsymmetric confocal cavity ($R_1 \neq R_2$, and $L = R_1/2 + R_2/2$) is not stable. For a symmetric confocal cavity the frequencies of all the high-order transverse modes are split into two families, either degenerate with the TEM_{00} mode frequency (in the case of $m + n$ = even integer), or located at the middle of the interval between the TEM_{00} mode frequencies (in the case of $m + n$ = odd integer). A confocal cavity is also stable and useful when the laser beam axis is not overlapped with the cavity axis, as shown in Figs. 2b and 2c. In the cases of Figs. 2b and 2c, the FSR is $c/4L$ and the frequencies of all the higher-order transverse modes are degenerate with the fundamental mode frequency. Thus, it is less demanding to match the mode of the input laser beam to the eigenmode of a confocal cavity. The penalty comes at high finesse when one discerns small asymmetry in the response function because one is not *exactly* at the special confocal condition.

The retroreflected field depends on the laser beam alignment into the cavity. In Fig. 2a the reflected field is the superposition of the directly reflected input field to the cavity and the field stored in the cavity that leaks out of the cavity back toward the source. In Fig. 2b only the stored field that leaks out of the cavity is present in the retroreflected field. In Fig. 2c there is no retroreflected field. This presence and critical-alignment dependence of the retroreflected field have practical importance in optical feedback problems back to the laser source and servo system implementation. The resonant frequency in Fig. 2c depends on the alignment as shown:

$$\frac{\Delta f}{f} = -\frac{\Delta l}{l} \approx \frac{x_1^3}{R_1^4} \Delta x_1 , \qquad (11)$$

where l is the round-trip optical length of the cavity, x_1 is the distance shown in Fig. 2c, and Δx_1 is the change in x_1. Care must be taken in the case when the absolute frequency is critical.

Nonconfocal stable cavities are also often chosen as the optical frequency references. By appropriately choosing the radii of the mirror curvatures and the mirror separation, the frequencies of (a few lowest) higher-order transverse modes can be moved far enough away from the frequency of the TEM_{00} mode such that they will not interfere with the error signal derivation near the resonance. When this design is implemented, the misalignment of the laser beam will only reduce the optical power coupled into the cavity and not importantly change the resonance frequency of the cavity. (The local heating-induced mirror deformation could result in the dependence of resonance frequency on the intracavity optical power; see [17].) Note that the cavity mode resonance frequency structure must also be considered in the vicinity of the optical sideband frequencies if rf methods are used to derive the error signal.

5.2.1.2 Mechanical Stress, Necessary Vacuum, etc.
Since the stability of the resonance frequency of a Fabry–Perot cavity is essentially determined by the stability of the optical length between two mirrors, attention has to be paid to stabilize this optical length. To appreciate this sensitivity, a change in the mirrors' optical separation by $\lambda/2$ (~300 nm for visible light) corresponds to a frequency change of one FSR. In the usual laboratory case, this will be something in the range of 100 MHz to about 5 GHz, while the attainable stability and linewidth can be easily in the hertz range. Thus, the stability of the spacer material is very important. Materials with a low thermal expansion coefficient such as fused silica, Zerodur [18, 19], and ULE [20] are often used. The structure of the cavity should be designed to avoid any longitudinal forces since that is the most sensitive dimension. The fabrication stresses in the material should be released properly. Even the residual stress in the thin layer on surfaces due to final machining or grinding should be released either by annealing or etching with hydrofluoric acid (HF). Using the same material for the spacer and mirror substrates enables one to use the optical contact technique to attach mirrors to the spacer without any adhesive. In the case that the optical contact method cannot be used, three or four

small dots of low-vapor-pressure adhesive [21] can be used to connect the mirror cylindrical surface to the spacer end surface. We have found that ultrasonic soldering of indium (In) metal between the mirror cylindrical surface and the spacer is another effective and reasonably stable mounting method. Of course, compared to adhesive, the vacuum compatibility of the indium metal approach is vastly superior.

To avoid fluctuation of the refractive index of air ($n_{air} - 1 \approx 2.7 \times 10^{-4}$ at 1 atm) due to air pressure changes, it is absolutely necessary to have a secure hermetic seal, and a good practice is to keep the cavity under vacuum. The vacuum dramatically reduces the influence of the laboratory's acoustic noise. The vacuum and enclosure also act as a low-pass filter to shield the cavity spacer from the temperature fluctuations. The effective coupling time constant can be about 1 day, provided a proper material with low thermal conductivity is used to attach this spacer to the vacuum enclosure. Radiative transfer is reduced by polishing the interior surface of the housing.

Another important but more subtle environmental disturbance is the vibration of the optical table or the floor of the building. A typical number for the vertical acceleration due to optical table vibration is about 1 milli-"g" $\approx 10^{-2}$ m/sec^2 with a peak Fourier frequency near 30 Hz. It arises from the superposition of mechanical vibrations from a large number of motors, each slipping behind half the line frequency by a different amount and each gently shaking the world. The vibration can deform the shape of the cavity, thus changing the resonance frequency of the cavity. Another coupling is from the vertical acceleration working through the Poisson's ratio of the spacer ($\sigma \sim 0.3$ for most materials) to "extrude the toothpaste" in response to the changing vertical force. In at least these ways [22] the vibrations cause modulation of the reference cavity's geometry and hence the laser frequency. The major part of the vibration consists of one or several discrete Fourier frequency components and generates a number of FM sidebands extending for several kHz. In addition, part of the vibration is a stochastic process with low-frequency components that broadens the laser linewidth, again typically by several kHz.

To estimate this vibrational influence, let us consider a stable glass–ceramic spacer with a length L and a uniform cross-section $W \times H$. Suppose the cavity is constructed such that it stands on one of the spacer's ends, as shown in Fig. 3a. In the earth's gravity field the relative length change is

$$\frac{\Delta L}{L} = -\frac{\rho g L}{2Y}, \qquad (12)$$

where $g \approx 9.8$ m/sec^2 is the earth's gravitational acceleration, and $\rho = 2.5 \times 10^3$ kg/m^3 and $Y = 90$ GPa are typical values for density and Young's modulus of compression, respectively. For a cavity of length $L = 0.3$ m, the optical frequency change is approximately 20 MHz/g in the visible range. Using this cavity vertically

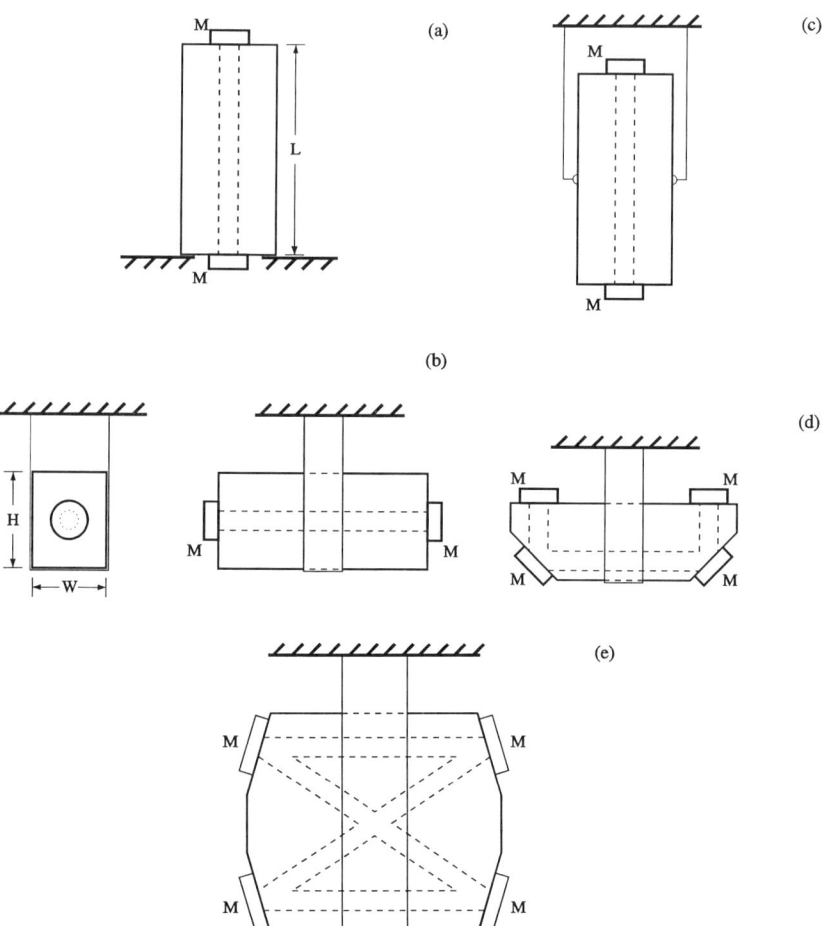

FIG. 3. Several mounting methods for reference cavities. See detailed discussion in text. (a) The spacer stands on one of its ends. (b) The spacer is suspended horizontally from the vacuum enclosure by a strap or ribbon. (c) The spacer is suspended vertically, mounted near the middle of its length. (d,e) Rectangular or ring-shaped cavities designed to reduce the sensitivity to vertical acceleration noise. The ring cavity is designed with zero net area to reduce the frequency noise due to vibration-induced rotations.

on a typical optical table, the resultant laser linewidth cannot be narrower than ~20 kHz due to the vibration. If we want to obtain a linewidth of 20 Hz, the vibration of the optical table has to be controlled to the 1 micro-"g" level if this vertical cavity mounting were used. Alternatively, the cavity can be mounted horizontally as shown in Fig. 3b. The relative length change in this case is given by

$$\frac{\Delta L}{L} \approx \frac{\sigma \rho g H}{2Y}, \tag{13}$$

where σ is Poisson's ratio. We can see that the frequency sensitivity to vibration has been reduced significantly but not enough. Clearly, one will try to use symmetry—mounting the cavity in its midplane as shown in Fig. 3c—and other mechanical tricks to reduce these perturbations. On the other hand, one should be able to design a suitable geometry (e.g., the ones shown in Fig. 3d,e) such that cavity length is immune to vertical acceleration by utilizing Poisson's ratio to compensate the vertical length and the horizontal length changes.

We have also had some good success using an active servo system to approach the ideal "vibrationless" working environment by a vertically active servo [23]. Unfortunately, using the airleg suspension technique to obtain soft support exacerbates tilt-induced problems. The cavity and indeed the entire tabletop sitting on top of the flexible connection with legs is a guaranteed converter from horizontal ground motion into table tilt. We have become accustomed to seeing the table's legs supporting the tabletop from below. How vastly more clever it would be to support the tabletop in its midplane or, better, the plane containing the percussion center, or even to hang the tabletop from above on shock cords. Of course, some people will try to use tilt sensors and servo-stabilize the tabletop. At present, by feedback we are back to the submicroradian domain from near milliradian table excursions, driven by the 6-second microseism accelerations. This level gives rise to a ±10-Hz excursion of the beat between two cavity-locked lasers. It is fair to say that, with a proper design of the reference cavity and electronics, it is relatively easier to stabilize the laser frequency to a subhertz level relative to the cavity resonance frequency than it is to prepare the cavity absolute stability at the same level, as needed for use in optimally locking the laser frequency to an absolute frequency, for example, an ultra-narrow atomic transition.

5.2.1.3 Tunability vs. Stability

In some applications the laser frequency needs to be tuned over some range, say several free spectral ranges. This can be realized using any means to change the round-trip optical length of the cavity. A typical method is to mount one of the cavity mirrors on a piezoelectric transducer (PZT). The high thermal expansion coefficient ($\sim 1 \times 10^{-5}$/K) of PZT degrades the cavity thermal stability. Due to the hysteresis and "creep" of the PZT, the cavity frequency cannot be predicted precisely according to the applied voltage. Thus, additional frequency marks are needed if precision scanning is required. Furthermore, the PZT material is less stable than spacer material, and thus introducing PZT usually degrades the cavity stability, even with some thermal compensation.

To maintain the stability of the cavity, an alternative method can be used to provide the needed tuning. This method uses a double-pass or even a quadruple-pass broadband acoustooptic modulator (AOM) to shift the laser frequency relative to the cavity resonance frequency (Fig. 4). Tuning the rf frequency applied

Top View

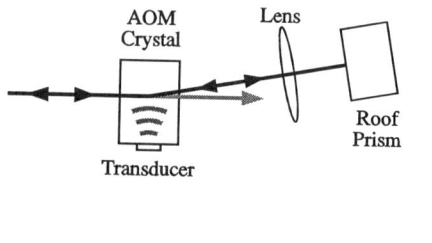

(a)

Side View

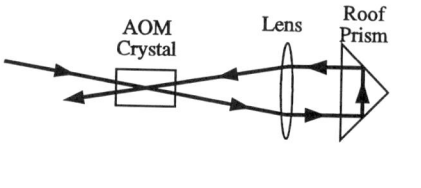

(b)

FIG. 4. Double-passed AOM. The frequency of the output beam is shifted twice by the frequency applied to the AOM. Quadruple pass can be realized using a curved mirror to retroreflect the output beam of the double-pass configuration. The output beam, which is now overlapped with the input beam spatially, can be separated using a polarization method (a linear polarizer in combination with a Faraday rotator or a $\lambda/4$ waveplate). Notice that the efficiency of some AOMs depends on the polarization, and some AOM crystals have birefringence.

onto the AOM scans the laser frequency offset produced by the AOM, but since this shifted frequency is held to the cavity's fixed resonance frequency by the servo system, the change in frequency ultimately appears on the laser itself. This method does not degrade the stability of the cavity. It also provides a one-to-one mapping between the rf frequency and the laser detuning frequency (more exactly, a 1:2 or 1:4 mapping). With a programmable rf frequency synthesizer, one can achieve precise tuning of the laser frequency relative to the cavity resonance frequency. The low efficiency of the multipass AOM is usually not a serious problem because one needs to send only about 100 µW optical power to the reference cavity. The tuning range of this method is limited by the bandwidth of the AOM used. If the tuning range needs to be more than one FSR of the cavity, a microcomputer-controlled system could unlock the laser from the cavity, move it to the next mode, and relock it to the cavity when the scanning limit is reached.

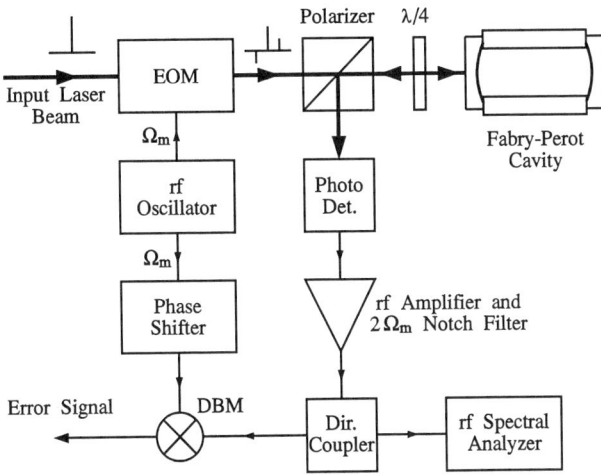

FIG. 5. Block diagram for optical frequency discriminator (Pound–Drever–Hall scheme) to derive optical frequency/phase error signal from a Fabry–Perot reference cavity.

An interesting type of pressure-sensitive IC resistance bridge has become available commercially, with full-scale sensitivity of 0–1 atm, for example. Electropneumatic valves with a roughly proportional response and a several-hundred-Hz bandwidth are also available. It is interesting to wonder if a closed-loop pressure controller would be able to tune the cavity resonance via the variable air index. Or perhaps the reference interferometer would be sealed and we would use the pressure to compress the cavity (length) in a controlled manner. The Joule–Thomson cooling may make mischief: we should prefer the optoelectronic approach for its clarity.

5.2.1.4 Error Signal Derivation When a Fabry–Perot cavity is used for the optical frequency reference, the error signal could be derived from either the transmitted beam or the reflected beam. It can be derived at dc or at radio frequency (tens of kHz to hundreds of MHz). Hils and Hall discussed the transmission signal in great detail [24]. Hänsch and Couillaud described a dc method to derive the signal in the reflected beam [25]. A more sophisticated method is to derive the error signal from the reflected beam at radio frequency, as described by Drever et al. [26, 27]. This method not only avoids all the technical difficulties of working at dc but also increases the signal-to-noise ratio (S/N) in the error signal by choosing the sideband frequency at a quiet place in the spectrum where the laser amplitude noise is small, ideally near the shot noise limit.

Figure 5 shows a schematic of this method. A polarization beam splitter and a $\lambda/4$ waveplate are used to separate the reflected beam. This combination can be either replaced by a polarization beam splitter and a Faraday rotator or, in the case of a confocal cavity, omitted. After passing the electrooptic phase modulator (EOM), which generates frequency modulation (FM) sidebands on the input laser field [28, 29], the resultant laser field can be written as

$$E_{\text{in}} = E_0 \exp\{-i\Omega t - i\beta \sin(\Omega_m t)\}$$
$$\approx E_0 \{J_0(\beta)e^{-i\Omega t} + J_1(\beta)e^{-i(\Omega+\Omega_m)t} - J_1(\beta e^{-i(\Omega-\Omega_m)t}\}, \quad (14)$$

where β is the modulation index, Ω_m is the modulation frequency, and J_i is the ith-order Bessel function. In Eq. (14) we are assuming a small modulation index β. The detected rf photocurrent of this input field vs. detuning Δ is

$$I_{\text{photo}}(t) \approx \eta_{\text{photo}} P_{\text{in}} \frac{2T_{\text{in}}\sqrt{R_r}\, J_0(\beta)J_1(\beta)}{1 - \sqrt{R_r}}$$

$$\times \left(\left[\frac{1}{1+\left(\frac{\Delta-\Omega_m}{\gamma}\right)^2} - \frac{1}{1+\left(\frac{\Delta+\Omega_m}{\gamma}\right)^2}\right]\cos(\Omega_m t)\right.$$

$$\left. + \left[\frac{2\Delta/\gamma}{1+(\Delta/\gamma)^2} - \frac{(\Delta-\Omega_m)/\gamma}{1+\left(\frac{\Delta-\Omega_m}{\gamma}\right)^2} - \frac{(\Delta+\Omega_m)/\gamma}{1+\left(\frac{\Delta+\Omega_m}{\gamma}\right)^2}\right]\sin(\Omega_m t)\right), \quad (15)$$

where η_{photo} is the responsivity of the photodiode in units of A/W, P_{in} is the input optical power, T_{in} and R_T are the power transmission of the input coupler mirror and the round-trip power reflectance of the cavity, respectively, and γ is the linewidth of the cavity. In deriving Eq. (15) we used the condition $\Omega_m > \gamma$. By changing the phase of the reference channel at the doubly balanced mixer (DBM), one can either detect the absorption signal, $\cos(\Omega_m t)$, or the dispersion signal, $\sin(\Omega_m t)$. Figure 6 shows these two signals. The dispersion is used for locking laser frequency to the cavity resonance frequency. The error signal is approximately linear vs. detuning when $|\Delta| < \gamma$, and it has the correct sign when $|\Delta| < \Omega_m$. For the purpose of designing a servo system, one is perhaps more interested in the transfer function of this cavity and the Pound–Drever–Hall scheme [27] when the dispersion signal is used. With zero detuning the transient response [9, 23], as shown in Fig. 7, is

$$I_{\text{photo}} \approx \eta_{\text{photo}} P_i \frac{4T_i J_0(\beta)J_1(\beta)}{1-\sqrt{R_r}} \Delta\varphi e^{-\gamma t} \quad (16)$$

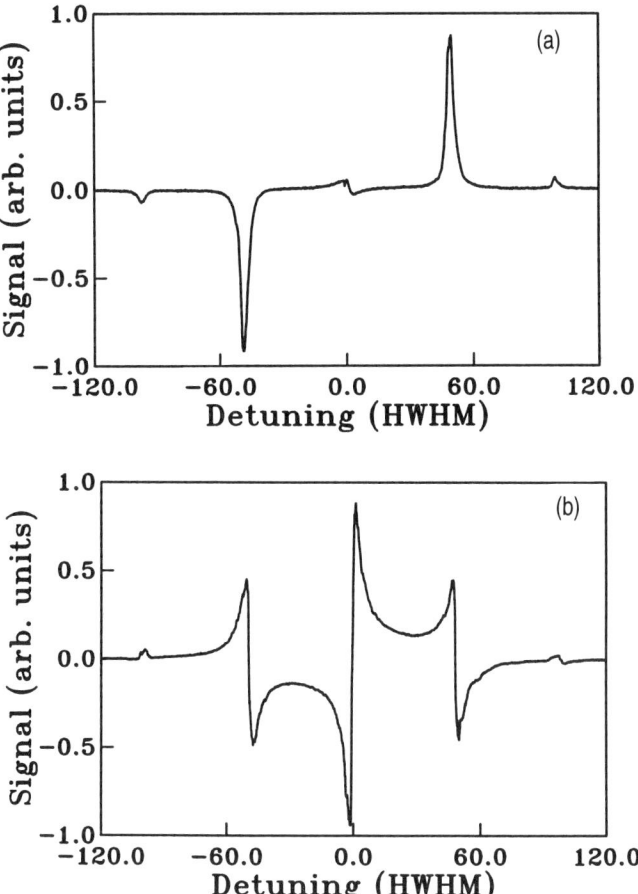

FIG. 6. Measured steady response of optical frequency discriminator vs. detuning. The modulation frequency is ~48 times the cavity linewidth (HWHM). The detuning is shown in units of cavity linewidth (HWHM). (a) Absorption term. (b) Dispersion term.

for a small step phase change $\Delta\varphi$ at $t = 0$, and

$$I_{\text{photo}} \approx \eta_{\text{photo}} P_i \frac{4T_r J_0(\beta) J_1(\beta)}{1 - \sqrt{R_r}} \frac{\Delta\Omega/\gamma}{1 + (\Delta\Omega/\gamma)^2} [1 - e^{-\gamma t}] \quad (17)$$

for a small-step frequency change $\Delta\Omega$ at $t = 0$, while the (small-signal) steady-state response [9, 23], as shown in Fig. 8, is

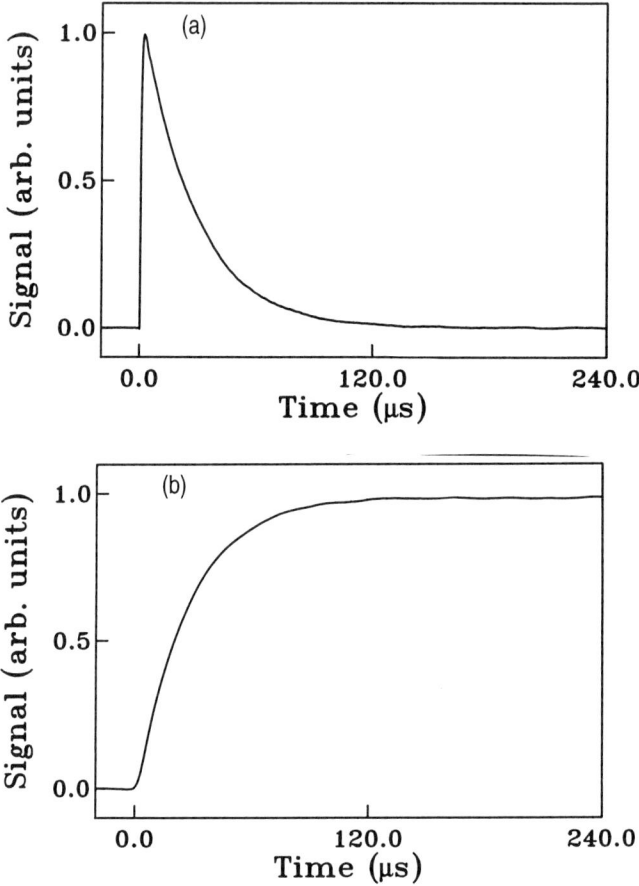

FIG. 7. Measured transient response of optical frequency discriminator when dispersion signal is used with zero detuning. Cavity linewidth (HWHM) is 6 kHz. The modulation frequency is 12.5 MHz, with a modulation index of ~0.9. (a) Response to a step-phase disturbance. (b) Response to a step-frequency disturbance.

$$I_{\text{photo}}(t) \approx \eta_{\text{photo}} P_i \frac{T_i\sqrt{R_r}J_0(\beta)J_1(\beta)}{1-\sqrt{R_r}} \frac{4\Omega_\alpha/\gamma}{\sqrt{1+(\Omega_\alpha/\gamma)^2}} \alpha \sin\left[\Omega_\alpha t + \tan^{-1}\left(\frac{\gamma}{\Omega_\alpha}\right)\right]$$

$$= \eta_{\text{photo}} P_i \frac{T_i\sqrt{R_r}J_0(\beta)J_1(\beta)}{1-\sqrt{R_r}} \frac{4/\gamma}{\sqrt{1+(\Omega_\alpha/\gamma)^2}} \alpha\Omega_\alpha \cos\left[\Omega_\alpha t - \tan^{-1}\left(\frac{\Omega_\alpha}{\gamma}\right)\right] \quad (18)$$

for phase modulation $\alpha \sin(\Omega_\alpha t)$ or frequency modulation $\alpha\Omega_\alpha \cos(\Omega_\alpha t)$.

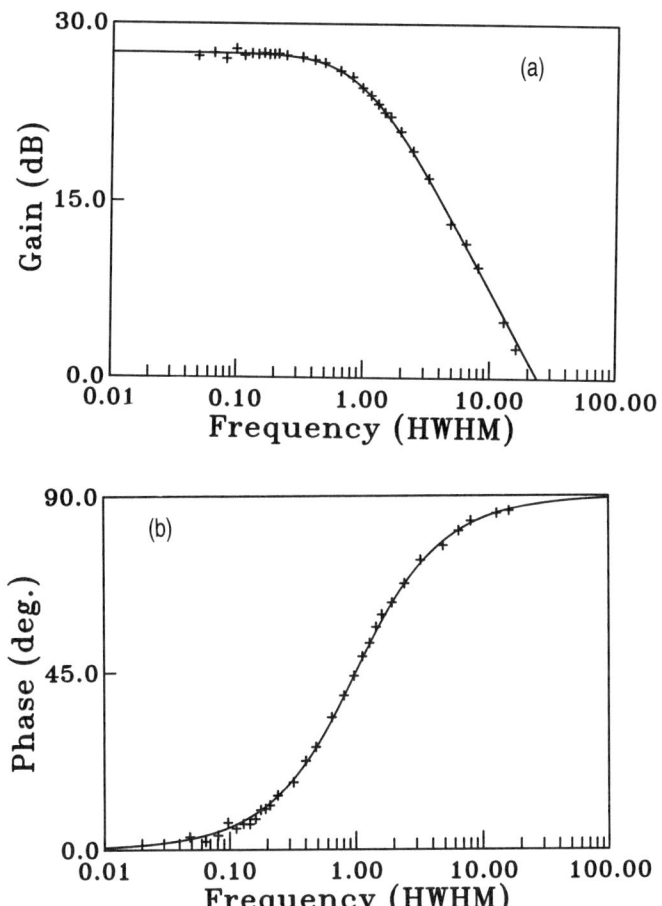

FIG. 8. Steady-state response of Pound–Drever–Hall scheme when dispersion signal is used with zero detuning. Cavity linewidth (HWHM) is 6 kHz. The modulation frequency is 12.5 MHz, with a modulation index of ~0.9. Crosses are the measured data, while the solid curves are theoretical curves according to Eq. (18). (a) Relative gain to FM noise. (b) Phase delay to FM noise.

Equations (16) and (17) show that the error signal is proportional to the optical phase change within the time $1/\gamma$ and is proportional to the frequency error when the time is longer than $1/\gamma$. Taking the full advantage of this error signal with a modulation frequency $\Omega_m > \gamma$, the ultimate bandwidth of a properly designed servo system is only limited by the time delay in the system, not by the cavity linewidth. With increasing bandwidth, of course, more shot noise is obtained, but this is ordinarily completely unimportant. Basically, at the highest frequencies the

Pound–Drever–Hall discriminator is acting as a phase (rather than frequency) detector. Thus, the extra measurement noise is written as a small laser phase noise, which is of little importance, as can be seen from Eq. (4) and the associated discussion. See also the example in Section 5.5.2.

5.2.2 Atomic Transitions

Atomic (molecular, ionic) transitions can also be used as optical frequency references for laser frequency stabilization. The (first-order) Doppler-free spectroscopic methods [30] are often exploited to reveal the homogeneous linewidth. Laser trapping and cooling [31–33] are also used to suppress Doppler effects (to all orders). Both trapped ions and the neutral atomic fountain approaches can provide the necessary interaction time for the extra-narrow linewidth (on the order of Hz). The main limit of this kind of reference is that S/N in the needed bandwidth for stabilizing the laser frequency directly is usually limited by the number of available atoms and/or nonlinearity of atom–field interaction, that is, saturation. Another limit is that typical system designs with the extra-narrow linewidth atomic transition lead to update rates ~1 sec^{-1}. Therefore, the laser linewidth is usually first narrowed by locking it to a stable Fabry–Perot cavity. The atomic transition is then used, only in a narrower bandwidth, to correct the "slow" change of the cavity resonance frequency.

To obtain the best optical frequency resonances, FM saturation spectroscopy (Fig. 9) is commonly used. This method provides a high S/N and directly produces a dispersion error signal for frequency locking. For a more detailed discussion, the readers are referred to the related chapter [34]. The environmental disturbances (stray electric and magnetic fields, collisional broadening and shift, etc.) need to be controlled such that the inherent limit of the stability is not degraded. In some cases, attention is needed for frequency shift due to radiation forces acting on the atomic absorbers [35]. In the visible and near-infrared regions, Te_2, I_2, and Br_2 are a few of the commonly used molecules for optical frequency stabilization. Work has shown the attractive feasibility of intracavity saturated absorption spectroscopy of vibrational overtones of molecules such as HCN and C_2H_2 [36], C_2HD [37], and CH_4 [38]. The frequency gap between the reference and desired frequencies could be covered by using AOM or EOM techniques to provide the necessary frequency offset.

5.2.3 Reference Laser and Optical Phase Locking

A frequency-stabilized laser can also serve as the optical frequency/phase reference. Heterodyne detection gives a beat note in the radio frequency region for processing and control while preserving the interesting optical frequency/phase difference between the reference laser and the laser to be stabilized. Using the

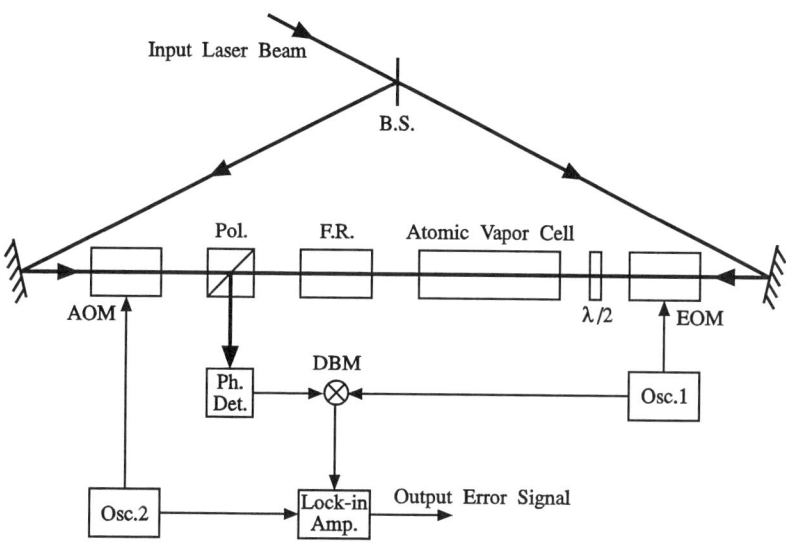

FIG. 9. Block diagram of saturation FM spectroscopy. See reference [34] for detailed discussion.

frequency discriminators or phase discriminators developed for rf applications, the test laser could be either frequency or phase locked to the reference laser. Since the optical phase locking methods are able to put most of the test laser power within the linewidth of the reference laser, even in the presence of measurement noise, they may be regarded as being superior to frequency locking [9]. The main application of using a stable laser as a reference is to add precise tuning (rf frequency precision) in a broad range (up to ~1 THz) [39, 40] and/or to "amplify" the stable laser's power without degradation of its optical frequency/phase stability.

5.3 Transducers

Laser frequency errors can be corrected by stabilizing the laser cavity's optical length. On the other hand, one can correct this frequency error after the light has already left the laser cavity by changing the optical path (changing the phase) and/or adding (or subtracting) a frequency to the laser output (changing the frequency). Either of these methods requires one or more transducers but offers unique advantages. Here we discuss some of the commonly used transducers and their properties. The main specifications of the transducers include their dynamic ranges, bandwidths, and optical insertion losses for intracavity application.

Piezoelectric transducers (PZT) are widely used as the transducers to change the laser cavity's optical length. The change in length of a PZT is proportional to the applied electric field. For an He–Ne laser with sealed mirrors, a PZT tube can be used to stretch or compress the glass tubing that is the spacer of the two mirrors. For a laser with unsealed mirrors, for example, a dye laser, one of the cavity mirrors can be mounted on a PZT disk or tube. The dynamic range of PZTs is large, up to ~0.1% of the disk thickness. Multilayer PZT stacks provide more range and/or less required voltage. (PZT stacks should be preloaded whenever possible to reduce hysteresis and to minimize compliance at mechanical interfaces.) A typical PZT has a bandwidth ranging from a few kHz to a few tens of kHz. To obtain the largest possible bandwidth of a PZT-mounted mirror, the mass of the mirror should be reduced to a minimum. (Buy a smaller mirror instead of cutting the thickness of an existing mirror. This helps avoid distortion of the mirror figure. Cementing of the mirror to the PZT offers another danger of distortion. We typically now use a hard wax rather than epoxy for this connection.) In mounting the PZT, one should try to provide absorption of the back-running acoustic wave generated by the reaction force to the PZT. The PZT driver should be able to provide sufficiently large current to charge the PZT capacitance rapidly.

A larger dynamic range for slow excursions can be realized by using galvo motor–driven intracavity Brewster plates. This is usually needed because of the requirement of a large-frequency scanning range rather than just for frequency stabilization. When a galvo motor is used, a suitable magnetic shielding (preferably, multilayer shielding) is needed to prevent pickup of the ambient magnetic field at the line frequency. A quiet galvo motor driver is also important.

When more bandwidth is needed than a PZT can provide, an intracavity EOM can be used [41]. The change in the refractive index of the EOM is proportional to the applied electric field [42]. Correspondingly, this changes the optical length of the laser cavity and in turn the laser frequency. The insertion loss comes from both bulk absorption of the crystal and the scattering loss of the surfaces. Due to their small bulk absorption, $NH_4H_2PO_4$ (ADP)-type crystals are most commonly used for intracavity EOM applications. Insertion loss of a pair of ADP crystals with a Brewster cut could be <1%/pass, including two fused silica windows sealing the cell. Another approach we have used to protect the crystal faces is a pair of BK-7 Brewster windows optically connected to the intracavity crystal with a suitable viscous index-matching liquid. Optically appropriate gels are now available for the fiberoptics industry, but we have not yet tried circulating 50 watts of intracavity optical power through such a material! For a typical EOM, optical phase change sensitivity is about $\lambda/800\,V$ in the visible range, which shifts the laser frequency about one free spectral range (FSR) for 800 V applied on the EOM in a ring cavity laser. Taking a CW ring dye laser as an example, the FSR is about 200 MHz, and the fast frequency fluctuation is about 1–10 MHz, depending upon the quality of the flowing jet stream. Thus, the driver for the EOM should have about a 100-V dynamic range. The capacitance between the electrodes of a typical EOM is some tens of pF. Certainly, a broadband driver is needed for the EOM. The bandwidth is usually limited by the driver's slew rate and delay time instead of by the response of the EOM crystal. The drawback of the intracavity EOM approach is its optical insertion loss, which could reduce the output power of a low-gain laser dramatically or even bring the laser under the lasing threshold. Due to the small dynamic range of the EOM, a combination of the PZT and EOM is often used to provide broad bandwidth and sufficient tuning range. For large tuning ranges, an intracavity galvo-mounted Brewster plate is additionally used, as noted earlier.

If one chooses to suppress the laser frequency noise outside the laser cavity, the process does not introduce any intra-laser-cavity loss. Although the external AOM and EOM used to suppress the frequency noise do indeed have losses and so reduce the useful optical power somewhat after the frequency noise is suppressed, this method probably gives more power than the intracavity method for a low-gain laser. In some cases it is difficult to use the internal method because of crystal birefringence or the nonavailabitity of an appropriate place within the laser resonator.

An AOM uses Bragg scattering to add the applied rf frequency to (or subtract it from) the laser frequency [42]. To retain laser beam pointing stability, a double-pass AOM could be used at the cost of less useful optical power (~50%) but increased sensitivity. The fundamental limitation of this transducer is the propagation time for the acoustical wave traveling from the transducer to the region of interaction in the AOM crystal. This ~300 nsec minimum time delay usually limits the servo bandwidth. One pitfall of using an AOM is the requirement of a voltage-controlled oscillator (VCO) that drives the AOM. In some commercial VCOs there is a built-in capacitor at the control voltage input to reduce rf leakage. This capacitor can easily delay the response of the VCO. For most VCOs the capacitor can be removed. A fringe benefit of using an AOM is that it can be used as a transducer for the amplitude stabilizer at the same time (of course, the bandwidth is still limited by the time delay).

An EOM can be used outside of the laser cavity to correct optical phase with a much smaller inherent time delay. The fundamental change produced by an EOM is a change in its optical length that alters the phase of the laser beam passing through. As the crystal driver, an integrator will respond to a step frequency request with a step frequency correction by changing the optical phase (Pockel's cell voltage) at a constant rate (although the driver eventually will be saturated or the crystal will break down).

Hall and Hänsch [43] used a combination of AOM and EOM units to construct an external laser frequency stabilizer with an electric interferometer topology to deal with the time delay in the AOM. A frequency error signal is first sent to the EOM via an integrator. After the propagation time delay of the AOM, the AOM takes over the correction job and the error signal sent to the EOM driver is canceled (and the accumulated output voltage is allowed to slowly relax). As will be discussed shortly, a similar setup used at JILA reduced a commercial dye laser linewidth from ~0.5 MHz to <0.5 Hz when locking the laser frequency to a Fabry–Perot cavity, and put >97% of optical power into the linewidth of a reference laser when locking the optical phase of the dye laser to a reference He–Ne laser.

Other methods of correcting the laser frequency can be used. They include changing the plasma discharge current of an He–Ne laser, changing the injection current of a diode laser, and using thermal expansion of the laser cavity. The designer has to take the transducer's unique properties into consideration. Very often several transducers have to be used in the system, with their individual responses smoothly tailored together in order to accomplish the task of laser frequency stabilization.

5.4 Loop Filter

While the typical error signal derived from a Fabry–Perot cavity with a Pound–Drever–Hall scheme is on the order of 1 V/MHz, the sensitivity of an intracavity EOM in a ring-cavity dye laser is about 0.1 MHz/V. It is obvious that gain is needed to use this error signal for the purpose of laser frequency noise suppression. As mentioned before, the loop filter (usually an electronic amplifier) is used to convert the error signal into the driving signal for the transducer(s). This loop filter provides necessary gain and phase compensation to obtain optimal system performance. In the system with multiple-references and/or multiple-transducers, this loop filter also coordinates the crossovers among the references and among the transducers.

Design of an optimized loop filter ordinarily requires consideration of a number of factors and often involves some compromises. While such a detailed servo loop filter design is certainly beyond the scope of this chapter [44], here we will discuss some guidelines for its application in laser frequency stabilization. In addition to guaranteeing the system stability, the design goal of a loop filter is to provide enough gain to suppress the laser frequency/phase noise, ideally down to the level limited by the shot noise, based on the best available knowledge of the derived error signal.

As mentioned at the beginning of this chapter, the laser frequency noise with technical origin has large modulation indices at lower Fourier frequencies, compared to the laser linewidth. The loop filter should be able to provide large enough gain at these Fourier frequencies. Because the slope of the gain versus the Fourier frequency is fixed by the loop stability requirement (at least near the unity gain frequency), one would like to push the unity gain frequency as high as possible.

The limit on the obtainable unity gain frequency in a servo system is the time delay from a perturbation acting on the laser until the action taken by the transducers begins to correct the noise. Although ultimately this time delay is limited by the finite speed of light, the main contribution of this time delay usually comes from the servo electronics. The existence of this unavoidable time delay limits the achievable performance of the whole system and makes the loop filter design more challenging (and more fun). There are several methods that could be used to analyze a servo system with a time delay. The proper mathematical description of such a system is differential–difference equations [45], which could be solved, at least numerically. Alternatively, a closed form of the stability criterion for such a system could be readily obtained using Cauchy's theorem in most cases [9, 46]. This stability criterion is useful for analysis of the achievable result of a given system. With ever-increasing computing power and availability of specialized software for analysis and simulation [47], the quantitative response of such a system could be obtained with much less efforts. For the present purposes, a reasonable criterion is to choose

$$f_{0\,\mathrm{dB}} \approx \frac{1}{2\pi\,\tau_{\mathrm{delay}}}, \tag{19}$$

where $f_{0\,\mathrm{dB}}$ is the unity gain frequency, and τ_{delay} is the system time delay. This choice will correspond to ~1 rad of delay-generated excess phase at the unity gain frequency.

After the unity gain frequency is determined according to the time delay of the system, the next step is to obtain as much gain as possible at the Fourier frequencies where the noise modulation indices are large. (For most cases one can just get merely enough gain at these Fourier frequencies.) The Bode plot [48] is conveniently used here for analysis purposes. The simplest loop filter would be an integrator. The corresponding −6 dB/octave slope at unity gain frequency guarantees system stability since the phase shift of 90° plus the 1 rad delay-generated phase are securely below 180°. The cost of this easy stability is a reduction in the achievable gain at lower frequencies. A −12 dB/octave at unity gain frequency causes the system to be unstable. Thus, a −9 dB/octave at unity gain frequency can be recommended [49]. This has the virtue that (in the absence of phase shift due to time delay) the shape of the dynamic settling curve is unchanged when the gain is changed, so that only the time scale changes. At about one decade below the unity gain frequency, it is usual to begin an even steeper slope in order to accumulate sufficient gain to suppress large-amplitude FM noise inputs at lower frequencies due to vibrations, temperature drifts, etc. For the fastest loops, where time delay is controlling, we usually cross unity gain with a slope of −6 dB/octave. Indeed, a frequency-flat "proportional" response about 3 dB below unity gain can

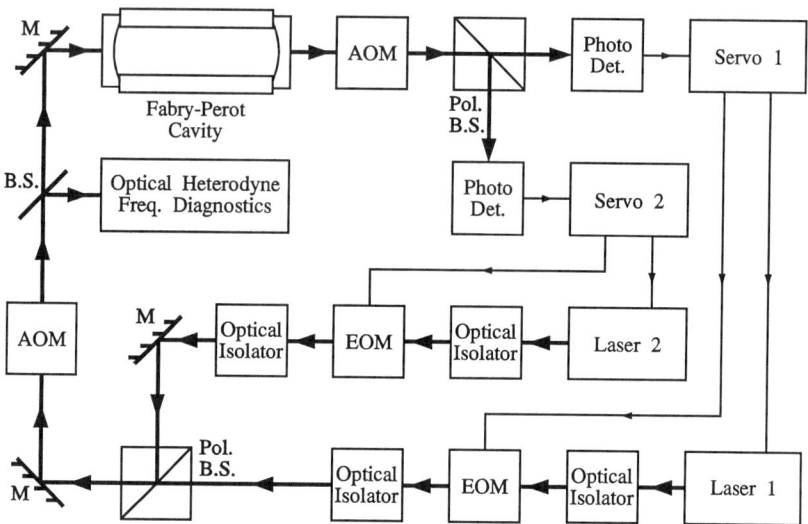

FIG. 10. Block diagram for He–Ne laser cavity locking. See text for discussion.

be added to benefit the transient response, extending an octave or more in both directions around the unity gain frequency.

More sophisticated designers might include some "smart" parts in the loop filter to further improve system performance. For instance, introduction of a clamped "bypass" integrator [27] does not increase the fast channel time delay, nor does it change the transient response of the system significantly, but it offers more gain when the system approaches equilibrium. We often use self-adaptive clamps, such as two series diodes connected with a reversed pair across a "bypass" integrator. When the error signal becomes small, this clamp gradually removes itself from any active role in the circuit, thus leading to a larger gain at low frequencies.

5.5 Design Examples

With increasing availability of high-performance optoelectronics, it might not be appropriate to recommend a "one-design-fits-all" circuit schematic that could become obsolete sooner than we expect. Instead, we discuss a couple of laser phase/frequency stabilization systems.

5.5.1 Short- and Long-Term Frequency Stability of an He–Ne Laser System

Figure 10 shows the block diagram of an experimental setup [50]. The purpose of this experiment is twofold. The first is to test the frequency stabilization

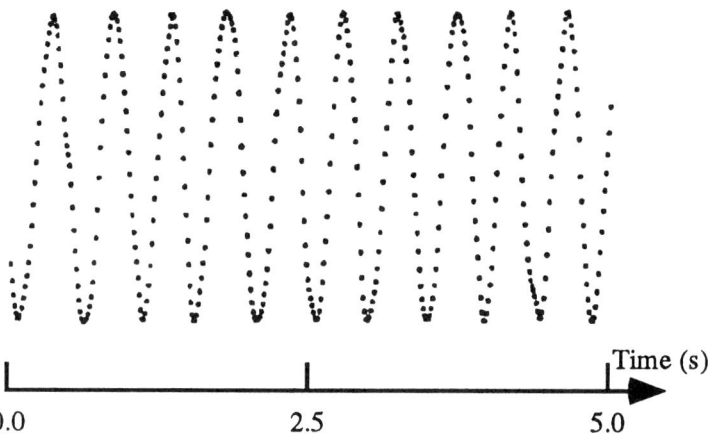

FIG. 11. Heterodyne signal of two He–Ne lasers independently locked to the adjacent longitudinal mode of a Fabry–Perot cavity. The detected heterodyne signal (~250 MHz) has been down-converted to ~2 Hz by a phase-stable frequency synthesizer.

electronics. This was accomplished using two He–Ne lasers locked to the adjacent longitudinal modes of a Fabry–Perot cavity. In the observed heterodyne frequency, the cavity resonance frequency change due to the environmental disturbances is reduced by a factor of FSR/f_{laser}, so that we were able to concentrate on the frequency-locking scheme. The dielectric mirrors were optically contacted to the polished ends of the Zerodur spacer [19]. The spacer was suspended in a vacuum chamber. The pressure of the chamber was maintained to less than 10^{-6} Pa by using an ion pump. The temperature of the chamber was stabilized to ~1 mK. The time constant for heat transfer from the chamber to the spacer was ~24 hours. The free spectral range and the full linewidth of the cavity were 250 MHz and 12 kHz, respectively. Discharge current, a PZT tube cemented to the laser tube, and heating tape bonded to the laser tube constituted a three-stage transducer for the two identical He–Ne lasers. Two Faraday isolators for each laser and two AOMs were used to reduce the possible optical feedback. Two orthogonally polarized laser beams were combined spatially using a polarization beam splitter. Another polarization beam splitter was used to separate the transmitted beams from the cavity for two independent servo systems. The heterodyne signal was further down-converted to ~2 Hz, as shown in Fig. 11. Fitting this waveform resulted in a linewidth of 50 milli-Hz for this beat signal. The limit of this linewidth resulted from the polarization method used to separate the transmitted beams. This method is sensitive to a change in system birefringence, especially the birefringence of highly reflective cavity mirrors.

The second purpose of this experiment was to test the long-term stability of this system. This test was accomplished by measuring the frequency difference of one

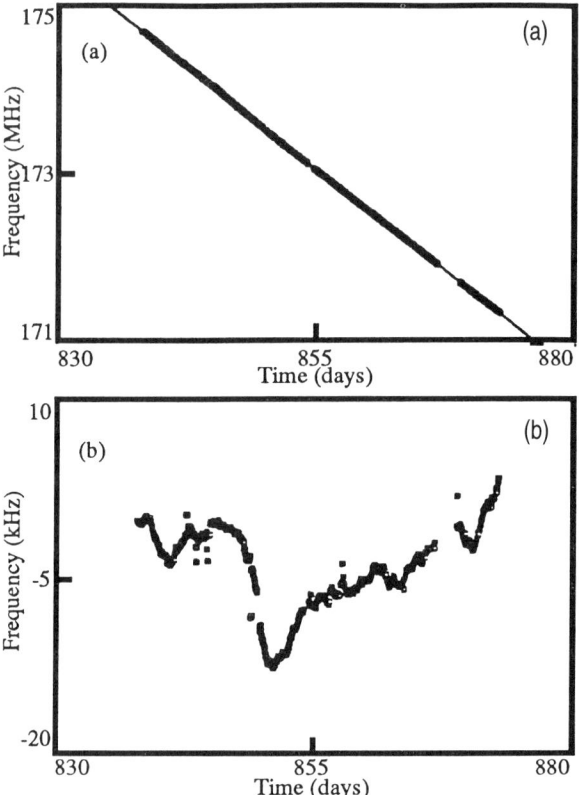

FIG. 12. (a) Zerodur [19] cavity frequency drift. (b) Residual frequency after the linear term 1.1 Hz/sec is removed from (a).

He–Ne laser locked to the Fabry–Perot cavity described above and another He–Ne laser locked to a hyperfine transition of iodine molecules. This experiment is also very useful for other precision measurements such as the Kennedy–Thorndike experiment [51]. The resultant heterodyne frequency is shown in Fig. 12. Even in a high-vacuum and stable-temperature environment, there was still a 1.1 Hz/sec linear frequency drift in this particular Zerodur [19] cavity. This material aging is related to both the material components and the processing history of the spacer under test [52]. Under a similar experiment, a different material, ULE [19], showed a 0.25-Hz/sec linear frequency drift [23].

5.5.2 Frequency and Phase Locking of a Commercial Dye Laser System

It usually requires a fast servo system to stabilize a dye laser's frequency [9]. Figure 13 shows an EOM/AOM combination of transducers together with an

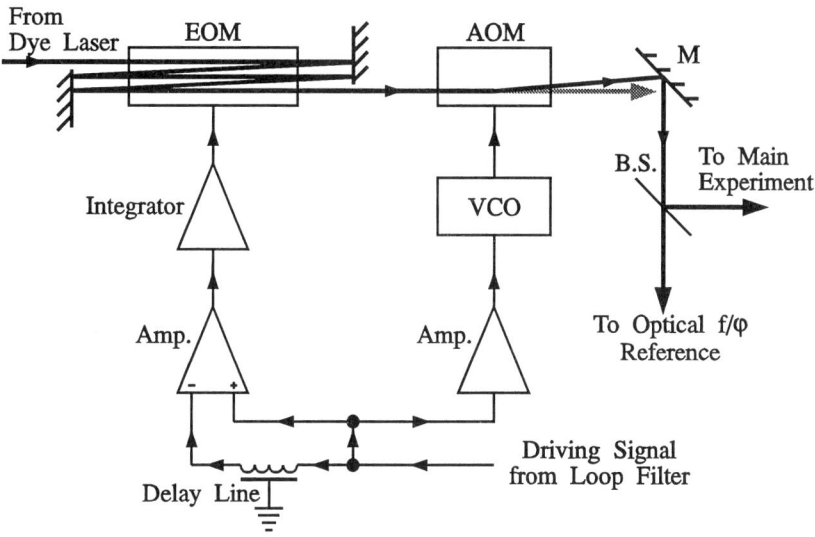

FIG. 13. Block diagram of EOM/AOM combination with an electrical interferometer for external laser frequency/phase correction.

electrical interferometer used to correct the laser frequency/phase outside the laser cavity. Efforts were made to reduce the time delay in the EOM channel. Notice that it is the time delay in this fast channel that sets the achievable unity gain frequency. The importance of the reduction in this time delay cannot be overemphasized, because this time delay cannot be compensated. For fast servo loops it cannot be ignored. The laser beam passed through the EOM five times, so that the sensitivity of the EOM was increased to 3.18 nm/V. This enabled us to construct an accurate EOM driver with a ±25-V range and a 10-nsec delay time using an extra-fast operational amplifier and GHz-bandwidth rf transistors. An electrical interferometer was used to cancel the error signal presented to the EOM driver at just the moment when the acoustic wavefront in the AOM reached the interaction point [43]. Operationally this means adjusting the stimulation-to-optical response delay of the AOM by displacing it along the direction of the sound wave propagation to match the preselected electrical delay line (400–600 nsec).

Figure 14 shows a frequency stabilization system for a commercial dye laser used in UV spectroscopy of laser trapped/cooled Na atoms [53]. A commercial dye laser (Coherent 699) [19] gives about 1 W of output power at $\lambda = 570$ nm with a 0.45-MHz linewidth when it is locked to its own reference cavity (Fig. 15). The spectroscopic feature ($5p$ state of Na atoms) has a natural linewidth of 450 kHz (FWHM) at $\lambda = 285$ nm with hyperfine splitting on the order of 10 MHz. In addition to laser linewidth narrowing, this experiment requires that the laser

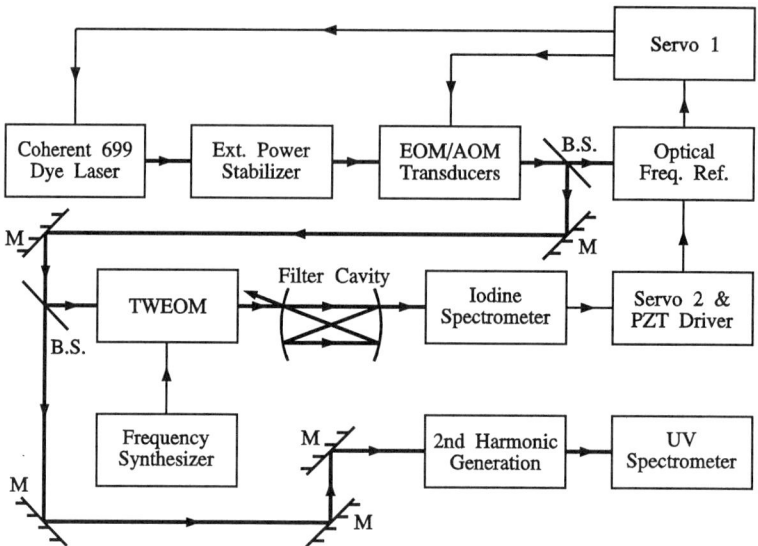

FIG. 14. Block diagram for external frequency stabilization of a commercial dye laser.

frequency be stabilized to an absolute reference frequency with precise tuning capability. We chose to accomplish these requirements in two steps: (1) to stabilize the laser frequency to an additional tunable Fabry–Perot cavity such that the laser linewidth can be narrowed relative to the cavity; (2) to stabilize the cavity resonant frequency with a tunable offset frequency relative to a nearby hyperfine transition of iodine molecules. These two steps were accomplished before the $\lambda = 570$ nm light was sent to the nonlinear optical crystal to generate second harmonics for the measurements.

In Fig. 14 the external power stabilizer used an EOM/polarizer combination and exhibited shot noise–limited performance for Fourier frequencies <1.5 MHz, with 85% of the input power available as useful power output. We used the Pound–Drever–Hall scheme, as shown in Fig. 5, to derive the error signal from an additional Fabry–Perot cavity, The spacer of this Fabry–Perot cavity is made of Zerodur [19] for thermal stability. The cavity has an FSR of ~330 MHz and a linewidth of ~250 kHz (FWHM). The cavity is hermetically sealed, with one of its mirrors mounted on a PZT stack for tuning the resonance frequency. Sidebands at 12.5 MHz were generated by an EOM with a modulation index of ~0.9 to derive the error signal. The EOM/AOM transducers with electrical interferometer topology were the ones used previously, as shown in Fig. 13. A much slower integrator tuned the laser's own reference cavity to make it track off any dc frequency error. In this way we could guarantee the laser beam pointing stability even though we only used a single pass through the AOM.

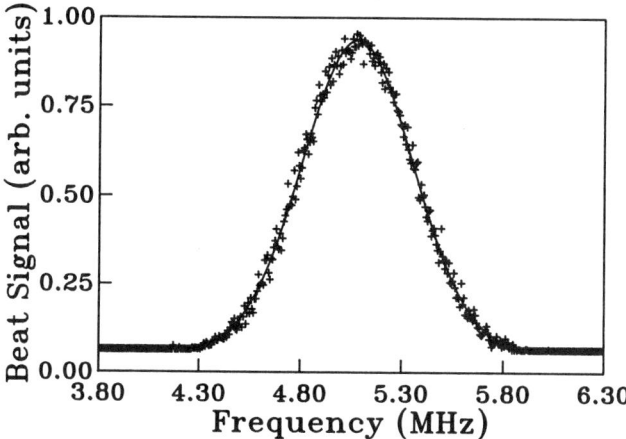

FIG. 15. Heterodyne signal between a reference He–Ne laser ($\lambda = 612$ nm) and the commercial dye laser, which is locked to its own reference cavity. The reference He–Ne has a linewidth of 10 kHz (FWHM). The crosses are measured data, while the solid curve is a Gaussian profile fit to the data.

The unity gain frequency of the servo loop (Servo 1) was ~4 MHz. Figure 16 shows the frequency error at different gain settings [9]. The top reference level was the reference for 125-kHz detuning. With the correction of the intentionally introduced saturation, this reference trace gave −3 dBm. The bottom trace was taken at the proper gain setting, which showed that the frequency noise was limited by the measurement noise (shot noise). The resultant linewidth of the laser was 0.49 Hz (FWHM) relative to the cavity.

Slow changes in the resonant frequency of this Fabry–Perot cavity were corrected by locking it to a hyperfine component of an iodine molecule (I_2) transition. To achieve a broad tuning range to cover the gap between the available I_2 transitions at $\lambda = 570$ nm and the Na transitions (at $\lambda = 285$ nm after the doubler), we used a traveling wave EOM (TWEOM) to generate FM sidebands on a portion of the light (modulation index $\beta \approx 1.8$). A confocal filter cavity was locked on the desired first-order sideband. The transmitted light was sent to a conventional FM saturation spectrometer (see Fig. 9). The error signal was used for a servo system (Servo 2 in Fig. 14) to keep the reference cavity frequency at the center of the hyperfine transition with a frequency offset. The residual error signal of the I_2 locking system showed that the servo system performance was limited by the achievable S/N for the chosen iodine transition. The long-term frequency uncertainty was estimated to be ~5 kHz. Tuning the driving frequency of the EOM with a programmable rf synthesizer, we effectively tuned the laser frequency relative to the iodine hyperfine transition with rf precision. The rest of the light (~0.5 W) was

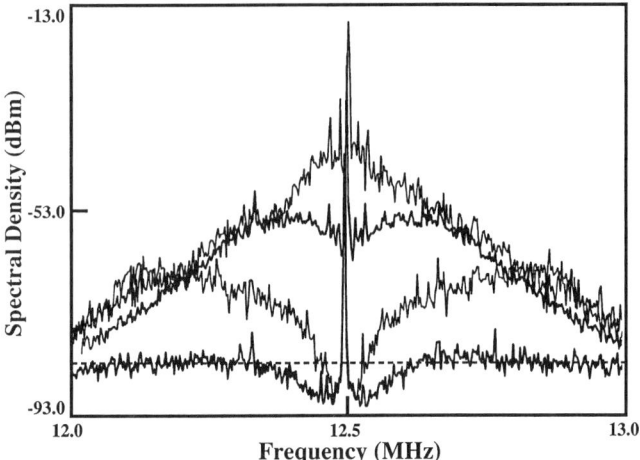

FIG. 16. Result of external dye laser frequency locking to a Fabry–Perot cavity. The resolution bandwidth is 1 kHz. With the proper gain setting, no intrinsic laser noise remains; even the shot noise below 100 kHz has been written as an equivalent FM onto the laser. The dashed line shows the shot noise level at 1-kHz resolution bandwidth. The dc-based error signal results when this rf spectrum is "folded" and displaced by ~12.5 MHz in the rf balanced mixer. Servo gain increases 10 dB for each descending curve.

sent to a ring buildup cavity containing an $NH_4H_2AsO_4$ (ADA) nonlinear optical crystal. Thus, we reliably generated several mW of frequency- and intensity-stabilized tunable light at $\lambda = 285$ nm for spectroscopy measurements.

Figure 17 shows an experimental setup for optical phase locking of the same dye laser onto a reference He–Ne laser. Optical phase locking can add precise tunability to the stable reference laser and/or "amplify" the stable reference laser's power, as mentioned in Section 5.2.3.

The external power stabilizer and EOM/AOM transducers with electrical interferometer topology were the same as in Fig. 14. The dye laser beam was combined with the reference He–Ne laser beam ($\lambda = 612$ nm) by a beam splitter for heterodyne detection. An avalanche photodiode (APD) detected the heterodyne signal between these two lasers. After being down-converted with the output of an auxiliary rf synthesizer, the heterodyne signal was sent to an emitter-coupled-logic (ECL) phase/frequency detector working at 45 MHz with a quartz oscillator as its local reference. This auxiliary rf synthesizer could tune the dye laser frequency relative to the reference He–Ne laser frequency while keeping the optical phase locked. After the servo loop parameters were optimized for this setup, the residual phase fluctuation was limited within ±0.65 rad, and exhibited a Gaussian distribution, as shown in Fig. 18. Further data processing showed that the variance of the

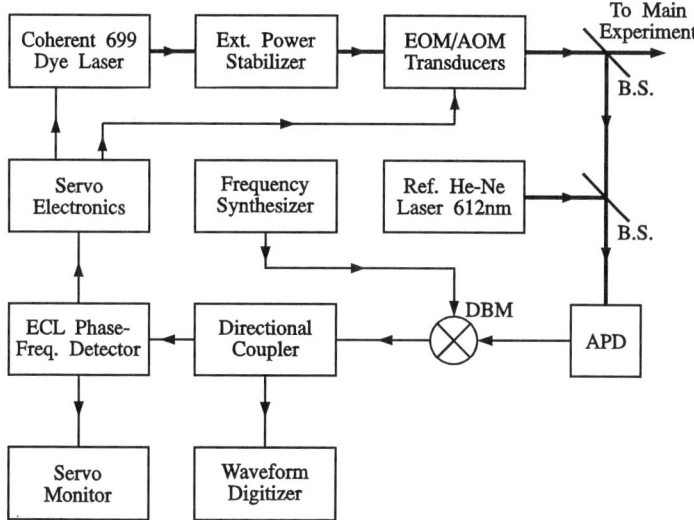

FIG. 17. Block diagram for external optical phase locking of the dye laser to a reference He–Ne laser at λ = 612 nm.

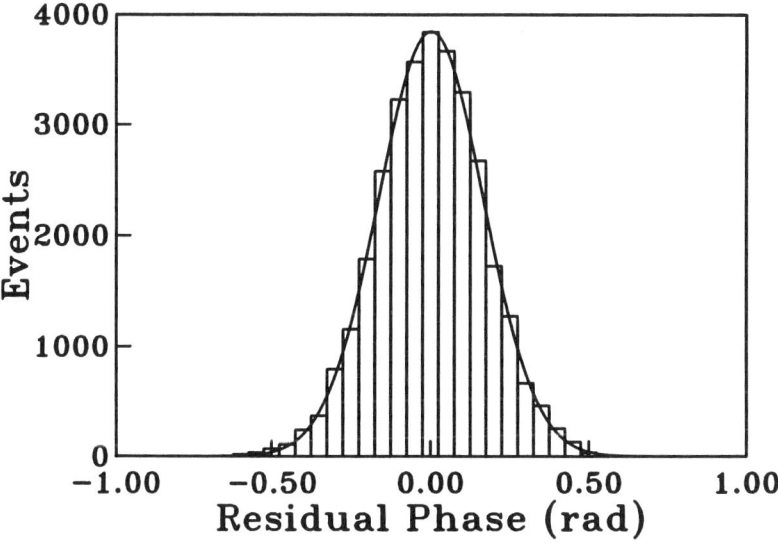

FIG. 18. Histogram of residual optical phase fluctuations after the dye laser was phase locked to reference He–Ne laser. Solid curve is the Gaussian distribution fit to the data with $\varphi_{rms}^2 = 0.028$.

residual phase fluctuation is ~0.028 rad^2, which implied that more than 97% of the dye laser power was put into the linewidth of the reference He–Ne laser [9].

5.6 Summary

In this chapter we presented the topic of laser frequency stabilization using a servo system. We discussed optical frequency references and transducers in reasonable detail. We presented some guidelines for loop filter design and emphasized the importance of the time delay in the system design. We showed two different laser phase/frequency stabilization systems along with their experimental results. We hope that the readers can benefit from the discussions here and design servo systems to fulfill their own laser phase/frequency stabilization requirements with newly available analysis tools and high-performance optoelectronics devices. Over the next several years, we expect to publish several detailed examples of these laser stabilization principles, particularly as applied to Ti:sapphire, diode-pumped Nd:YAG, and semiconductor diode laser systems.

References

1. Schawlow, A., and Townes, C. (1958). *Phys. Rev.* **112**, 1940.
2. See Fox, R. W., Zibrov, A. S., and Hollberg, L. (1996). "Semiconductor Diode Lasers," Chapter 3 in this volume, and the references therein.
3. Middleton, D. (1948). *Q. Appl. Math.* **5**, 445.
4. Middleton, D. (1949). *Q. Appl. Math.* **7**, 129.
5. Middleton, D. (1951). *Philos. Mag.* **42**, 689.
6. Middleton, D. (1952). *Q. Appl. Math.* **9**, 337.
7. Middleton, D. (1952). *Q. Appl. Math.* **10**, 35.
8. Elliott, D. S., Roy, R., and Smith, S. J. (1982). *Phys. Rev. A* **26**, 12.
9. Zhu, M., and Hall, J. L. (1993). *J. Opt. Soc. Am. B* **10**, 802.
10. Born, M., and Wolf, E. (1980). *Principles of Optics*, Pergamon, New York.
11. Kogelnik, H., and Li, T. (1966). *Appl. Opt.* **5**, 1550.
12. Siegman, A. E. (1986). *Lasers*, University Books, Mill Valley, CA.
13. DeVoe, R. G., and Brewer, R. G. (1984). *Phys. Rev. A* **30**, 2827.
14. Anderson, D. Z. (1984). *Appl. Opt.* **23**, 1238.
15. Rempe, G., Thompson, R. J., and Kimble, H. J. (1992). *Opt. Lett.* **17**, 363.
16. Hauck, R., Kortz, H. P., and Weber, H. (1980). *Appl. Opt.* **19**, 598.
17. Uehara, N., and Ueda, K. (1995). *Appl. Opt.* **34**, 5611.
18. Schott Glass Technologies Inc., Duryea, PA. See also note [19].
19. Mention of a commercial product is for technical communication only. It does not imply endorsement, nor does it suggest that other products are necessarily less suitable for the application.

20. Corning Inc., Corning, NY. See also note [19].
21. For example, Torr Seal, Varian Associates, Lexington, MA. See also note [19].
22. Landau, L., and Lifshitz, E. (1959). *Theory of Elasticity*, Pergamon Press, Oxford.
23. Zhu, M., and Hall, J. L. (1992). "Short- and Long-Term Stability of Optical Oscillators," in *Proceedings of the 1992 IEEE Frequency Control Symposium*, Hershey, Pennsylvania, May, p. 44.
24. Hils, D., and Hall, J. L. (1987). *Rev. Sci. Instrum.* **58**, 1406.
25. Hänsch, T. W., and Couillaud, B. (1980). *Opt. Comm.* **35**, 441.
26. Pound, R. V. (1946). *Rev. Sci. Instrum.* **17**, 490.
27. Drever, R. W. P., Hall, J. L., Kowalski, F. V., Hough, J., Ford, G. M., Munley, A. J., and Ward, H. (1983). *Appl. Phys. B* **31**, 97.
28. Bjorklund, G. C. (1980). *Opt. Lett.* **5**, 15.
29. Hall, J. L., Hollberg, L., Baer, T., and Robinson, H. G. (1981). *Appl. Phys. Lett.* **39**, 680.
30. See, e.g., Letokhov, V. S., and Chebotayev, V. P. (1977). *Nonlinear Laser Spectroscopy*, Springer-Verlag, Berlin.
31. Phillips, W. D., ed. (1983). *Laser-Cooled and Trapped Atoms*, National Bureau of Standards, Special Publication No. 653, Washington, DC.
32. Meystre, P., and Stenholm, S., eds. (1985). "Feature on the Mechanical Effects of Light," *J. Opt. Soc. Am. B* **2**, 1705 ff.
33. Chu, S., and Wieman, C., eds. (1989). "Feature on Laser Cooling and Trapping of Atoms," *J. Opt. Soc. Am. B* **6**, 2019 ff.
34. Bergquist, J. (1996). "Doppler-Free Spectroscopy," in *Experimental Methods in the Physical Sciences: Atomic, Molecular, and Optical Physics*, Volume 29B, Academic Press, New York, and the references therein.
35. See, e.g., Grimm, R., and Mlynek, J. (1989). *Appl. Phys. B* **49**, 179.
36. de Labachelerie, M., Nakagawa, K., Awaji, Y., and Ohtsu, M. (1995). *Opt. Lett.* **20**, 572.
37. Ye, J., Ma, L.-S., and Hall, J. L. (1996). *Opt. Lett.* **21**, 1000.
38. Suzumura, K., and Sasada, H. (1995). *Jpn. J. Appl. Phys.* **34**, L1620.
39. Kourogi, M., Nakagawa, K., and Ohtsu, M. (1993). *IEEE J. Quantum Electronics* **29**, 2693.
40. Brothers, L. R., Lee, D., and Wong, N. C. (1994). *Opt. Lett.* **19**, 245.
41. Helmcke, J., Lee, S. A., and Hall, J. L. (1982). *Appl. Opt.* **21**, 1686.
42. Yariv, A. (1971). *Introduction to Optical Electronics*, Holt, Rinehart and Winston, New York.
43. Hall, J. L., and Hänsch, T. W. (1984). *Opt. Lett.* **9**, 502.
44. See, e.g., Dorf, R. D. (1989). *Modern Control Systems*, Addison-Wesley, Reading, MA.
45. See, e.g., Bellman, R., and Cooke, K. (1963). *Differential–Difference Equations*, Academic Press, New York.

46. Whittaker, E. T., and Waston, G. N. (1927). *Modern Analysis*, Cambridge Univ. Press, Cambridge.
47. For example, Matlab and SimuLink, from MathWorks Inc., Natick, MA. See also note [19].
48. Bode, H. W. (1975). *Network Analysis and Feedback Amplifier Design*, Krieger Publishing, Huntington, New York.
49. Hall, J. L. (1985). *Lectures on Laser Frequency Stabilization for Scientific Applications*, University of Colorado, unpublished.
50. Salomon, C., Hils, D., and Hall, J. L. (1988). *J. Opt. Soc. Am. B* **5**, 1576.
51. Hils, D., and Hall, J. L. (1990). *Phys. Rev. Lett.* **64**, 1697.
52. Bayer-Helms, F., Darnedde, H., and Exner, G. (1985). *Metrologia* **21**, 49.
53. Zhu, M., Oates, C. W., and Hall, J. L. (1993). *Opt. Lett.* **18**, 1186; (1993). Erratum in *Opt. Lett.* **18**, 1681.

6. PULSED LASERS

Michael G. Littman and Xiao Wang

Department of Mechanical and Aerospace Engineering
Princeton University
Princeton, New Jersey

6.1 Introduction

Pulsed lasers are versatile tools for the scientist and engineer that have played an important role in the development of modern optical physics. The first laser, invented by T. H. Maiman, was a pulsed laser [1]. In this laser a ruby crystal (chromium-doped sapphire) served as the gain medium, which was excited by light from a pulsed flashlamp that surrounded the crystal. Two parallel faces of the crystal were coated with silver, which served to trap photons emitted along the optic axis so that they would pass through the crystal many times before escaping. The laser operated at room temperature and produced pulses of light with a wavelength of 694 nm (near infrared).

Pulsed lasers have proven to be especially helpful in the study of transient phenomena that occur in time intervals as short as femtoseconds, or as long as milliseconds or beyond. The shortest-duration laser created to date produced pulses of six-femtoseconds duration, a time that corresponds to only three optical cycles [2]. Because of their high peak power, pulsed lasers have also been useful in helping researchers study and apply effects that scale nonlinearly with increasing optical field strength. Pulsed lasers have been operated that have achieved enormous peak powers of as much as 10^{18-19} W/cm^2, which corresponds to a field strength strong enough to strip ground-state electrons from atoms.

In comparison with continuous-wave (CW) lasers, pulsed lasers generally are easier to operate, are more flexible, have higher peak powers, and cover a larger part of the electromagnetic spectrum. Pulsed lasers are also easier to build, because the high gain of the laser medium that is typical of pulsed lasers means that intracavity laser optics can be of relatively low optical quality and lossy. Thus, such inexpensive optics as aluminized mirrors, uncoated lenses, and polarizers are usable in the construction of pulsed lasers. For these reasons, many researchers choose to use pulsed lasers even when CW lasers might appear to be preferable. At present, pulsed lasers are widely used in areas ranging from atomic and molecular spectroscopy to remote sensing, from fiberoptic communications to laser medicine.

In this chapter we provide an introduction to a wide variety of pulsed lasers. For an in-depth discussion of lasers in general, the reader is referred to A. Siegman's excellent text [3]. The chapter begins with a section that provides a brief overview of pulsed lasers and introduces some of the basic ideas that have led to the many different forms of pulsed lasers that now exist. This section can be viewed as a guided tour to pulsed lasers, and includes discussion of their operating principles, their attributes that might be useful in a variety of applications, as well as a brief survey of the types of lasers that are commercially available. The next section is a buyer's guide, where a range of issues are discussed that are relevant to the selection of specific lasers for specific applications. The last section gives a step-by-step guide to the building of pulsed tunable dye lasers of varying levels of performance. Pulsed tunable dye lasers are representative of most pulsed lasers, with the added feature that they are especially forgiving of imperfect optical elements and operating conditions. The robustness of pulsed tunable dye lasers derives from the very high gain (e.g., 1000 per pass) of the dye medium. For this reason they are ideal for exploration by the novice.

6.2 Pulsed Lasers

6.2.1 General Concepts

The LASER (**L**ight **A**mplified by the **S**timulated **E**mission of **R**adiation) is a quantum-mechanical device. There are three main elements in the typical laser: (1) a gain medium that has the capacity for light amplification along an optic axis, (2) an exciting energy source that raises the gain medium to a higher internal quantum state, and (3) a cavity that feeds back (i.e., resonates) light along the optic axis so as to efficiently extract energy stored in the gain medium. When the gain medium is excited by the energy source, states of higher energy become more occupied (i.e., populated) than states of lower energy, thereby creating a population inversion. (Population inversions are unusual in nature because quantum systems in thermal equilibrium follow a Boltzmann distribution, in which case lower-energy states are preferentially populated relative to higher-energy states.) When population inversion is maintained in the gain medium, light at or near a resonance frequency corresponding to a transition between the higher- and lower-energy states in the gain medium will be amplified when passing through the gain medium, by the process of stimulated emission. (If the medium were not inverted, light at or near the resonance frequency traveling in the medium would be attenuated by the complementary process of stimulated absorption.) If the cavity gain exceeds the cavity loss, the device will be self-sustaining and form what is generally referred to as a laser oscillator.

Gain media can be excited in numerous ways, including: (1) by electron collisions such as those that excite atoms in a gaseous discharge lamp, (2) by chemical reaction in which product molecules are formed in an excited state, or (3) by direct optical excitation due to absorption from an intense light source such as a flashlamp or laser. When the gain medium is optically excited, it is usually the case that the exciting light will be of higher energy (i.e., shorter wavelength) than the laser transition. So, for example, the flashlamps used in a ruby laser flood the crystal with light in the blue and near-ultraviolet portion of the spectrum, while the laser itself operates in the near-infrared.

There are two possible ways to achieve a pulsed output from a laser: (1) excite the gain medium for a brief but finite duration, or (2) allow for the circulation of light in the gain medium only for a finite time. The first method makes use of a pulsed excitation source. The second makes use of an intracavity optical switch. Depending upon how they are operated, optical switches can allow for either Q-switching, discussed later in this chapter, which provides for the rapid dumping of energy stored in the gain medium, or mode locking, which is capable of providing extremely short-duration optical pulses and is a subject treated in another chapter of this volume.

6.2.2 Ruby Laser

The first laser, built by Maiman in early 1960, used a ruby crystal as its gain medium [1]. Figure 1a shows the cavity structure of a ruby laser. Ruby, which is composed of Cr^{3+} ions embedded in sapphire host crystal, is a gain medium representative of the so-called "three-level system." A schematic energy diagram is shown in Fig. 1b. A time history of the laser output and the flashlamp output is shown in Fig. 1c. The electrons in the Cr^{3+} are first pumped by a pulsed flashlamp from ground states to a band of high-energy states. These states then relax rapidly through nonradiative processes into the first excited state of Cr^{3+} ion, which is long-lived. The laser transition, $\lambda = 694$ nm, occurs between the first excited electronic state and the ground state, and in the process the ground state becomes repopulated, thereby destroying the population inversion. Thus, the laser is pulsed, its width determined by the on time of the flashlamp. Since normally all Cr^{3+} ions are in the ground state at room temperature, in order to achieve the condition for population inversion it is necessary to excite more than 50% of the Cr^{3+} ions. Thus, ruby lasers require intense pumping, and as a result they are no longer widely used.

6.2.3 Q-Switching

Shortly after the invention of the ruby laser, the method of Q-switching was introduced by R. W. Hellwarth to create a high-intensity burst of light [4]. Q is the quality factor of a resonator, and it is a measure of its energy storage ability. For a Q-switching operation, a population inversion is built up during a pumping period that in the ideal case matches the excited state lifetime of the laser medium.

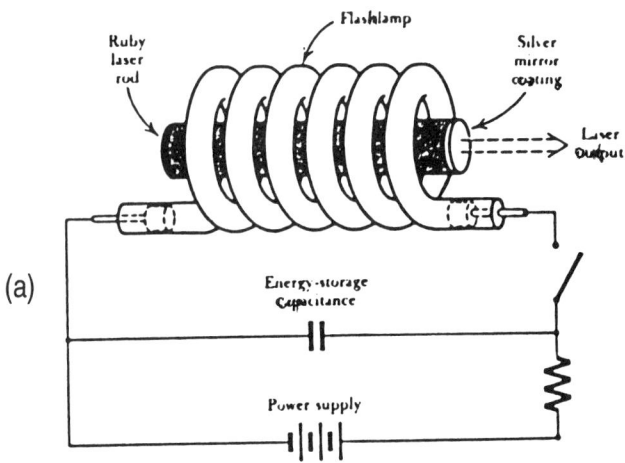

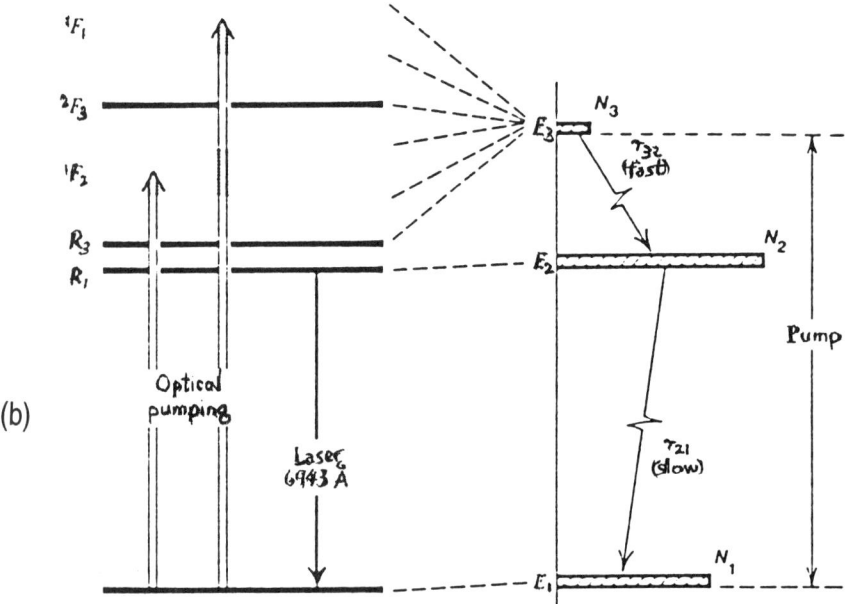

FIG. 1. The ruby laser: (a) physical configuration; (b) energy level diagrams (on right is schematic idealization; on left is real physical system); (c, opposite) time history of laser and flashlamp output. From Siegman [3]. Reprinted with permission from University Science Books.

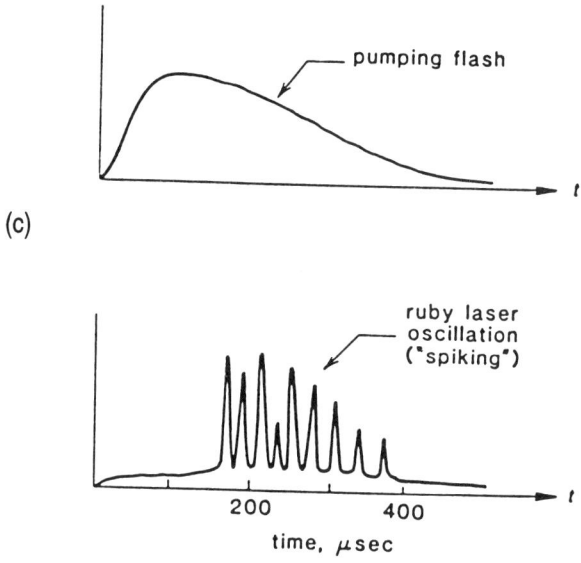

(c)

Fig. 1c

(For the ruby laser this time is approximately 3 msec.) During this pumping period the laser cavity Q is kept at a low value by keeping closed an intracavity optical switch in order to prevent circulation of light within the cavity. At the end of this pumping period, when the upper level is heavily populated, the cavity Q is suddenly switched from a low to a high value by opening the optical switch. Laser action then builds up rapidly, beginning with random noise light due to spontaneous emission. (Excited quantum systems decay spontaneously by the emission of light when there are no external driving fields. Spontaneous emission allows laser oscillators to be self-starting. If there was no spontaneous emission, it would be necessary to inject light into a laser cavity to get it oscillating. Spontaneous emission is distinct from stimulated absorption and stimulated emission in that these latter two processes require external driving fields.) Spontaneous light that falls in a narrow range of angles defining the optic axis of the gain medium is reflected by the mirrors and coherently amplified by the stimulated emission process. In a very short time, corresponding to a few passes through the gain medium, most of the energy stored in the medium is extracted. (The stored energy in the gain medium falls to zero typically in several nanoseconds.) For most solid-state laser media, this duration is much shorter than their natural (spontaneous emission) decay time. Q-switching is thus used to create a burst of light that is briefer and more intense than one would obtain without an optical switch.

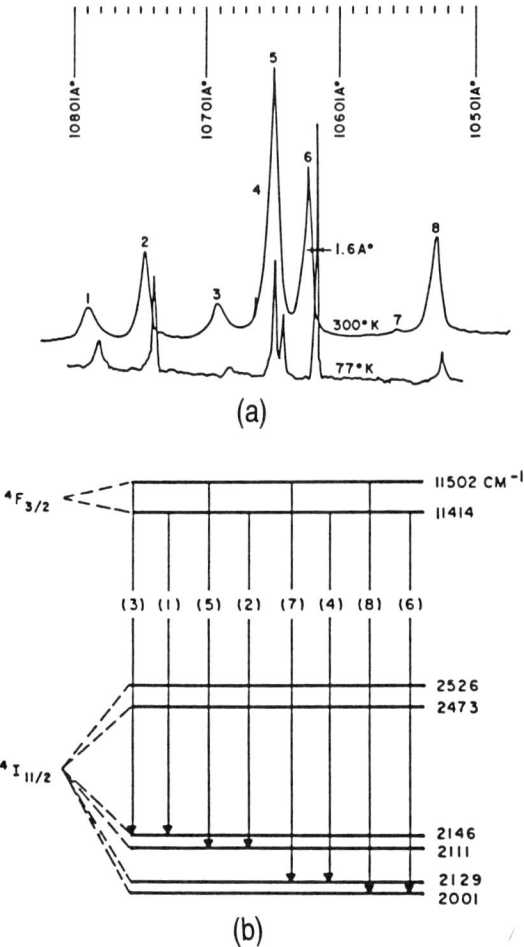

FIG. 2. Nd:YAG laser: (a,b) energy level spectrum and diagram; (c, opposite) energy level diagram idealization. From Geusic *et al.* [5]. Reprinted with permission from the American Institute of Physics.

If a laser medium has a long excited state lifetime, the population in the upper state can be built up gradually to a high level during that time, which can be accomplished using a modest pumping source. Therefore, Q-switching is easier to achieve in laser media with long excited state lifetimes, such as ruby. Other factors, such as the gain of the laser medium and the saturation fluence level of the medium, also influence the energy storage in the medium, and thus the efficiency of Q-switching,

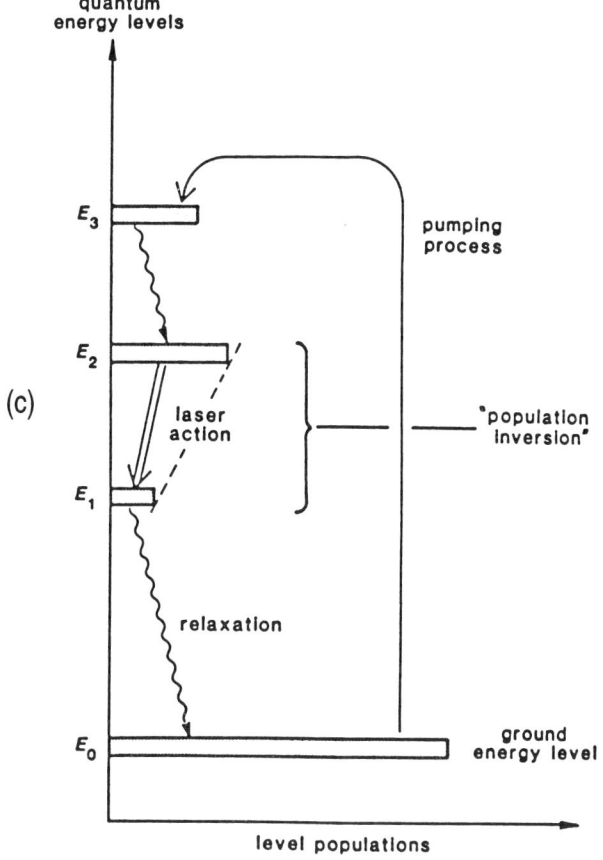

Fig. 2c

6.2.4 Nd:YAG Laser

Another laser in the solid-state laser family, the neodymium-doped yttrium aluminum garnet (Nd:YAG) laser, was first demonstrated by Geusic *et al.* in 1964 [5]. The Nd:YAG laser has been used extensively in a variety of applications, and especially for those requiring high peak powers. It is a four-level laser system (see Fig. 2), where the lower laser level is not the ground state. Therefore, in contrast to the Cr^{3+} ion that is used in the ruby laser, the Nd^{3+} ion is much easier to invert. The YAG host crystal is hard and has high thermal conductivity and good optical quality. The combination of easy pumping of the Nd^{3+} ion and a robust YAG host results in an excellent laser material that has high gain and low threshold for laser operation. Furthermore, the excited state lifetime of Nd:YAG is 230 μsec, which can be easily matched by the duration of a typical xenon flashlamp; therefore, Q-switched operation can be easily realized in the Nd:YAG laser. Commercially available Q-switched Nd:YAG laser oscillator-amplifier systems can generate pulses that are several nanoseconds long, with 1–2 J of energy per pulse, and operate at repetition rates of up to tens of hertz. (Higher-frequency operation at these power levels leads to cracking of the laser crystal due to thermal stress.)

Although the output of the Nd:YAG laser is in the infrared, at a wavelength of 1.06 μm, by means of frequency doubling, tripling, or quadrupling using nonlinear crystals, laser pulses can be obtained in the visible (532 nm) and ultraviolet (355, 266 nm), with pulse energies of up to hundreds of millijoules achievable from "tabletop" light sources. Harmonic generation can be very efficient, especially at high power, where as much as 80% of light at the fundamental can be frequency shifted to the visible and near-ultraviolet. Crystals are not usable for wavelengths shorter than about 200 nm because of direct optical absorption by the crystal material itself. However, high-order harmonics can be generated by using nonlinear effects in rare gases, which do not absorb readily. These have been used to produce pulsed soft x-rays at wavelengths as short as 7.3 nm [6].

Another laser that uses the Nd^{3+} ion is the Nd:glass laser, where glass is used as the host material. Although glasses have excellent optical quality, they are usually less homogeneous and have lower thermal conductivity compared to the crystalline hosts. Glasses are used mainly because they can be made in large sizes; therefore, more energy can be stored and extracted. Nd:glass lasers are used to generate high-peak-power light pulses that are applicable to thermonuclear fusion.

6.2.5 Cavity Mode and Injection Seeding

In steady state an optical cavity can only resonate light of discrete wavelengths that satisfy the following condition: the cavity length is equal to an integral number of half-wavelengths of the circulating radiation. These wavelengths define the axial or longitudinal cavity modes. Most pulsed lasers operate in quasi-steady state. If the gain bandwidth of the laser medium is narrower than the spacing between the adjacent axial modes, single-mode laser operation can be realized. It is more often, however, that the laser transition linewidth is far larger than the spacing between the adjacent cavity modes. In this case, the transient laser cavity usually will support several modes simultaneously, unless an additional mode selection mechanism is used in the cavity that forces one mode to have higher gain than all the others. In a Q-switched Nd:YAG laser, for example, when the intracavity optical switch is first opened, the laser builds up from spontaneous emission, and many axial modes within the laser gain bandwidth will build up at the same time, so that the output pulse contains many axial modes. Mode-beating in the pulse results in an undesirable irregular temporal and spatial profile, and large near-random fluctuations from pulse to pulse. This unsteadiness is a typical drawback of multimode pulsed lasers.

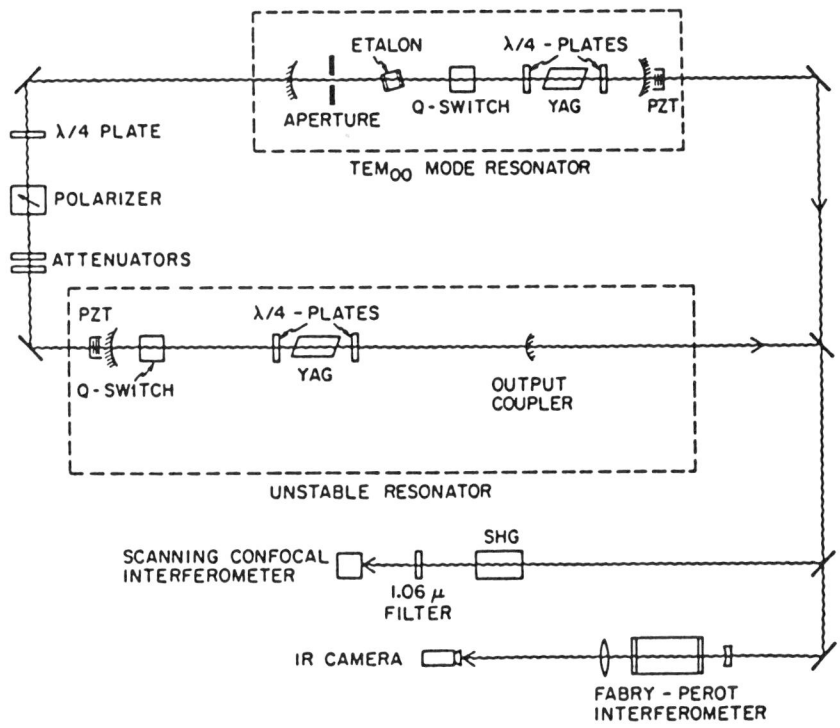

FIG. 3. Schematic of an injection-seeded Nd:YAG laser. From Park *et al.* [7]. Reprinted with permission from the IEEE.

Intracavity Fabry–Perot etalons can be used to reduce the number of modes, but they introduce large insertion loss and therefore reduce the output power. Injection seeding, applied by R. L. Byer [7] for mode selection in Q-switched Nd:YAG lasers, is a better method. Figure 3 is a schematic diagram for injection seeding of a Nd:YAG laser. For injection seeding, an external optical field with a bandwidth narrower than the mode spacing of the high-gain cavity and a field intensity much higher than that of the spontaneous emission noise field is fed into the high-gain laser cavity. If the frequency of the injected field matches an axial mode of the high-gain cavity, the injected photons will "seed" the cavity and be preferentially amplified. This will eventually deplete the gain in the medium, so that the other modes cannot be amplified, and only one mode will oscillate. The injection field is usually obtained from a single-mode CW laser operating at a frequency that falls within the gain bandwidth of the laser medium. Since there is no insertion loss in the cavity, injection seeding does not reduce the output energy of the laser. With injection seeding, Q-switched laser pulses with Fourier-transform-limited spectral linewidths have been obtained. Injection seeding also stabilizes laser operation significantly, reducing the time jitter and power variations from pulse to pulse. Injection-seeded Nd:YAG lasers are commercially available but are significantly more expensive.

6.2.6 Dye Lasers

Organic dye in solution also makes an excellent laser material. The first laser using this medium was a rhodamine 6G dye laser pumped by a frequency-doubled ruby laser, as demonstrated by P. Sorokin and J. Lankard in 1966 [8]. The upper state lifetimes of dyes are typically in the nanosecond range, so that a dye laser pumped by pulses nanoseconds or longer usually produces a pulse width equal to the duration of the pump pulse. A major advantage of dye lasers is their continuous tunability, which is a consequence of the large number of overlapping excited states in the heavy dye molecules. Tuning is achieved by using frequency selecting elements (for example, gratings) in the laser cavity. A grating can be used to feedback light selectively at the desired frequency for recirculation and amplification. By mechanical adjustment of the grating angle, the frequency for oscillation can be adjusted. Tunability is an important property that makes dye lasers popular light sources in laser spectroscopy. Figure 4 is a schematic diagram of a commonly used grating-tuned dye laser [14]. There are many different laser dyes available for operation in a broad range of optical wavelengths. Together, a collection of as few as 10 dyes can cover the spectral range from near-UV to infrared. Harmonics of a Q-switched Nd:YAG laser at $\lambda = 532$ nm and $\lambda = 355$ nm are especially well suited for the pumping of dye lasers to obtain frequency-tunable pulses in the nanosecond range. Short-duration gas discharge lasers such as XeCl excimer lasers at $\lambda = 306$ nm are also very useful in pumping dye lasers.

A disadvantage of dye lasers is that dyes can have short operational time due to bleaching of the dye molecules by the pump light source. Bleaching can require the replacement of dye solution after times as short as 20 minutes for the most fragile dyes. Dyes like rhodamine 6G are very robust, and they typically do not need replenishment for extended periods of time. Fragile dyes also have limited shelf life: sometimes less than one day after they are mixed. Another drawback is that dyes and solvents are sometimes environmentally hazardous.

6.2.7 High-Repetition-Rate Lasers

In some applications, high sampling rate, rather than high peak power, is needed. Therefore, a laser capable of providing a high repetition rate output pulse train is required. This can be realized by continuously pumping the laser medium and repeatedly Q-switching the laser at repetition rates up to several thousand pulses per second. For example, an acoustooptic Q switch can be used in an

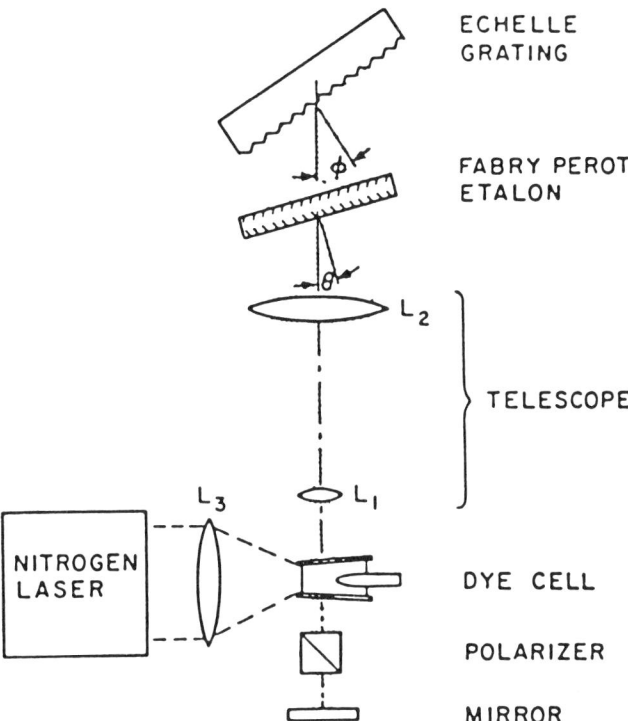

FIG. 4. Schematic of a narrow-band tunable dye laser. From Schawlow [14]. Reprinted with permission from John Wiley & Sons Inc.

Nd:YAG laser to provide a train of pulses at low energy per pulse and up to a 20 kHz repetition rate, with 10–20 watts average output power.

6.2.8 Gas Lasers

Another important class of pulsed lasers uses gas as the laser medium and a high current electron discharge as the pumping source. Examples are the nitrogen (N_2) laser, excimer lasers, and the CO_2 laser. Nitrogen lasers typically generate pulses that are several nanoseconds long, with up to several hundred millijoules of energy. The lasing wavelength is in the near-UV, at 337 nm, and it is suitable for pumping of dye lasers operating in the range of 400 to 800 nm. N_2 lasers were common pumping sources in the 1970s.

Excimer lasers, typically using mixtures of halogen and rare gases as the gain medium, are unique in that the laser media (excimer molecules) are bound only in the excited state. Since the molecule flies apart the moment it emits light, there are no ground state absorbers of light. As a consequence, population inversion is naturally accomplished as soon as the excited state is created. The laser wavelength depends on the kind of laser medium used, ranging from 193 nm (argon fluoride) to 351 nm (xenon fluoride). Typical pulse characteristics are: pulse duration 10–20 nanoseconds; pulse energy several hundred millijoules per pulse; pulse repetition rates up to 200 Hz. Because excimer lasers can conveniently produce high-intensity radiation in the UV region, they are widely used in industrial and medical applications. Like N_2 lasers, excimers are also used to pump dye lasers.

Excited CO_2 radiates at $\lambda = 10.6$ μm and is another important gas laser. CO_2 lasers have very high efficiency, can be operated with pulse lengths ranging from nanoseconds to and milliseconds, and have an average power up to several kilowatts. Aside from applications in scientific research, CO_2 lasers are widely used in industries where high power densities are frequently needed for material processing.

6.2.9 Tunable Solid-State Lasers

During the last decade, progress has been made in the search for tunable solid-state laser materials in an attempt to find better alternatives to tunable dye lasers. Today, the most important laser material in this class is titanium-doped sapphire crystal (Ti:Al_2O_3), introduced by P. Moulton [9]. The main advantage of Ti:sapphire crystal is its huge laser gain bandwidth, from 670 to 1100 nm. It also has a very large gain cross-section that is half that of Nd:YAG at the peak of its tuning range. As a host crystal, sapphire is extremely hard and has high thermal conductivity and excellent optical properties. The absorption band of Ti:sapphire centers at 500 nm; therefore, it can be pumped effectively by argon, copper vapor, and frequency-doubled Nd:YAG lasers, as well as rare gas flashlamps. Laser cavities developed for dyes are perfectly suitable for Ti:sapphire, with the dye cell replaced by Ti:sapphire crystal.

The upper laser level lifetime of Ti:sapphire is only 3.2 μsec. This presents some difficulties for traditional flashlamp pumping for Q-switch operation in a Ti:sapphire laser, because it is extremely difficult to make xenon flashlamps that operate with such short pulse duration. Therefore, in the flashlamp-pumped Ti:sapphire laser, the lamp duration is usually about 10 μsec. Although flashlamp pumped and Q-switched Ti:sapphire lasers are commercially available, they are not as efficient as a Nd:YAG laser. The usual configuration uses a Q-switched Nd:YAG laser to pump the Ti:sapphire laser. Using this pumping method and injection seeding, Fourier-transform-limited pulses that are several nanoseconds long and have several hundred millijoules of energy per pulse have been produced [10]. The Ti:sapphire laser can also be mode-locked when pumped by an intense CW laser such as an argon ion laser, and because of its large bandwidth it can be used to generate extremely short laser pulses in the tens of femtoseconds range.

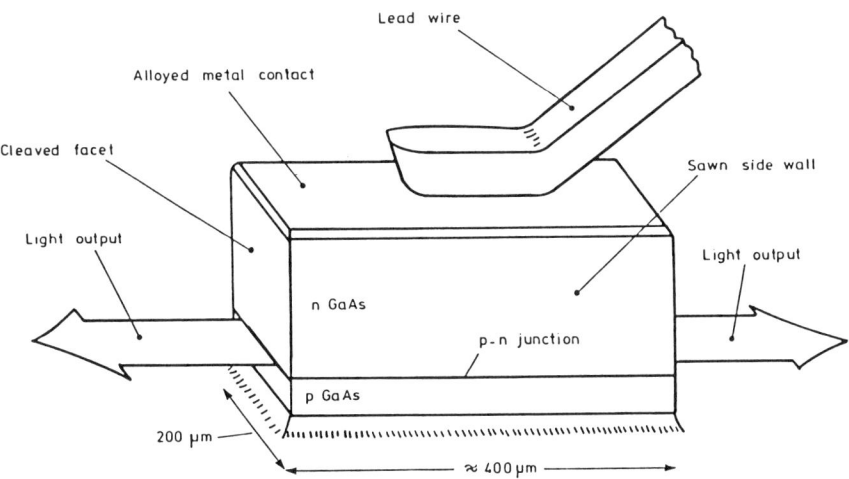

FIG. 5. Schematic of a GaAs semiconductor laser. From Thompson [11]. Reprinted with permission from John Wiley & Sons Limited.

6.2.10 Semiconductor Lasers

The semiconductor laser was invented by Basov *et al.* in 1961 [11], and is rapidly becoming one of the most widely used laser systems. A conducting p–n junction emits light; it also exhibits gain and can be converted into a laser if the emitted light is fed back to the junction. Figure 5 is a schematic diagram of a GaAs diode laser. Semiconductor lasers are significant in that they produce light with an efficiency that surpasses almost all other laser systems. Laser diodes are compact, relatively cheap, and have long operating lifetimes. They are commercially available at wavelengths ranging from around 800 nm (GaAlAs) to 10 µm (Pb-salt). GaInAsP laser diodes that emit light at 1 to 1.7 µm are particularly important for optical communication. Diodes that emit coherent blue light have been demonstrated recently [12]. Since the density of optical data storage is inversely proportional to the square of the wavelength of the light used, approximately four times higher data density can be realized if a blue laser diode is used instead of the near-infrared laser diode that is used in the current storage system. Semiconductor lasers are currently a subject of very active ongoing research.

Compared to other laser systems, diode lasers are relatively easy to operate. They only require a power supply with several volts, and typically less than 1 A of current output. (Although currents up to hundreds of amperes may be required for high-power diodes.) Pulsing of the laser diode output is extremely easy: it can be accomplished by directly modulating the driving current. The repetition rate can easily be adjusted from dc to the GHz range, although frequency chirping is usually associated with this kind of direct modulation, because the frequency of the laser also depends on the driving current. Laser diodes are available from CW single-frequency diodes that put out 150 mW of power, to bar stacks that produce up to 1.5 kilowatts peak power with 400-μsec pulses. A diode laser's frequency can be tuned by changing either the temperature or the diode current, or by using an external cavity that is similar to that used in tunable dye lasers.

Diode lasers have become popular pumping sources for solid-state lasers. Since the GaAlAs diode's wavelength can be tuned to match one of the absorption bands of the Nd:YAG crystal, GaAlAs diode can be used to pump an Nd:YAG laser. In the future, the bulky flashlamp-pumped solid-state lasers may well be replaced by compact diode-pumped setups. The laser wavelength of the GaAlAs laser also matches the lasing range of the Ti:sapphire crystal, so that it is possible to use a GaAlAs diode laser to injection seed the Ti:sapphire laser.

6.3 Buyer's Guide

In this section we explore issues frequently encountered when buying a pulsed laser and develop a checklist to aid the decision making process.

Each application places different requirements on the laser system. It is important to know, therefore, what a specific application demands, and to make sure the laser system that is purchased meets all those requirements. Of the many factors, the following are especially important: peak power, energy per pulse, repetition rate, pulse duration, wavelength, tunability and linewidth, transverse mode structure, pulse stability, and longitudinal mode structure. Let us examine them in more detail.

- *Peak power and energy per pulse.* Researchers studying such nonlinear effects as multiphoton ionization or pumping of atoms, harmonic generation in gases, laser-initiated plasma, soft x-ray generation, and relativistic behavior of atoms need lasers with high peak power. Researchers studying linear processes such as photon absorption where the detected signal scales up with energy per pulse, regardless of the peak laser power, do not require pulsed lasers. But pulsed lasers are often used for this purpose because of their simplicity, flexibility, and ease of use.

- *Repetition rate.* In laser-induced chemical reactions, to avoid creating nonlinear effects and unwanted reactions, laser intensity must be kept low so that the signal will be weak. In this case, one may want to sum up a large number of events to get a good signal-to-noise ratio. Therefore, a high-repetition-rate laser is preferred. In many cases the repetition rate of the laser is limited by power supply capacity. For some lasers such as the Nd:YAG, however, the repetition rate is limited by the laser material. At high energy per pulse (e.g., 1000 mJ), an Nd:YAG crystal may crack at a repetition rate in excess of 20 Hz.

- *Pulse duration* can be determined by the properties of the laser medium (e.g., lifetime), the duration of the pump source, or switching elements. To study transient effects, for example, to measure the lifetime of an excited state, one has to make sure the laser pulse duration is shorter than the duration of the phenomenon to be studied.
- *Wavelength.* Solid-state lasers such as Nd:YAG and semiconductor lasers operate in the infrared. To obtain visible or ultraviolet light from high-power solid-state lasers, one can use harmonic generation methods, although conversion efficiency is high only at power densities close to the damage limit of nonlinear crystals. Gas lasers such as N_2 and excimer lasers operate in the ultraviolet. Dye lasers and Ti:sapphire lasers are useful when light is needed in the near-infrared to the near-ultraviolet. The needs for specific wavelengths are quite varied. For example, in laser surgery, light is needed at a variety of wavelengths because biological tissue responds differently at different wavelengths. In molecular spectroscopy, light is needed at specific wavelengths to excite specific molecules.
- *Tunability and linewidth.* In applications such as laser spectroscopy, it is important that the laser be tunable in the wavelength range of interest. In high-precision measurements, the laser linewidth has to be narrow enough to be able to distinguish the features of interest. Dye lasers can be tuned over a large spectral range by adjusting intracavity frequency-selective elements and by changing the dye. The pump laser must be selected with attention to the needs of the gain medium, such as absorption bandwidth, excited state lifetime, and bleaching effects. The Ti:sapphire laser covers a large spectral range in the near-infrared. Its coverage can be further increased by frequency doubling or tripling using nonlinear crystals.
- *Transverse mode.* Many dye lasers and some solid-state lasers produce light in the lowest-order transverse mode, where the spatial intensity distribution in the beam cross-section is a Gaussian function. In some high-power solid-state lasers such as the Nd:YAG, the presence of higher-order transverse modes can lead to hot spots which can damage optics. If the application requires tight focusing of the beam, the fundamental transverse mode output is preferred.
- *Pulse stability.* This describes the variation from pulse to pulse of peak power, pulse energy, or pulse position in time relative to a marker pulse or external trigger pulse. Smaller variation means more reproducible events and therefore is preferred.
- *Longitudinal mode structure.* In lasers with either a single longitudinal mode or many longitudinal modes, the temporal profile of the output beam tends to be stable, and often can be approximated by a Gaussian function in time. In lasers with a few longitudinal modes oscillating simultaneously, beating be-

tween modes creates large variations in the temporal profile of the output pulses. This mode-beating can produce some unexpected results. For example, if the pulse-to-pulse variation in energy is 10%, the variation in energy of the second harmonic may be much greater than the 20% that one would expect if these pulses were smooth Gaussian profiles. For similar reasons, multimode lasers also require much higher damage threshold optics (typically by a factor of five) than one would guess by looking at the power specifications of typical lasers.

The next thing to consider is ease of laser operation and reliability. One wants a laser that is well-made, highly reliable, and requires minimum maintenance. Many lasers that are commercially available are operational nightmares. Be careful. Before buying a laser, one should also know the laboratory requirements that the installation will require. Here are some questions one should ask.

- *Is the laser well-designed and easy to operate?* Solid state lasers are usually easy to operate, but there is a large variation between different manufacturers. For example, some lasers are easy to turn on and off, while others may require complicated procedures. In some cases lasers may need an hour of warmup time before they are operational. Well-built lasers also have many clever devices built in to avoid accidental damage of the laser rod and optics by the user. It is well worth the extra cost to get those safeguards.

- *Installation.* Are there necessary facilities in the laboratory for laser operation? High average power lasers require special powerlines. They may also need cooling water to operate, or exhaust hoods to remove or deal with dangerous substances.

- *Environmental impact.* Dye laser operation involves dealing with organic solvents and dyes, some of them known to be carcinogenic. Therefore, safety precautions in handling the dyes and solvents and provisions for proper disposal of the waste material are necessary. Excimer lasers use gases that are highly toxic and sometimes flammable. Extreme care must be taken to prevent leaks in the circulation system, and the waste gas should be disposed of properly.

- *Safety features.* Are there shutters built into the laser? What about emergency shutoff? Are there high-quality laser safety goggles available for the laser?

- *Other equipment.* A stable laser table is a must. High-damage-threshold laser optics are needed for high-peak-power lasers. One also has to prepare the mounts, power meters, detectors, spectrometers, and other accessory equipment. Make certain that you can test the operating specifications in your laboratory. Frequent examination of the specifications can ensure proper laser operation, early damage detection, and savings in maintenance and repairing cost.

6.4 Builder's Guide

6.4.1 Introduction

This section describes how to design, assemble, and operate various forms of nanosecond-duration pulsed tunable dye lasers. We provide a step-by-step guide to building these dye lasers, from the simplest superradiant laser to a high-performance single-mode dye laser. Many of the basic concepts we introduced in the first section are discussed in detail here in a practical setting. It is our hope that by going through this exercise the reader can understand the issues that are representative of those that come up in the design of almost all lasers.

All of the dye lasers to be discussed in this section are pumped by direct optical excitation. This means that another light source (usually another laser) is used to create population inversion in the gain medium (i.e., an organic dye in solution) as required for laser action.

The pumping light source can be any of several different lasers, including a nitrogen laser (3371 Å), a frequency-doubled Nd:YAG laser (5320 Å), or an XeCl excimer laser (3080 Å). The only requirements are that the pumping laser must be of a shorter wavelength (i.e., higher frequency) than that of the dye laser, and that the gain medium is able to absorb light at the pumping laser wavelength. All of the examples presented in this section are obtained using a relatively inexpensive and low-powered nitrogen laser having an output pulse energy of about 1 millijoule in a 10-nanosecond pulse. The pump pulse is in the form of a rectangular-shaped beam of area equal to roughly 1 cm^2. This beam is focused using spherical and cylindrical lenses to a line on the face of a high-quality optical cell. (The cell and lenses must be made of a material that is transparent at the pump laser wavelength. For 3371 Å it is necessary to use a material such as fused silica (quartz)). This beam of light having a cross-section in the form of a line is absorbed in a fraction of a millimeter by the organic dye solution in the cell. Penetration depth and excited dye volume height are shown in Fig. 6. The peak power of the pump laser is about 1 mJ/10 nsec/1 cm^2 or 100 W/cm^2, and it operates at a pulse repetition rate up to 100 pulses per second.

Given that this section concentrates on the laser as a practical tool, most of the discussion is concerned with the hows and whys of various optical configurations and advice on how to select optical elements. The discussion deals with issues that must be addressed and understood if one is to obtain a useful instrument. The practical bent also means that there are forms of the instrument that are not ideal in the sense that laser designers who are purists would not likely consider them as desirable. However, if the goal is to obtain a light source that is matched to one's needs, then there is no reason to build a better laser than one has to. This is especially true with respect to aspects such as spectral width and beam quality. In fact, in many applications the lower-quality instrument may be preferable in that

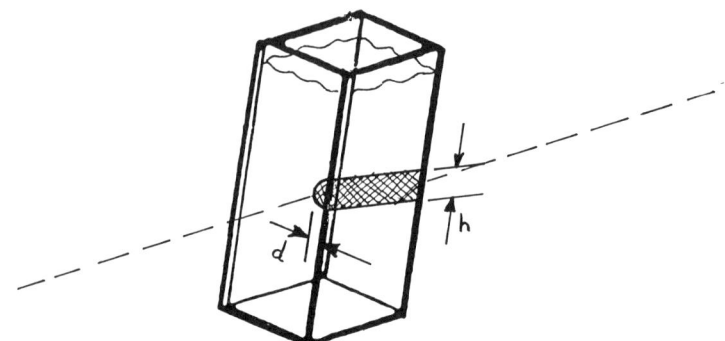

FIG. 6. Excited dye volume in a liquid-filled cell. Penetration depth and excited volume height are shown. (The cell is shown rotated 10° clockwise about the pumping laser propagation axis, and the long axis of the excited dye volume is rotated 3° clockwise about the pumping laser propagation axis. The cell is square in cross-section, and the central axis of the exit light from the excited volume lies in the horizontal plane.)

it might have other properties that are beneficial. For example, a lower-quality tunable laser might be more efficient, or have a broader tuning range, or have less amplitude jitter from pulse to pulse. Some of the lasers described in this section can be constructed out of parts that are extremely inexpensive and readily available in many laboratories.

6.4.2 A Laser Without Mirrors: Amplified Spontaneous Emission

It is surprising to many that a laser does not require mirrors, or a cavity, or feedback, in order to operate. A laser requires only that light be amplified by the process of stimulated emission. Amplification will occur in an optically active medium anytime that the population is inverted, meaning that there are more quantum particles (e.g., atoms, molecules, ions, electrons, electron–hole pairs) in an excited state than there are in a lower state. The system is "inverted" in the sense that a quantum system in thermal equilibrium would not exhibit such a feature. Inversions require effort to create, although they are known to exist in nature under special nonequilibrium conditions. (One frequently noted case involves the inversion of ammonia vapor in the atmosphere of Jupiter, which is excited by strong microwave emissions from the planet's surface.)

Evidence of laser action can be obtained from observing the spatial distribution of light emanating from a nonspherical volume of excited quantum particles. For example, if the excited volume has a long aspect ratio, let us say an elongated cylinder in the shape of a pencil, then spontaneous light emitted at one end of the pencil in the solid angle subtended by the remainder of the pencil-like volume will be amplified (see Fig. 6). In contrast, light emitted to the side will not be amplified to the same extent because the pathlength through the medium is shorter. Recall that the curve of growth in intensity is exponential as described by Beer's law. The effect of all of this is that the light from the volume in total will be bunched along the long axis of the volume. As one turns up the population inversion, this bunching effect becomes more pronounced. Axial amplification of spontaneous light is known as *Amplified Spontaneous Emission*. It is equivalent to single-pass lasing, and the bunching or collimation of light is an indicator of laser action. The greater the bunching, the higher the gain of the medium. (If the medium is weakly interacting due to the low density of scatterers or low oscillator strength or equal populations in the upper and lower states, then the light emitted will be spherically symmetric far away from the volume. The light emitted will simply be the sum of the light from each independent voxel (i.e., volume element). This is the expected distribution from a dye medium that is not very strongly pumped. If the medium is strongly interacting but not inverted (i.e., absorbing), then a dark beam will be observed along the long axis of the pencil-shaped volume, that is, the light along the axis will be absorbed and preferentially scattered to the side.

To observe single-pass lasing in the laboratory is easy if one uses a highly stable and fluorescent dye such as rhodamine 6G. This dye is one of the most efficient UV converters known, which is also why it is used as an attention-getting coloration for highway reflectors. (This dye was also the first one used by Peter Sorokin, the IBM researcher who invented the dye laser. He selected it on its fluorescent properties as listed in a handbook of fluorescent materials. To this day it is the most commonly used dye in lasers.) One simply fills a 1 cm × 1 cm × 5 cm quartz spectrophotometer cuvette (i.e., dye cell; dimensions are not critical, although the flatness of the inner face and the sharpness of the inner frontal corners is critical; high-quality dye cells can be obtained from Nippon Scientific Glass or Helma) with a $1-5 \times 10^{-4}$ molar solution of rhodamine 6G in ethanol. Other solvents such as methanol or water are also usable as long as one is able to achieve an adequate concentration of dye so that the pump light is absorbed completely in a depth of less than 1 mm as it enters the cell.

The purity of the solvent is not important when working with rhodamine 6G, although it is important when working with other dyes that are at risk of breaking down following exposure to the UV light from the pump laser. The breaking down or "bleaching" effect of UV light is a problem with many dyes. This is the same effect that causes coloring agents in paints and dyes to fade in sunlight. For dyes that are prone to fade such as the Oxazine IR dyes, one must take care to use very pure (reagent grade) solvents and be prepared to replenish the dye solution periodically.

The pump light is focused into the cell using spherical and cylindrical lenses. The dye cell is slightly rotated (5 to 15°) from the vertical about the propagation axis of the pump light so that reflections from the faces do not contribute to the lasing action. This is needed because we are concentrating on single-pass lasing. In later forms of the laser we will continue to use this geometry. Unwanted reflections can interfere with the action of feedback elements as used in later forms

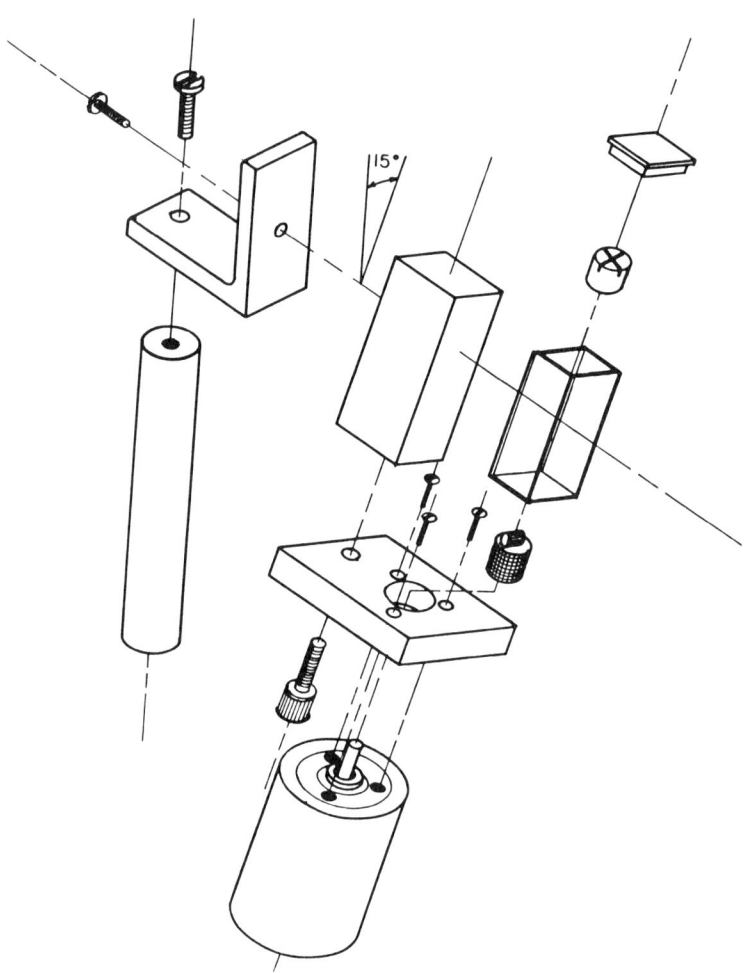

FIG. 7. Dye cell mounting at a 15° angle to the vertical.

of the laser and are to be avoided at all costs. It is useful to rotate the line axis of the focused beam in the same direction as the tilt of the dye cell but by an amount that is less by a factor roughly equal to $(1 - 1/n)$, where n is the index of refraction of the dye solution. The cell rotation and its mounting is shown in Fig. 7. The rotation of the axis of the focused beam allows for light coming from the long axis of the excited dye volume to be refracted in such a way that the ASE lies in the horizontal plane (i.e., the plane of the tabletop). Take the extra time that it takes to get the axis right. Optical misalignment out of the tabletop plane will cost you time in the long run.

An He–Ne alignment laser on the tabletop is a very useful tool in this regard. For example, to determine the correct angle to rotate the line focus of the pump laser in order to obtain table-planar operation, a planar He–Ne alignment beam can be sent into the cell. As it passes through the dye medium it will leave a streak that can be used as an indicator of the correct angle. After the laser is operating you can make additional angle corrections by adjusting the cylindrical lens rotation angle.

The dye cell is stirred with the help of a magnetically coupled stirrer in the cell. In stable dyes such as rhodamine, stirring is necessary for pulse repetition rates in excess of 10 Hz. In dyes that are bleached by the UV pump light stirring is always needed. Since most of the pump energy is converted to heat following the pulse (even the best dye lasers have an efficiency of at most 30 or 40%), it is necessary to gently swirl the dye or to flow it. This brings in fresh dye for each shot and makes the excitation volume more optically uniform. If one does not do this, the beam may appear distorted. This is the same effect that causes optical distortion if one observes an object along a path in air over a hot source such as a radiator or a stove.

A word about safety. These types of dye laser do not appear to be very intense until one reaches pulse energies on the order of 10–100 mJ per pulse. Because of this, many researchers get sloppy and allow primary beams and reflections to bounce around the laser laboratory. This is to be avoided at all costs. The average power of these lasers is indeed low, but peak power can be very high. Eye damage is the result of a mixture of peak power and average power. One shot directly in the eye from a 100-µJ dye laser is enough to burn a hole in your retina. A 100-µJ pulse in 1 cm^2 when observed on a white surface looks relatively dim, even though its peak power is 10 kW/cm^2. One reason for this perceptual error is that the eye integrates light over roughly a 40-msec period (a 25-Hz scan rate) in conditions of normal room illumination. These nanosecond dye laser pulses are much shorter in duration than the eye's integration period, so that when we compare them to other light sources that are on for the entire integration period the laser light can look unimpressive. Do not be fooled. The same warning is relevant when one works with lasers outside our normal spectral response. An 8000- or 4000-Å laser may not look very bright since our retinal photosensitivity is low at these wavelengths; however, if one gets this light in the eye it can cause a great deal of damage. Eye damage due to UV may not show up right away. We recall an incident at MIT where a fellow using a quartz microscope and a mercury vapor light source (a UV source) burned his retina. He did not know it until several hours after exposure, by which time he was in a great deal of pain.

Laser goggles are recommended when working with high-powered lasers. For the UV, inexpensive UV goggles may be obtained from a variety of sources (e.g., the Cole–Palmer Instrument Company). Check the absorption factor when obtaining laser goggles. Goggles are rated by their absorption. For example, a goggle might absorb 99.9% of the incident light in the UV, which corresponds to an optical density (OD) of 3.0. Goggles for the dye lasers are more problematic because dye lasers operate at wavelengths over the entire visible spectrum. It may be necessary to have many sets of goggles for different ranges. Goggles with very high optical density can be very expensive, often in excess of a thousand dollars each. Also, do not get goggles that make the laboratory appear so dim that it increases your chances of having another type of accident. This is especially important if one works near rotating machinery or around high-voltage instruments. In fact, the single greatest hazard in a laser lab is the high-voltage power supply used to run most lasers. Laser power supplies are lethal.

Be reasonably careful not to let the UV light hit your skin. UV in high intensities can cause burns and is carcinogenic. The 3371-Å light in the nitrogen laser is not very dangerous in terms of skin damage, however. Shorter-wavelength UV light can be very dangerous. Particularly the fourth harmonic of the Nd:YAG at 2660 Å, which is sometimes used in pumping UV dyes, is in the dangerous wavelength band for producing skin cancer.

The procedure for obtaining single-pass lasing is straightforward. Turn on the pump laser to low power. Make sure that the magnetic stirrer is operating. The UV beam can be made visible by letting it hit a white business card. Business cards appear "whiter than white" because of an organic dye (stilbene) that they are soaked in. This is the same whitener that is used in laundry detergent to make laundered clothes appear bright. (The clothes are actually being dyed.) These whitening agents absorb UV light and reradiate in the visible range, where we can see it. Focus the light to a line, as discussed previously. Observe the dye light emitted from the excitation volume on each side of the dye cell. Do this by using white cards placed about 10 cm away from the dye cell at right angles to the optic axis of the laser and on either side of the cell. Notice that the UV pump light is absorbed as soon as it enters the dye medium. Also look at the lenses and the cell face to see if you see a bluish-white streak where the UV has passed through the transparent material. If you have selected, in error, a cell or lenses made of pyrex or some other type of glass, then the UV beam will be visible as it goes through the material. Quartz or fused silica will not have this bluish-white appearance.

Turn up the UV pump laser power. This can be accomplished, for example, by turning up the operating voltage on the pump laser. Watch the beams on the white cards as the power is increased. You will notice that the light emitted along the optic axis of the dye laser will become progressively brighter and more bunched along the long axis of the cell. This is evidence of the single-pass lasing that was discussed earlier. You will likely also observe another bright spot due to a reflection off the cell wall. This spot will appear more well-defined and might even be very bright. It is the result of two passes through the gain medium. The gain in the case of single-pass lasing can be very high. Given that a 100-µJ pulse of rhodamine 6G light (energy of about 3 eV per photon) contains about 2×10^{14} photons, depending on the aspect ratio the gains may be as large as 10^{10} or higher. (In order to calculate the exact gain, it is necessary to consider the factors that determine the probability that a spontaneous photon win be emitted and will fall in the solid angle defined by the geometry. When the gains are so high, the likelihood that many spontaneous photons contribute to the final emission is high, and as a result the level of coherence of this light source is low.)

To convince yourself that the gain is high, try turning down the pump laser so that the dye laser output is only weekly bunched along the axis, and then bring another business card close to one side of the dye cell so that some of the reflected light is bounced back in the direction of the cell. You should notice that on the screen on the other side of the cell a spot due to this small amount of reflected light will be visible. The laser is able to amplify any light that passes through it during the time that the population is inverted, even if it comes from a diffuse source.

Under some circumstances vertical diffraction "stripes" may be observable in the dye laser beams on each of the screens. This will be especially pronounced if the dye concentration is high. A high dye concentration means that the penetration into the cell is minimal. The shorter the penetration, the larger the diffraction angle from the confined emitting volume. Think of this as the diffraction from a slit where the diffraction angle in radians is roughly the wavelength divided by the slit width. Given that the screen is 10 cm away from the cell and that the wavelength of the dye laser is about 0.5 μm, a penetration depth of 50 μm will result in stripes of separation equal to 10 cm × 10^{-2} radian or 1 mm. Thus, if one observes multiple stripes separated by more than 1 mm it means that the penetration depth is very small and that the dye concentration is probably too high.

This discussion about the penetration depth also suggests the need for a high-quality dye cell. Since all of the "action" in the dye laser is happening in a very narrow region just behind the front face of the dye cell, it should be obvious that the surface quality of the inner face of the dye cell needs to be high (at least $\lambda/4$ or better) and that the inner corners of the dye cell need to be sharp and without curvature.

A few comments about the laser wavelength are in order. If the output of the dye laser is analyzed as a function of the pump power, it will be noted that the spread in wavelength will reduce dramatically as the pump power is increased. This occurs as the result of gain narrowing in which the central wavelength of the band for which the gain is high will be preferentially amplified over wavelengths that are in the "wings" of the gain profile. This is a consequence of the exponential curve of growth for intensity and the fact that the gain factor, which is in the exponent, is frequency-dependent.

Many dyes other than rhodamine 6G can be used to obtain light from this single-pass laser. In fact, the first tunable dye laser was tuned by changing dyes and solvents and by modifying dye concentrations. (Peak wavelengths of a given dye can be affected by choice of solvent, and concentration.) Rhodamine dyes operate in the red–yellow–green range, Oxazine dyes work in the IR–red range. Coumarin dyes operate in the blue–near-UV range. DCM operates in the blue–green range. Other specialty dyes are available for use in certain narrow bands. A good source of laser dyes and information is Eastman Kodak (see Fig. 8). Sometimes it is possible to mix dyes to get coverage in a hard-to-reach band. This was the case in the 7500-Å range, but this problem range has been covered by the emergence of a new laser material, titanium:sapphire, which uses a titanium ion embedded in a sapphire host crystal. This outstanding tunable solid-state material has virtually eliminated the use of the dye lasers in the spectral region to the red of 7500 Å.

In general, one wishes to select dyes that absorb strongly at the wavelength of the pump laser. Stated another way, one should select a pump laser wavelength that best matches the absorption properties of the dye that one wishes to use. Organic dyes have a broad spectral range of usefulness because they are quite heavy (molecular weights of 500 are typical) and as a consequence have many modes of oscillation. This high degree of freedom is what allows them to be used as a tunable medium. Lighter and simpler molecules have a regular quantum line and band structure that has many "holes" in the absorption/emission spectrum. Atoms are even worse, in that they have a spectrum that is typically sparse. Thus, a helium–neon laser can oscillate at only certain wavelengths that correspond to optical transitions in neon. A carbon dioxide molecular gas laser similarly will oscillate at only certain wavelengths corresponding to transitions in the carbon dioxide molecule. Of course, since it is a molecule, there are a large number of nearby lines corresponding to quantized rotations of the molecule. Heavy molecules like organic dyes have so very many modes of oscillation and vibration that their spectrum is smooth over a wide range, and it is for this reason that they are useful as a tunable gain medium.

The strongest "spring" in the organic dye molecule is the carbon–carbon double bond. One vibration of this bond is about 1200 cm^{-1} of energy. The decaying electronically excited dye molecule will typically drop down to a lower electronic state in the lowest unoccupied vibrational state (usually the first excited vibrational state). This vibrational state is at an energy that is 1200 cm^{-1} below that corresponding to the peak absorption energy. Thus, rhodamine 6G, which has a peak emission at 6000 Å (16,666 cm^{-1}) has a peak absorption at 5500 Å (18,000 cm^{-1}). Thus, we see that, while it is possible to use UV to pump this dye, it would be far more efficient to use light at 5500 Å since this corresponds to the peak of dye absorption. UV pump lasers are common in many laser labs because most of the fluorescent dyes will absorb the light. A single pump source such as a nitrogen laser or an XeCl excimer laser can be used to cover the entire visible spectrum merely by changing dyes. Atomic and molecular spectroscopists especially take advantage of this feature. Sometimes the bleaching effects of a UV pump make it

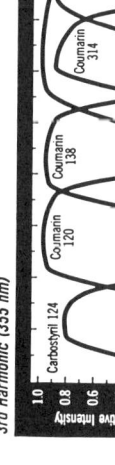
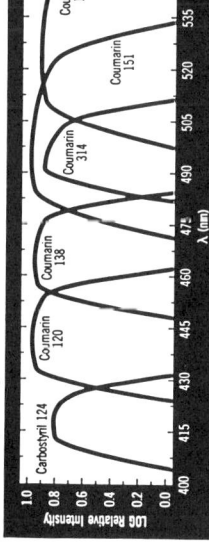
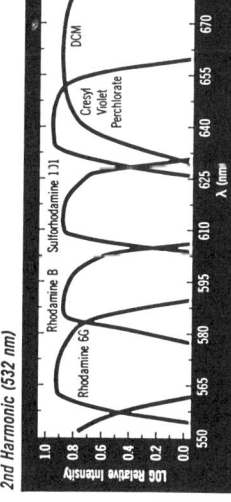
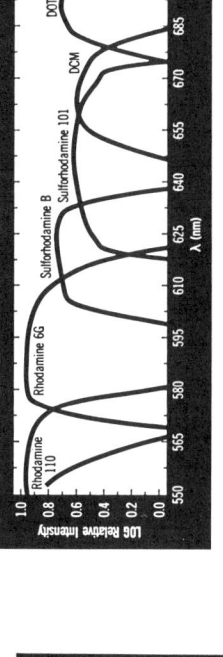

FIG. 8. Dyes for use in dye lasers: nitrogen and Nd:YAG laser pumping sources. From Eastman Kodak [8]. Reprinted with permission from the Eastman Kodak Company.

necessary to use the light from one dye laser to pump another. This is especially useful with the fragile IR dyes that are easily bleached by the UV pump light.

An excited dye cell is usable as a light amplifier. It can amplify light from a pulsed laser oscillator or from a CW laser. The acronym MOPA refers to "master oscillator/pulsed amplifier." MOPAs are used routinely in many laboratories where many millijoules of tunable light are required. The usual configuration uses one or two pulsed amplifiers of the side-pumped geometry. In many cases, the last stage of amplification is a longitudinal amplifier cell. For pumping using a laser with a rectangular output beam such as many excimer lasers, it may be convenient to incorporate a Bethune cell in the last stage of amplification. The Bethune cell is a cleverly bored-out prism created by Don Bethune, formerly of the IBM Research Laboratory. Dye flows through the cylindrical bore. (Bethune cells are available commercially from the Santa Ana Laser Company.) Light from a pump source having a rectangular beam shape, typical of high-powered excimer lasers, is internally reflected radially to excite the dye from several sides. The dye concentration is lower in this stage, so that the pump light can penetrate further into the dye volume. This kind of amplifier is needed because, as the peak power of the dye laser light increases, damage to optical elements becomes more of a problem. Longitudinal amplifiers are used to obtain hundreds of millijoules of tunable dye light. The concept of increasing the size of the pulse amplifier as the intensity builds in each stage is routine in the laser field. The rule of thumb is to keep the dimensions as small as possible given the constraints imposed by the strength (in this case, the optical damage power threshold) of materials.

Side-pumped pulsed amplifiers are also useful for increasing the power of CW lasers. One configuration that is sometimes used involves a tunable CW dye laser that is pulse-amplified by a series of several side-pumped pulsed amplifiers. The gain of a chain of amplifiers can be very high. Unfortunately, this type of tunable oscillator–pulsed amplifier system is very expensive, large, and difficult to maintain.

6.4.3 The Advantage of Mirrors: Multiple Passes Through the Gain Medium

It is usually the case that one wishes to have a laser light source that has many desirable properties. Properties such as high power, good beam shape, low beam divergence, narrow spectral width, tunability, and spatial/temporal coherence are often required. In the superradiant laser discussed in the previous section, many of these properties are absent.

The use of mirrors to recirculate the amplified light for further amplification is extremely useful in helping to obtain a laser that has many of the beneficial properties mentioned above. Recirculation may be accomplished in a standing-wave cavity, as shown in Fig. 9, or in a ring cavity [3]. Historically, the standing-

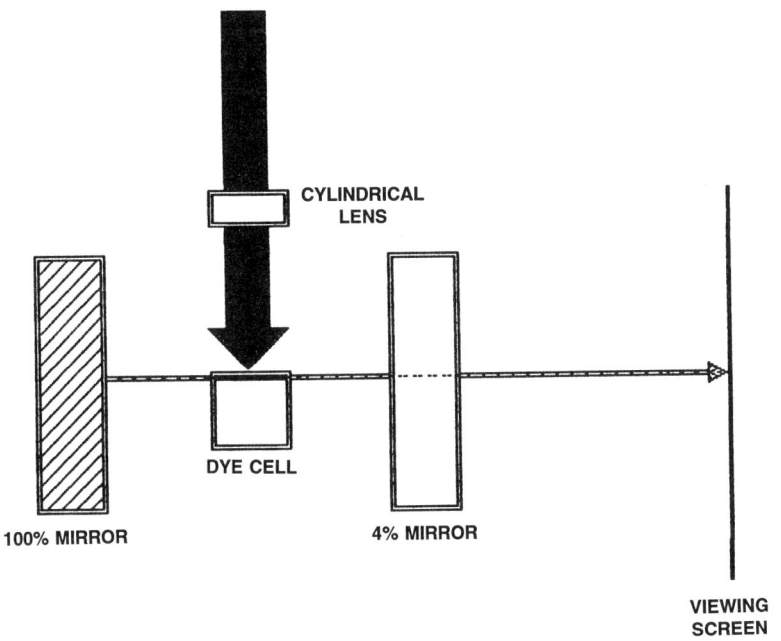

FIG. 9. Standing-wave laser cavity.

wave cavity has received the greatest use, and it is the type of cavity that will be considered here.

Let us consider the excited dye cell of the superradiant laser in the previous section when it is placed between two mirrors, one fully reflecting and one partially reflecting and separated by 30 cm. Since these two mirrors are assumed to be aligned perfectly, the gain medium plus reflectors may be replaced conceptually by a series of dye gain cells. Light travels at a speed of roughly 30 cm/nsec, so that the 10-nsec duration of the excited medium translates into a series of 10 gain cells of 1 cm pathlength and separated by 30 cm. In essence we have formed a chain of dye gain cells that is 300 cm long. The gain of each cell does not have to be anywhere as high as it was in the superradiant laser of the previous section in order to extract light efficiently, as long as the exponential curve of growth of the dye light is given the opportunity to build to the same value as before.

The beam divergence of this multiple-pass laser is seen immediately to be much less than the single-pass laser. The single-pass laser beam divergence was set by geometry, as is the multiple-pass version, except that the multiple-pass version has a much smaller aspect ratio and therefore a much smaller divergence.

Since each stage of the amplifier chain is pumped less than the laser in the previous version and since the solid angle is smaller, the number of spontaneous photons that initiate the laser to oscillate is greatly reduced. In the previous case, many "seed" photons were amplified, so that the output was in the form of a collection of many amplified incoherent photons. In this case, we have fewer "seed" photons that are amplified to a greater extent, with the result that the coherence of the multiple-pass laser oscillator is far better than that of the single-pass laser. Note that, since the laser is still operating on many wavelengths, coherence is not high. Coherence of macroscopic measure (i.e., lengths of centimeters or times of nanoseconds) is only possible if we additionally add narrow spectral filters to the laser cavities. By coherence we mean that the phase of the electromagnetic wave at another point in space or time is known given that we know it at a point we use as an origin. Coherence is a necessary ingredient of interference. Interference effects will be considered later in this section.

6.4.4 Intracavity Spectral Filters for Narrow-Band Operation and Tuning

When dye lasers were first invented, it was not appreciated that the laser frequency could be tuned with the help of intracavity optical filters. At first, the suggested method for tuning was to change dyes or the dye chemical environment (e.g., pH, solvent, temperature). In 1971, Strome and Webb discovered that the use of a spectrally dispersing intracavity prism would allow for a reduction in laser bandwidth [13]. It was also discovered that the operating wavelength could be changed by adjusting the angle of the dispersing prism. An aperture in the cavity also helped to reduce the linewidth. In essence, the laser became an *active* prism spectrometer. The years that followed this discovery led to many cavity configurations that included an enormous variety of intracavity tunable optical filters whose purpose was twofold: to narrow-band the regenerative oscillator and to tune it. Among many intracavity elements, researchers used Lyot (birefringence) filters, prisms, gratings, and interferometers. In this chapter, we consider only the use of reflecting diffraction gratings and interferometers for narrow-banding and tuning.

Diffraction gratings are desirable as narrow-band filters because they are highly dispersive. A grating, if used properly, has a $d\lambda/d\theta$ that may be an order of magnitude larger than that of a typical prism. Also, since they are reflective, it is possible to use gratings far into the UV. Their use, however, does introduce some difficulties. One problem is that diffraction gratings are easily damaged by high-powered laser beams. Another is that they are quite lossy in comparison to other filters, such as birefringent filters and prisms. For these reasons, gratings are not used often in lasers of low gain (gain per pass must exceed loss per pass, if the laser is to oscillate, i.e., regenerate), and they are not used in oscillators in which the intracavity power is high. What defines high power, however, is not easily stated because it depends on the geometrical configuration in which the grating is used in the laser.

The diffraction grating obeys the grating following equation:

$$m\lambda = d(\sin \theta_i + \sin \theta_d),$$

where d is the grating period, θ_i and θ_d are the incident and diffraction angles as measured with respect to the grating normal, m is the order of diffraction, and λ is the wavelength. In the special case where $\theta_d = -\theta_i$, the diffracted light is retroreflected. At this wavelength, the grating acts like a plane mirror at right angles to the optic axis of the laser. For other wavelengths, the light is not retroreflected. If one places a small aperture or slit in the cavity, the off-axis light will not reach the gain medium and therefore will not be amplified. This configuration is known as the Littrow arrangement. For the side-pumped geometry that has been discussed so far, the narrowly defined gain medium in the form of a thin filament along the front face of the dye cell serves the purpose of both gain medium and slit. Thus, if one replaces the highly reflecting mirror in the laser oscillator of the previous section with a retroreflecting and dispersing grating, the laser will oscillate at the retroreflected wavelength. Tuning of this laser is then readily accomplished by rotating the grating relative to the surface normal, as shown in Fig. 10. The output from the laser can be either the exit beam off of the grating (the zeroeth-order reflection) or the beam exiting from the partial reflector. The beam through the partial reflector is usually used since its direction does not change as the grating is rotated. It should also be obvious that, since there are two exit beams, the design is wasteful of light.

This laser is impressive in its ease of construction and its efficiency (except for the extra exit beam that was just mentioned). It is not especially impressive with respect to its linewidth, however. The linewidth of this laser is typically larger than 0.2 nm. For comparison, the Doppler-broadened atomic absorption linewidth of sodium vapor at low pressures is 0.002 nm. The reason for the large bandwidth of the laser is that only a very small portion of the grating is being used for dispersion. The linewidth can be expressed as:

$$\delta\lambda \approx \lambda^2/L,$$

where λ is the laser wavelength, and L is the linear dimension of the portion of the grating that is being used for dispersion. The laser spot size is typically 1 mm in this kind of set up, which makes the linewidth about 0.2 nm for an operating wavelength of 500 nm.

It should be obvious from this discussion that, if one could increase the intracavity beam size so that the entire surface of a large (>5 cm) grating could be used, it might be possible to achieve a very narrow band source. In fact, this was accomplished by T. Hänsch in 1972 [14] by means of an intracavity beam-expanding telescope. This arrangement significantly narrows the bandwidth. However,

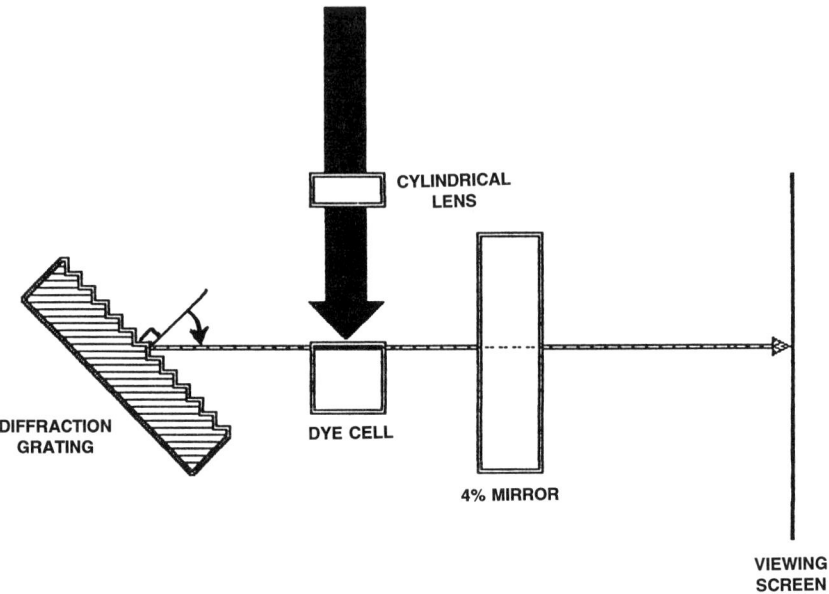

FIG. 10. Top view of a grating tuned dye laser.

using refractive optics in a laser cavity has some drawbacks: intracavity optics give rise to surface reflections that can compete with the feedback from the tunable filter; an intracavity optic has to be of high quality in order for it not to distort the spatial wavefront. High-quality optics are expensive. Refractive materials are also dispersive, which means that the beam expander can only be collimated at one wavelength. Other wavelengths will experience either focusing or defocusing, which has the result of broadening the laser linewidth, and it leads to a linewidth that changes as the laser is scanned in wavelength. Lastly, the requirement of exact alignment demands high-quality mounts and stages.

All of these problems are averted if the grating is used in grazing incidence, as shown in Fig. 11. In this case, the grating is held fixed and the tuning mirror is rotated about the grating. This configuration is known as the grazing incidence dye laser and it was developed in 1976–77 independently by Littman and Metcalf and Shoshan, Danon, and Oppenheim [15].

In the grazing-incidence configuration, the grating angle is steep enough that the intracavity optic beam illuminates the entire width of the grating. The grating diffracts the beam at an angle given by the grating equation. The diffracted light comes off as a streak. A mirror is used to retroreflect the light back to the grating, where it is rediffracted back to the dye cell. Only the wavelength that corresponds

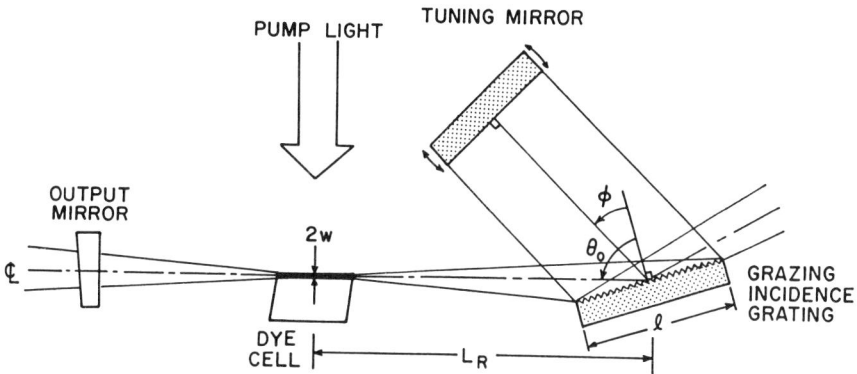

FIG. 11. Top view of a side-pumped grazing-incidence dye laser.

to the value given by the grating equation for specified angles θ_i and θ_d is reflected back to the dye cell for amplification. Other wavelengths are mapped to return angles that miss the dye gain medium and therefore are not amplified. This laser is tuned by varying the angle of the tuning mirror relative to the grating.

The major advantages of the grazing-incidence approach is that it is easy to implement with relatively inexpensive parts, it does not need careful focusing or alignment, and it allows for the cavity to be made very compact.

6.4.5 Single Longitudinal Mode Tunable Dye Laser

The above setup can be further improved to achieve single longitudinal mode operation (see Fig. 12). Unlike the previous ones, the new laser is end-pumped. The pumping light is focused into a dye cell by a spherical lens with a 75-cm focal length. The dye cell has a 2-mm pathlength and is polished on two sides. The end mirror is a fully reflecting mirror with a high damage threshold [16].

Since the dye cell's pathlength is only 2 mm, the gain per pass is small. As a result, the two-pass spot we used for alignment before is too weak to be useful. Therefore, we bring in an He–Ne laser beam pointed opposite to the dye laser output beam for alignment purposes. Care is taken to make sure that the alignment beam properly strikes the surfaces of the grating, end mirror, and tuning mirror. Alignment is completed when a single beam from the dye laser cavity is retroreflected back to the He–Ne laser. The pump laser is then focused onto the dye cell to a spot that is defined by the He–Ne laser. The pump light and the He–Ne light should form an angle of about 5°.

Note that at this time the angle of the tuning mirror corresponds to backreflection of the He–Ne wavelength (628 nm). Therefore, to make the laser oscillate, the tuning mirror is rotated to an angle that corresponds to a wavelength that is within

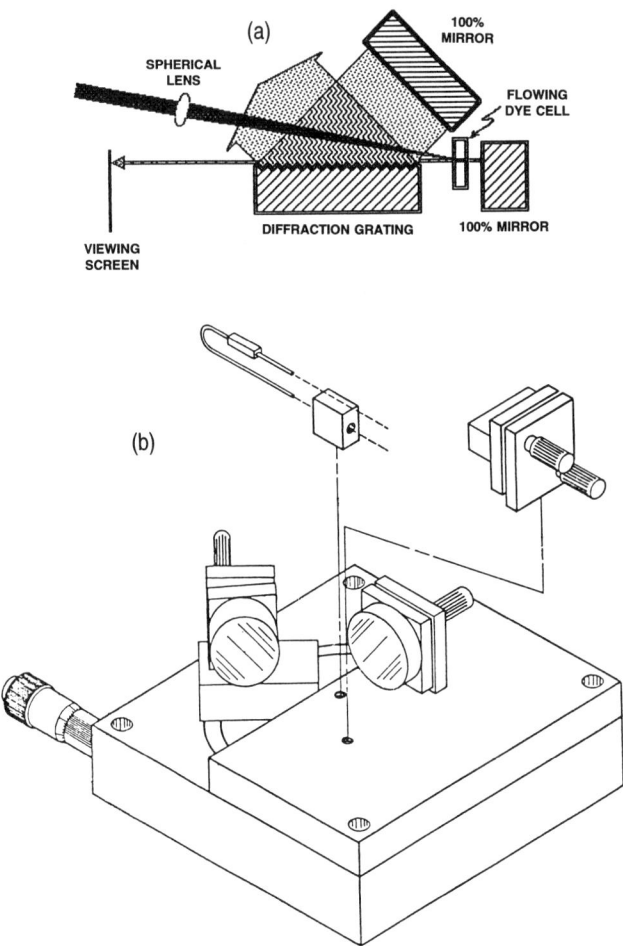

FIG. 12. Single-mode longitudinally pumped grazing-incidence dye laser: (a) top view; (b) exploded view.

the laser gain range. Once this is achieved, the laser will be oscillating, although it might be doing so in several longitudinal modes. Fine adjustment of the tuning mirror pitch angle, the position of the focus lens in the pump beam, and the grazing incidence angle of the grating is made while the output is monitored by a 20-GHz free-spectral-range Fabry–Perot etalon, until single-mode operation is achieved. The laser is continuously tunable by rotation of the tuning mirror about a unique pivot axis that is defined by the intersection of the surface planes of the grating, end mirror, and tuning mirror [17].

6.5 Summary

This chapter is not intended to be a complete review or full coverage of the subject of pulsed lasers. It is a practical guide to researchers who have had only limited exposure to the subject but are interested in building, using, and understanding lasers as tools for research. We first discussed the general concepts of laser operation and briefly introduced the readers to various kinds of pulsed lasers that are important in applications. A checklist was then developed for potential laser buyers. Finally, we provided a step-by-step illustration on how to build dye lasers with various levels of performance, so that the readers can get some hands-on experience with building lasers they can use and gain better understanding of the important issues in laser operation.

References

1. Maiman, T. H. (1960). *Nature* **187**, 493.
2. Fork, R. L., Brito Cruz, C. H., Becker, P. C., and Shank, C. V. (1987). *Opt. Lett.* **12**, 483.
3. Siegman, A. (1986). *Lasers*, University Science Books, Mill Valley, CA.
4. Hellwarth, R. W. (1961). In *Advances in Quantum Electronics*, p. 334, Columbia Univ. Press, New York.
5. Geusic, J. E., Marcos, H. M., and Van Uitert, L. G. (1964). *Appl. Phys. Lett.* **4**, 182.
6. Macklin, J. J., Kmetec, J. D., and Gordon III, C. L. (1993). *Phys. Rev. Lett.* **70**, 766; L'Huillier, A., and Balcou, P. (1993). *Phys. Rev. Lett.* **70**, 774.
7. Park, Y. K., Giuliani, G., and Byer, Robert L. (1984). *IEEE J. Quantum Electronics* **QE-20**, 117.
8. Sorokin, R. P., and Lankard, J. R. (1966). *IBM J. Res. Develop.* **10**, 162; and Eastman Kodak Laser Dye Technical Note and Catalog.
9. Moulton, P. F. (1982). Solid State Research Report No. DTIC AD-A124305/4, Lincoln Laboratory, MIT, Lexington, MA, pp. 15–21.
10. Rines, Glen A., and Moulton, Peter F. (1990). *Opt. Lett.* **15**, 434.
11. Basov, N. G., Krokhin, O. N., and Popov, Y. M. (1961). *JETP* **40**, 1320; see also Thompson, G. (1980). *Physics of Semiconductor Laser Devices*, Wiley, New York.
12. *Laser Focus World*, September 1992, p. 15.
13. Strome, F. C., and Webb, J. P. (1971). *Appl. Opt.* **10**, 1348.
14. Hänsch, T. W. (1972). *Appl. Opt.* **11**, 895; Schawlow, A. (1973). In *Fundamental and Applied Laser Physics*, M. Feld, A. Javan, and N. Kurnit (eds.), Wiley, New York.
15. Littman, M. G., and Metcalf, H. J. (1978). *Opt. Lett.* **3**, 138; Shoshan, I., Danon, N. N., and Oppenheim, U. P. (1977). *J. Appl. Phys.* **48**, 4495.
16. Littman, M. G. (1984). *Appl. Opt.* **23**, 4465.
17. Liu, K., and Littman, M. G. (1981). *Opt. Lett.* **6**, 117.

7. TECHNIQUES FOR MODELOCKING FIBER LASERS

Irl N. Duling III

Naval Research Laboratories
Washington, DC

7.1 Introduction

One of the more interesting developments in the past 15 years has been the proliferation of well-designed fiber components. This development has been driven by the interest in these systems by the telecommunications industry. The promise of fiber is high bandwidth and reliable communications.

It is fortunate for the fiber laser researcher that this is the case, for the large market has generated a number of fiber components that otherwise would be unavailable or would at least be much more expensive. Many of the techniques that will be discussed in this chapter use no specialized components at all and are therefore easily acquired from a number of vendors.

The chapter begins with a short discussion of general methods of handling optical fibers and of the methods of cavity construction available. These techniques would be used in any laser that might be constructed. There will then be the main body of the work on modelocking methods that have been used in fiber lasers up to this time. This will not be a comprehensive history, but more an overview of the more common or interesting techniques that have been employed and any distinguishing features of that particular technique. The references at the end of the chapter can lead the researcher to more detailed discussions of each method, but this chapter should allow him to choose the method most suitable to his ends and to get started in the experiments.

7.2 Cavity Building

There are a few attributes of fiber lasers that make them particularly forgiving in constructing a cavity. One of the major differences is that the gain medium (rare-earth doped optical fiber) can have an unsaturated gain per pass of over 1000 [1]. This from a material that in bulk lasers is considered to have relatively low gain per pass. In the fiber case, the length of the gain medium can be tens of meters, whereas the bulk laser is limited to around ten centimeters.

In a bulk laser, diffraction must be taken into account and the gain medium is limited by the Rayleigh length of the focus. In the case of a fiber there is no such limitation. The core of the fiber confines the light so that the gain can be of a length dictated by pump absorption or other requirements.

The high gain in the fiber lasers provides a unique situation among lasers. Bulk lasers generally fall into two categories that are roughly delineated by the size of the emission cross-section of the lasing transition. Lasers with a large emission cross-section are generally high gain per pass and show a large amount of gain saturation with the passage of a single pulse. The dye laser and the semiconductor diode laser fall into this category. On the other hand, lasers with a small emission cross-section, like the Nd:YAG laser, show little gain saturation from a single pulse and low gain per pass. In contrast to both of these situations, the fiber laser shows little single pulse saturation but a high gain due to the length of the gain medium. The laser can therefore tolerate rather large intracavity loss, but shows the characteristics of lasers with little or no gain saturation.

There are two other advantages of the fiber form. The first is that the researcher does not have to concern himself with the stability of the cavity or any form of mode matching, as might be necessary for a bulk laser. And secondly, the tightly confined mode and the long cavity lengths generally used means that even the small nonlinear coefficient of silica becomes significant. In fact, it becomes necessary for some of the modelocking mechanism to be discussed later in this chapter.

7.2.1 Splicing

As mentioned previously, the cavity is usually constructed completely of optical fiber. Since there are usually multiple components to connect, a decision must be made as to how they are to be connected. There are two main techniques: connectors or fiber splicing. The advantage of connectors is that the cavity can be easily rearranged and launching the light into test equipment (e.g., detectors, optical spectrum analyzers) is simplified. On the other hand, unless the connector is kept very clean and not overly used, the loss and backreflection at the connection will undoubtedly increase. The connections then become suspect for anything that might go wrong with the system (even when they are not at fault).

There are two types of connectors that could be considered: "physical contact" (PC) and "angled physical contact" (APC). The major difference is that APC connectors almost completely eliminate backreflections since the contacting interface is angled at about 10°. The only drawback is that the loss can be slightly higher as the manufacturing tolerances are a bit more stringent. In addition, coupling to test equipment can be a bit more difficult as the light comes out of the fiber at an angle.

Connectorized components can be preserved with a few basic precautions. Clean the connectors each time prior to mating, and periodically repolish the connectors to maintain their finish.

The alternative to connectors is fiber fusion splicing. A fully automated fusion splicer can be obtained for somewhere in the neighborhood of $25,000 and will produce splices routinely with as little as 0.1 dB of loss (the loss in dB = $10 \log_{10}$ (fractional loss)). The great advantages are that backreflection at the splice is essentially eliminated, the resultant connection is smaller, and there is no problem with maintaining the end surfaces. The drawbacks are that a small amount of fiber is lost as each splice is done (which will eventually make the component useless), it takes slightly longer to perform a splice (although it can be done in about 2 min), and the cost of the fusion splicer.

Care must be taken in all fiber lasers to minimize backreflections into the lasing mode from a variety of surfaces. These include the surfaces of the modulator, lenses, and the ends of the fiber itself. The surfaces are most often angled to send the reflection out of the cavity. The additional loss incurred by this technique is not a problem for most fiber lasers due to the high gain of the amplifier.

7.2.2 Polarization Controllers

One of the fundamental variables inside the fiber laser cavity is the polarization state of the light. This can be used to advantage, as in modelocking with nonlinear polarization rotation (Sections 7.3.4 and 7.3.5), but usually there is some polarization dependent component in the optical path which requires a specific state of polarization (SOP). In order to adjust the SOP, polarization controllers have been developed. The simplest controller is made by winding the fiber around a disk of appropriate diameter that lies in a plane including the fiber and can be rotated freely about the fiber. The stress-induced birefringence acts like a bulk waveplate to adjust the polarization state. Theoretically, it only takes two waveplates to adjust the polarization, but it is easier to use three: a quarter-wave, a halfwave, and another quarter-wave retarder. These polarization controllers are often called "flappers" or "Mickey Mouse ears" and can be purchased from a number of vendors or built by a local machine shop. The diameter of the disk and the number of turns of fiber dictate the order of the waveplate [2].

Other types of polarization controllers have been developed that squeeze the fiber or use electrooptic elements, but the least expensive method is the controller described above [3].

7.2.3 Faraday Mirrors

Often it is better to use a cavity that is all polarization-maintaining (PM) fiber and avoid the question of polarization control. In the case of an all-PM cavity, the hardest component to obtain is the polarization-maintaining gain fiber. A solution has been found to placing nonpolarization-maintaining components in a PM cavity, and that is by the use of a Faraday mirror [4]. This device consists of a 45° Faraday rotator and a high reflecting mirror at the end of a fiber pigtail. The operation of the Faraday mirror is to compensate for the birefringence of a fiber section by reflecting an orthogonal SOP after the first pass through the fiber section. This places the light that was on the slow axis of the fiber on the fast axis and that which was on the fast axis on the slow axis. The light retraces the evolution of the polarization in the second pass through the fiber, causing the birefringent effect to be "unwound." It is then possible to construct a device where the polarization state is maintained but contains components that are not polarization-maintaining. Such a device is pictured in Fig. 1. The light is launched into the non-PM section of fiber through a polarizing beam splitter, propagates through the standard fiber components, reflects from the Faraday mirror, and then retraces its path. Since the light was launched in a linear state, it returns to the polarizing beam splitter also in a linear state, but at 90° to the original state. The light then exits the polarizing beam splitter through the opposite port [5]. In this way the polarization state is stabilized.

7.2.4 Isolators

Many of the fiber lasers are built in a ring configuration to remove the need for mirrors and eliminate spacial hole burning. In order for the ring to operate unidirectionally, it is necessary to insert an isolator in the cavity. Isolators are available commercially for less than $500 due to their high-volume use in communications systems. The most common isolators available today are polarization independent (PI). Manufacturing tolerances have improved, but PI isolators can have significant temporal walkoff between orthogonal polarizations. While not a problem for a communications system operating at 10 Gb/sec or below, as the pulses get shorter the time difference can become significant. Early isolators had up to 7 psec of walkoff, while currently most isolators limit this to less than a few hundred femtoseconds. Even at this level, considering that some of the fiber lasers described here can produce pulses shorter than 100 fsec, the walkoff can cause pulse distortion or breakup.

The effect inside a nonlinear fiber laser cavity, rather than causing pulse distortion, is twofold. One, a pulse condition could be imagined where the pulse splits into two and thus generates a train of pulses at a high repetition rate. The second, more likely, result would be that the pulses would choose to run on one or the other axis so that no breakup occurs (since in a cavity with nonlinear switching this will be the lowest loss condition). This means that the isolator is operating in some sense as a system polarizer. As a result, it is common in passively modelocked lasers to use a single polarization isolator to determine the polarization state of the laser.

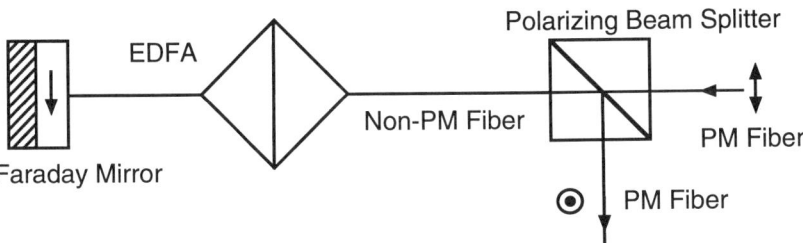

FIG. 1. Birefringence in the Er:fiber amplifier (EDFA) is compensated by conversion of the light to the orthogonal polarization at the Faraday mirror. The polarization is orthogonal at every point on the return path and reaches the beam splitter in the orthogonal linear state regardless of the amount or orientation of the birefringence in the EDFA.

The use of a single polarization isolator can have an added benefit. If the isolator is pigtailed with PM fiber and followed by a PM output coupler, the output polarization state of the laser will be fixed with no additional constraint of laser operation. This also allows insertion of polarization-dependent components (e.g., a lithium niobate modulator) in the laser cavity without the need for separate polarization controllers.

7.3 Modelocking

This section on modelocking methods is the largest and is organized to put the more classic active modelocking first (with a digression to cover frequency-shift modelocking), then semiconductor modelocking (which is similar to the old technique of modelocking with a saturable absorber), and then the more recent methods based on Kerr nonlinearity (loop mirrors and nonlinear polarization rotation). The presentation is geared more to practical considerations in constructing these lasers and less to an exhaustive history of the techniques.

7.3.1 Active Modelocking

One of the major advantages of active modelocking is the control that it gives the researcher over the characteristics of the laser. The external drive determines synchronization of the pulse train to an electrical reference that can then be used to trigger external phenomena. This has particular advantage when the laser is being used to study an electrical device, as in electrooptic sampling, or when the modelocked train must be modulated by an electrical signal, as in digital communications. It is possible to synchronize a passively modelocked laser to an external reference through feedback on the fundamental cavity frequency and the phase of

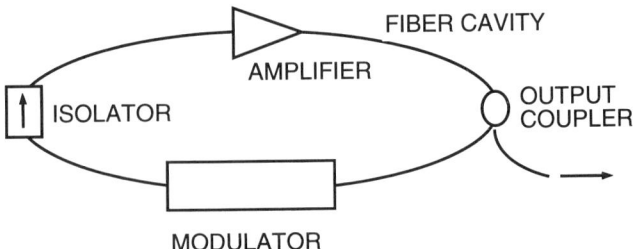

FIG. 2. The standard cavity for an actively modelocked fiber laser contains: an Er:fiber amplifier, an integrated optic modulator, an optical isolator; and an output coupler.

the pulse in the cavity [6], but this technique has not been shown to obtain synchronization of better than a few picoseconds.

Another area in which fiber lasers differ from their solid-state counterparts is that they are often operated harmonically modelocked. As a result, there are a large number of pulses circulating in the cavity. The modelocker defines windows in time where each of these pulses should reside. If passive modelocking is present in the laser, the pulses will not necessarily fill the time slots that they should and may appear as pairs of pulses in a single time slot, or there may be empty slots, or in the case of a nonsoliton fiber laser all of the energy may be in a single pulse.

7.3.1.1 Amplitude Modulation A standard actively modelocked fiber laser is illustrated in Fig. 2 [7]. The components are an Er:fiber amplifier, an isolator, a modulator, and an output coupler. There are also a number of optional components that could be included that are not shown: a tunable filter, an element for changing the cavity length (e.g., a thermally controlled section of fiber, a fiber wound PZT cylinder, an adjustable free space propagation section), dispersion compensation (e.g., dispersion compensating fiber, prism sequence), polarization controllers, and a polarizer.

The actively modelocked laser has the normal control questions of any of its solid-state counterparts. How do I match the cavity length and the drive frequency? What determines the pulsewidth? What determines the operating wavelength?

The drive frequency is easily matched to a multiple of the fundamental cavity mode spacing by looking at the output on an RF spectrum analyzer, as described in Section 7.4.3. Keeping the two frequencies matched is of greater concern. A laser that might have a cavity mode spacing of 1 MHz, operating at 10 GHz, and generating 1-picosecond pulses will require the drive frequency to be matched to the cavity harmonic to within a few hundred hertz. That corresponds to a few microns out of 200 meters. It is then obvious that small temperature changes or small operating frequency shifts will cause a change of this magnitude. In order to lock the frequencies, a variety of feedback systems have been developed. Adjusting the cavity length is accomplished by: a fiber wrapped around a PZT cylinder

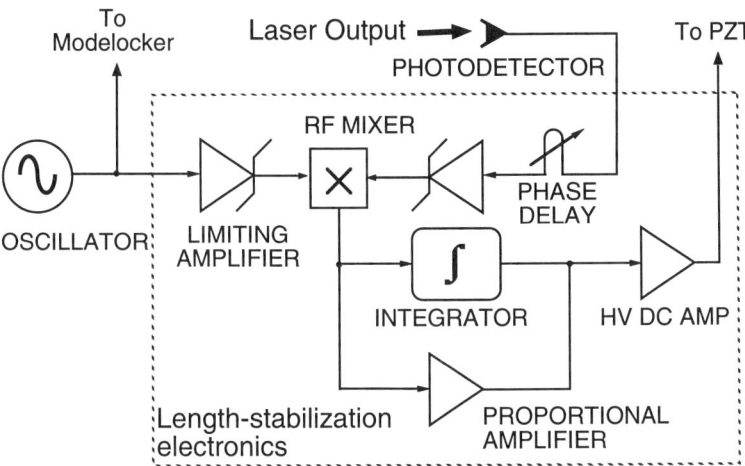

FIG. 3. Block diagram of a feedback circuit based on mixing of the modelocker drive with the modelocked pulse train. Limiting amplifiers are used to ensure equal amplitudes, and an integrator compensates for long-term drift.

[8], bulk optics and a translation stage, a coil of fiber in a thermal enclosure [9], and metal-clad fiber for resistive heating. The error signal is usually derived from the phase shift of the pulse train that occurs when the two frequencies are mismatched. A typical circuit is shown in Fig. 3. The error signal is generated by mixing the output of a photodiode detecting the pulse train with the drive frequency. It is then put through the feedback circuitry, which should include proportional, integral, and differential control. The integration takes care of slow variations in cavity length, while proportional control compensates for the fast component. The output of the feedback circuit is passed through a high-voltage amplifier and used to drive a fiber-wound PZT. Some groups have separated the fast and slow control mechanisms, applying proportional control to a PZT and integral control to a thermally controlled section of fiber [9].

An alternate and very effective error signal has been derived from the magnitude of the relaxation oscillation noise peak as seen on the RF spectrum analyzer. This technique has been implemented using a computer for feedback control [10].

One subtlety of active harmonic modelocking is that the laser should be driven on an odd harmonic of the laser cavity. Driving on an even harmonic allows two independent pulse trains to be generated. This case is characterized by production of pulses with alternating amplitudes as seen on the sampling scope.

The pulsewidth in purely active modelocking is determined by a number of limiting parameters depending on the regime of operation. In the low-power regime the pulsewidth will be given by the classic Kuizenga–Siegman equation:

$$\tau_p = \frac{(2 \ln 2)^{1/2}}{\pi} \frac{g_0^{1/4}}{\delta_1^{1/2}} \left(\frac{1}{f_m \Delta f_a} \right)^{1/2}, \quad (1)$$

where g_0 is the saturated single-pass gain, δ_1 is the modulation parameter, f_m is the modulator drive frequency, and Δf_a is the lasing bandwidth [11]. This is the dependence that for a 100-MHz modulation will give 50–100-psec pulses and for a 10-GHz modulation will give 5–10-psec pulses. Obviously, if there is a frequency filter in the laser the pulses can be lengthened. An example of the application of this equation to a fiber laser is given in reference [12].

When building a fiber laser out of non-PM fiber, the cavity incorporates a natural frequency filter. The fiber birefringence in conjunction with the intracavity polarizer forms an effective Lyot filter. Even for short lengths this can limit the optical bandwidth and broaden the pulses. At the very least it will prevent smooth tuning of the laser cavity. As implied, smooth tuning of the laser can be recovered if the cavity is constructed from all-PM fiber.

By using a wide bandwidth Mach–Zehnder modulator, a step-recovery diode can be used to generate pulses to drive the modulator. This allows shorter pulses to be produced than with sinusoidal modulation. The drive frequency in Eq. (1) is replaced by the highest frequency component of the step-recovery diode regardless of the repetition rate. In addition, harmonic modelocking can be done, where the modulator is driven at a multiple of the fundamental frequency of the cavity. In this way, thousands of pulses can be sustained in the cavity and repetition rates as high as 40 GHz have been produced [13].

It is often desirable to generate pulses shorter than predicted by Eq. (1). One solution would be to combine active modelocking and one of the passive modelocking techniques. As mentioned earlier in this section, this generally leads to pulse dropouts or double pulsing and is in general hard to stabilize. The exception is the work by Doerr [14], where the nonlinear switching was set such that it limited pulses from increasing in intensity due to its sinusoidal character. This system did, however, produce pulses that were longer than those normally obtained with this passive technique.

A more robust method for shortening the pulses in an actively modelocked fiber laser is to take advantage of the high-power amplifiers and dispersion-compensating fiber currently available. By increasing the intracavity power and reducing the cavity dispersion, it is possible to cause the normally Gaussian pulses of classic active modelocking to evolve into solitons. This technique is appropriate for pulses that can operate in the anomalous dispersion regime. The equation which applies for the soliton energy is

$$E_0 = 1.135 \tau P_1, \quad (2)$$

where τ is the soliton full width at half maximum, and P_1 is the soliton peak power, given by

$$P_1 = 0.776 \, \lambda^3 \, \frac{|D|A_{\text{eff}}}{\pi c n_2 \tau^2}, \tag{3}$$

where c is the speed of light in a vacuum, D is the fiber dispersion, A_{eff} is the effective area of the propagating mode, λ is the laser wavelength, and n_2 is the nonlinear refractive index.

As these equations indicate, the soliton will shorten as the energy increases and the dispersion decreases. For a limited amount of power from the amplifier, it is necessary to lower the dispersion to get shorter pulses. Also, decreasing the repetition rate will allocate more of the average power to each pulse, resulting in potentially shorter pulses. Obviously, the pulse will not shorten further than any bandwidth filter that might be present in the cavity. More on the effects of solitons in the cavity will be discussed in the context of passive modelocking techniques in Section 7.3.1.3.

7.3.1.2 Phase Modulation An alternative method of active modulation is to use a phase modulator. A number of systems have been produced that utilize either bulk or integrated phase modulators of various types. The laser follows the classic patterns for modelocking with a phase modulator, as has been described elsewhere [15].

An interesting class of phase-modulated lasers is characterized by intracavity powers high enough to cause the pulses to propagate nonlinearly (as solitons). These lasers, known as frequency-shift lasers, operate on the same principle as the sliding-guiding frequency filter of soliton-based communications [16].

The requirement for modelocking can be viewed as making the CW lasing condition have higher loss than the modelocked (pulsed) condition. If this fact is coupled with the knowledge that a soliton can recover bandwidth lost at a frequency filter through self-phase modulation, a system can be envisioned where the passband of the filter is shifted slightly as a function of time, so that linearly propagating light will eventually be blocked, but the soliton will be able to follow the shift by recovering the lost bandwidth after each filtering. To turn this concept into a laser, the filter is kept at a fixed frequency and all of the light in the cavity is shifted in frequency on each round trip. The frequency shift can be implemented by either an acoustooptic frequency shifter [17] or by a phase modulator [18]. This laser is an interesting hybrid between active and passive modelocking. It requires an external drive frequency, but there is no specified synchronization between the drive and the laser pulses.

One of the advantages of modelocking with an amplitude modulator is that it provides an electrical reference that can be synchronized to an external process, whether that is an electrical circuit under study or a modulator encoding the modelocked pulse train with data. In addition, by actively modulating the cavity, interpulse noise is reduced, and in harmonic modelocking the pulses will be evenly spaced within the cavity period.

7.3.1.3 "Soliton" Modelocking
To get shorter pulses than those predicted by classic Kuizenga–Siegman modelocking requires nonlinear propagation of the pulse train. As the internal cavity power is increased, the pulses will begin to shorten due to the soliton-forming process and will evolve from the Gaussian pulseshape characteristic of pure active modelocking to a pulse that is primarily hyperbolic secant in shape (the classic soliton shape). In fact, the pulse will tend to be hyperbolic secant in the wings of the pulse and very Gaussian in the center. For theory and experimental evidence refer to [19, 20]. The result of this increase in pulse power is that the pulses can now have durations below that predicted by Kuizenga–Siegman by up to a factor of five [20]. This is a very powerful technique and is well-behaved enough to be used even in lasers intended for communications systems.

7.3.2 Semiconductor Saturable Absorbers

Another method of modelocking that has been used to good advantage is that of passive modelocking using a solid-state analogue to a saturable absorber. The saturable absorber of choice has been the semiconductor. These saturable absorbers use real carrier generation to produce saturable absorption. The time response of the carriers has two components, causing short pulses to be generated (by the fast component of recovery) and the threshold for pulse forming to be low (by the slow component of recovery). Semiconductors have been used in bulk [21], as multiple quantum wells [22], and as saturable Bragg reflectors [23]

Modelocking is usually obtained in the band tail, resulting in pulses that are well-separated in the cavity and of durations down to 320 fsec [24]. The success of semiconductor modelocking has led to the development of turnkey commercial lasers that rely on this technique and to its use in addition to other modelocking mechanisms in the cavity to further organize or shorten the pulses.

A typical fiber laser modelocked with a semiconductor saturable absorber is pictured in Fig. 4. Most semiconductor modelocked fiber lasers have linear cavities due to the reflective nature of the saturable absorbers. Due to the low number of elements, it is possible to make the fundamental frequency higher than that in other modelocked fiber lasers.

7.3.3 Nonlinear Loop Mirror Modelocking

One of the first mechanisms used to do passive modelocking in fiber lasers was the Kerr effect. By using a nonlinear interferometer, the small differential phase shift experienced by pulses of different amplitude could be transformed into amplitude-dependent modulation in the cavity. If a Sagnac interferometer is used, the device is called a nonlinear optical loop mirror (NOLM) or a nonlinear amplifying loop mirror (NALM), depending on whether the amplitude imbalance of the counterpropagating light is from a coupler that is not 50:50 or from an

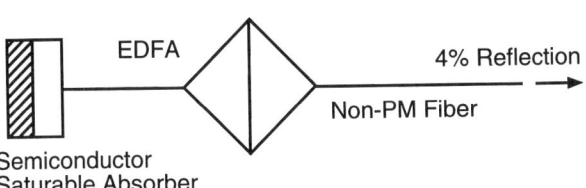

FIG. 4. The diagram of a semiconductor modelocked fiber laser. The semiconductor element can be bulk, a multiple quantum well, or a saturable Bragg reflector.

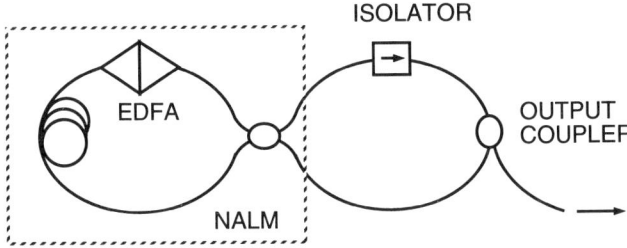

FIG. 5. The figure eight laser consists of a ring fiber laser modelocked by either a nonlinear amplifying loop mirror (shown here) or a nonlinear optical loop mirror.

amplifier included at one end of the loop. If the interferometer is implemented by the two birefringent axes of the fiber and a polarizer, then it is called nonlinear polarization rotation. This will be treated in Section 7.3.4.

If the output of an NALM is connected through an output coupler and an isolator to its input, it forms a figure eight laser (F8L) (see Fig. 5) [25]. The nonlinear switching characteristic [26, 27] of the NALM acts as a fast saturable absorber that shortens the pulses with each round trip until the pulse limit is reached. The cavity also generally includes two polarization controllers, one in the NALM and one in the loop with the output coupler. These polarization controllers must be set properly for the laser to operate in pulsed mode. The controller inside the loop mirror will provide a differential phase bias in the loop [28], and the other polarization controller compensates for the stress birefringence in the unidirectional loop and for any difference in the input and output polarization states of the loop mirror. In concept there should be a small amount of leakage through the loop mirror to aid in pulse initiation. In practice the proper setting is found by a systematic search starting from the point where the loop mirror is completely reflecting. The initiation of pulsing will be characterized by an instantaneous increase in the average power. In contrast to nonlinear polarization rotation, the nonlinear loop mirror can be constructed using all-PM fiber. This creates a very stable and environmentally insensitive laser, but the bias to initiate pulsing must be provided by uneven splitting at the central coupler of the laser.

Nonlinear switching of the loop mirror and the required suppression of the CW modes of the cavity causes the laser to exhibit bistability of the average power. The laser can operate at CW low average power and then with the introduction of a transient perturbation to the cavity it can snap into a higher average power pulsed mode. Most commonly the laser will enter this pulsed mode with more than one pulse in the cavity (the number of pulses dictated by the amplifier saturation conditions discussed in the previous section). The figure eight laser, and most other Kerr effect modelocked fiber lasers, will have a self-starting threshold (that point where the amplifier is pumped hard enough that an external stimulus is not necessary) that is higher than the double-pulse threshold. The condition for self-starting can be understood as that point where a strong noise burst (fixed phase relationship between the cavity modes) will exist in the cavity long enough to form a pulse. The lifetime of modal coherence can be measured by the width of the RF beat notes detected if the output is put on a detector and into an RF spectrum analyzer [29]. Experimentally the lifetime is affected by internal reflections in the cavity. It has been shown that for this reason a ring cavity will in general have a lower self-starting threshold since it would take two internal reflections to make a subcavity, in contrast to the linear cavity laser, where a single internal reflection and one of the end mirrors will make a higher-Q subcavity [30]. It is therefore possible to create a ring laser that is Kerr effect–modelocked that exhibits a single pulse self-starting threshold.

As in any laser, the repetition rate is dictated by the cavity length. In the case of multiple pulses in the cavity, it is possible to arrange that the pulses be spaced at even intervals around the cavity by internal [31] or external [32] feedback. The disadvantage of internal feedback is that the subcavity must be interferometrically stabilized relative to the main cavity. It has also been shown in polarization-switched lasers (to be treated in the next section) that the pulses in the cavity can self-organize at a repetition rate between 0.5 and 1 GHz by the action of the acoustic soliton interaction.

As a soliton travels down a fiber, the electrostrictive effect causes an acoustic wave to be launched. As it propagates out of the core, it first creates a repulsive force and then an attractive force. Lasers have been shown to be stable when operated in the region of attractive force that exists after the soliton [33].

7.3.4 Nonlinear Polarization Rotation Modelocking

As mentioned in the previous section, one way to look at nonlinear polarization rotation is as a nonlinear interferometer whose separate paths are the two birefringent axes of the fiber. If one axis of the fiber carries more light than the other, it will experience more phase shift due to the effect of the nonlinear refractive index. This differential phase shift between light on the two axes of the fiber will result in an intensity-dependent polarization state exiting the length of fiber. By adjusting the launch polarization state and angle with respect to the fiber birefringent axes and placing a polarizer at the other end of the fiber, an effective fast saturable absorber can be created [34]. In practice this is a very simple laser to construct. It has the same components as the actively modelocked laser shown in Fig. 2 but without the modulator. Two polarization controllers are usually included, one just after the single polarization isolator and one just before the isolator. Nonlinear switching occurs in the whole cavity, and strictly unidirectional operation reduces the effects of internal reflections in the cavity, lowering the threshold for the onset of pulsing, as will be discussed in the next section. This laser has also been implemented using bulk components for the polarization controller, isolator, and polarizer, which allows the light rejected at the polarizer to be used as the output. In this way the polarization controller can also give a variable output coupling and the laser can be optimized for high-output power [35].

In the soliton regime, the laser produces pulses down to about 100 fsec and exhibits the same transform-limited soliton output pulses as the F8L. If the laser is driven hard enough to produce multiple pulses, they will usually occur with arbitrary placement in the cavity. Multiple pulsing will be discussed more in the next section.

7.3.5 Stretched Pulse Modelocking

In some circumstances, the limitation on the energy per pulse imposed by soliton formation may be undesirable. In that instance, another fiber laser design modelocked by nonlinear polarization rotation has been developed. What has been called at various times an alternating dispersion laser, an additive pulse modelocked fiber laser or a stretched pulse modelocked fiber laser is essentially a passively modelocked ring laser (as described in Section 7.3.4) but operating in the net normal dispersion regime [36]. The most significant result is that, without the mechanism of dispersive wave formation in the laser, there is no bandwidth limitation and the pulse can increase arbitrarily in energy. The peak power will be clamped by the peak transmission of the nonlinear switching characteristic of the nonlinear polarization rotation, but the pulse is free to spread due to the net positive dispersion.

Net positive dispersion is obtained in the cavity by constructing a portion of the cavity with standard fiber components that have anomalous dispersion at the lasing wavelength, and the remainder of the cavity by fiber that has been dispersion-shifted to the point that it has normal dispersion. In all of the alternating dispersion lasers reported to date, the erbium-doped fiber was in the normal dispersion regime. This comes about as a consequence of attempting to draw high-efficiency erbium-doped fiber to make the most of limited pump light. The tight mode field confinement leads naturally to dispersion shifted waveguides.

The output of this laser is highly chirped, but pulses as short as 77 fsec have been generated using external compensation [36]. Where the output coupler is located will determine the sign of the pulse chirp as it leaves the cavity. Pulses from this laser are not transform-limited, but since there is no dispersive wave generated the autocorrelation trace is much cleaner and does not show additional energy in the wings.

7.3.6 Laser Operating Parameters

In designing a fiber laser, one of the driving operating parameters is the desired pulsewidth. The major determining factor in the pulsewidth is either total dispersion in the cavity or any effective frequency limit that might be present. In the case of a frequency filter in the cavity, the obtainable pulsewidth will be determined by the Fourier transform of the oscillating bandwidth. In that case, the pulselength will be limited by

$$\tau \Delta \nu = 0.3148 \tag{4}$$

for a hyperbolic secant pulseshape, where $\Delta \nu$ is the optical bandwidth of the pulse. The time bandwidth products for various pulseshapes are shown in reference [37].

It was realized that there is another frequency-dependent loss mechanism that can limit the oscillating bandwidth in soliton supporting fiber lasers [38]. The spectrum of the laser (particularly in passively modelocked lasers) displays sharp sidebands (see Fig. 6). These sidebands are found to be located where

$$\Delta \lambda_N = \pm N \lambda \sqrt{(2N/cDL) - 0.787 \, (\lambda^2/(c\tau)^2)}, \tag{5}$$

where $\Delta \lambda_N$ is the offset to the Nth sideband, and L is the cavity length [39]. The sidebands are a result of the periodic perturbation of the soliton by the elements of the cavity. When the soliton changes energy, it reshapes to satisfy Eq. (2) and in the process sheds energy in the form of a dispersive wave. Since the soliton propagates with a single propagation constant for all wavelengths, and the dispersive wave propagates according to the dispersion of the cavity, the dispersive waves generated on successive round trips of the cavity constructively interfere at the specific wavelengths given by Eq. (5). The location of the sidebands is then a direct measurement of the total dispersion in the cavity and can be used as a powerful diagnostic [38]. By plotting the order of phase accumulation versus relative frequency offset and fitting the result to a polynomial with an offset, the significant orders of dispersion can be extracted. This difference at zero frequency corresponds to the phase velocity differential between the soliton and linearly propagating light and in real units is a function of the pulsewidth [40]. In some instances, fitting the sideband spectra is the simplest way to determine the dispersion of the fibers in the cavity since amplifier fibers are rarely characterized in terms of dispersion. By measuring the total dispersion in the cavity and then adding

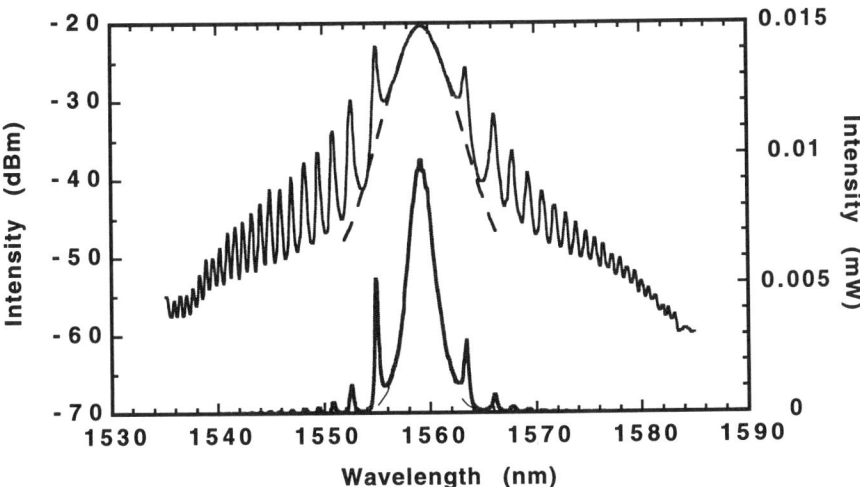

FIG. 6. Log and linear plots of the dispersive wave sideband structure on the spectrum of a passively modelocked fiber laser.

or removing a length of the desired fiber, the dispersion can be determined by the difference in total dispersion.

From the perturbation theory analysis of Gordon [41], the rate of energy loss from the soliton into the dispersive wave is proportional to the amplitude of the soliton spectrum at the location of the sideband. This results in an effective frequency filter whose width is dependent on the total dispersion in the cavity (seen as the product DL in the first, and usually dominant, term of Eq. (5)). By examining the results from many different fiber lasers, the pulsewidth is found to correspond reasonably with the condition that

$$\tau = \sqrt{L|\beta_{2\text{-ave}}|}, \tag{6}$$

or

$$\tau^2 = \frac{L\lambda_0^2}{2\pi c}|D|, \tag{7}$$

where $|\beta_{2\text{-ave}}|$ is the magnitude of the second-order dispersion in the cavity measured in psec^2/km, and L is the cavity length in km.

The theory as presented by Gordon [41] describes the effect of a sinusoidal perturbation to the pulse and resulted in only two sidebands. In actual systems, of course, the perturbations are not sinusoidal but can be Fourier decomposed into sinusoidal perturbations. This leads to the multiple sidebands that are observed. More than 60 sidebands have been observed in some systems [38]. Because the

(a)

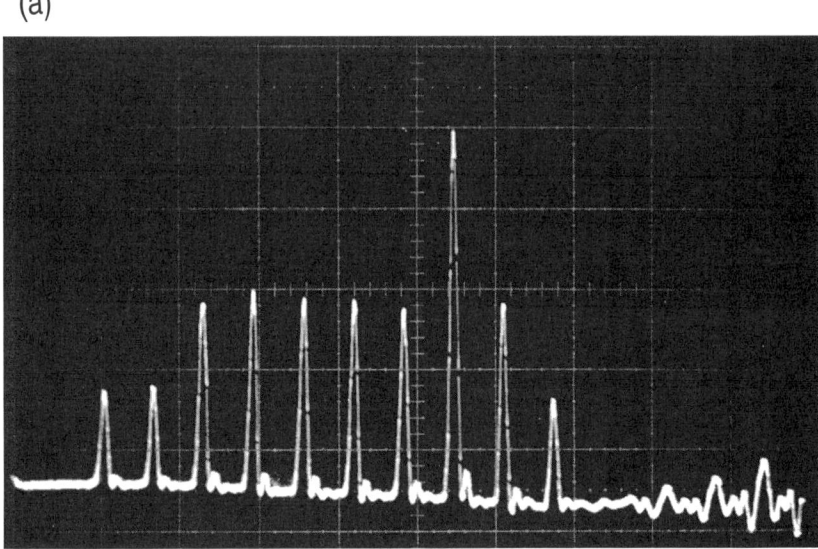

(b)

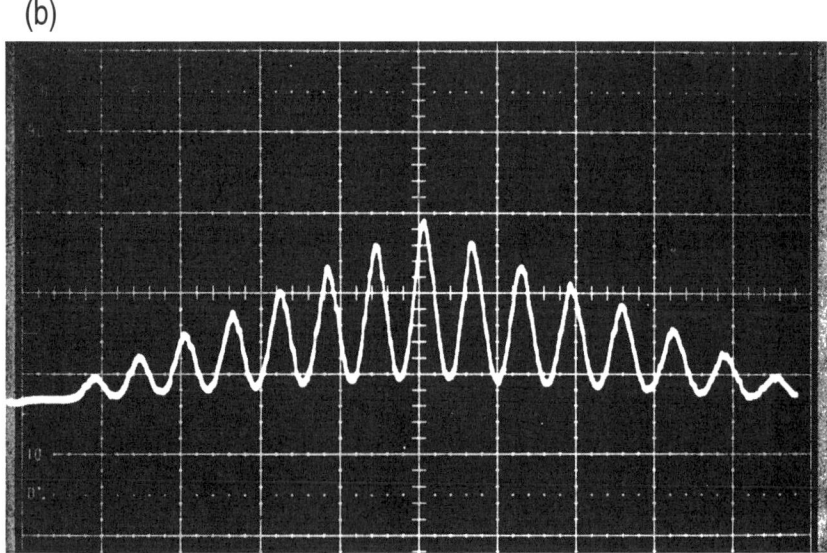

FIG. 7. A passively modelocked fiber laser can emit bursts of pulses repeated at the cavity round trip. These bunches show the quantized nature of the pulse energy (a). The autocorrelation of an equally spaced train corresponding to 8 pulses at 240 GHz (b).

exact form of the perturbations and the phase of their components are not known, it is difficult to predict the relative strength of the various order sidebands.

The result of frequency limitation is an effective way to control the time bandwidth product and produce very stable pulseshapes, but it also leads to a lower limit on the pulsewidth and therefore an upper limit on the energy of the pulses from the fiber laser. This leads to quantization of the average power from the laser. Since each pulse is limited in energy, that energy divided by the round-trip time of the cavity leads to an average power per pulse. In a laser there is an additional requirement on the average power that it must be sufficient to saturate the amplifier to the point where the gain is equal to the loss in the laser. This is the requirement for steady-state operation. If this saturation power is larger than the single pulse average power, the laser will produce multiple pulses in the cavity until there is sufficient average power to produce the required saturation.

When multiple pulses are present in the cavity, there are only weak mechanisms to affect their location relative to one another. The locations are most often random, but small etalons in the cavity can produce ordered bursts of pulses. An example of the quantization of the pulse energy and self-ordering of the train is evident in Fig. 7. Self-ordering in the neighborhood of 1 GHz has been observed and is attributed to the acoustic wave generated by the passing of the soliton in the fiber [42]. The saturation power in an erbium amplifier is dependent on the pump power. As a result, the number of pulses in the laser can be controlled. If the laser starts with multiple pulses, the number can be reduced by carefully reducing the pump power until only one pulse remains [33]. In practice the final few pulses will sometimes drop out together, but with careful adjustment of the pump power single-pulse operation can be achieved.

If the requirements of soliton peak power and pulsewidth dependence due to the sidebands is combined (Eqs. (3) and (7)), the result is that

$$P_1 = \left(\frac{1.552\lambda}{n_2 L}\right), \tag{8}$$

with the consequence that the peak power of the pulses from the laser is only a function of the cavity length. This is counterintuitive, because one would think that, if the dispersion in the cavity increased, the soliton would be required to have a higher peak power. But since the pulsewidth increases with the dispersion, there is no net increase in peak power.

The presence of sidebands indicates that there is energy coming out of the laser that is not a part of the soliton. The fraction of the average power contained in the dispersive wave can be as high as 30% [43]. It is possible to remove the dispersive wave inside the cavity by inserting a frequency filter of the appropriate width [44]. As the filter is reduced in width, the sideband amplitude will decrease with little increase in the soliton width until the output is limited by the filter itself.

7.4 Diagnostics

Finally, it may be useful to say a word about diagnostics. The fiber laser can exhibit a wide range of behaviors, from single pulses, to driven relaxation oscillations, to large noise bursts with bandwidths exceeding 80 nm. To determine which of these regimes is being generated it is helpful to have a wide range of test equipment to analyze the output.

In a normal solid-state laser, it is sufficient to have a fast photodiode with a sampling scope, and an optical spectrum analyzer to determine the laser wavelength (if it is a tunable source), and an autocorrelator. In the case of a fiber laser, it is helpful to also have a slow detector, a power meter, and an RF spectrum analyzer.

7.4.1 Autocorrelator

The use of the autocorrelator in addition to measurement of the pulsewidth will give valuable information on how clean the pulses are. Particularly in the actively modelocked lasers, the pulses will often exhibit broad wings and/or instabilities. As a detection that is sensitive to the square of the intensity, the autocorrelation is particularly sensitive to amplitude instabilities. In addition, if the laser is modelocked to a high repetition rate, the autocorrelator can be adjusted to display the cross-correlation between adjacent pulses. In a system that is operating well, the cross-correlation should be the same as the autocorrelation in amplitude and width. Degradation of either of those indicates either missing pulses in the pulse train (not true harmonic modelocking) or jitter between adjacent pulses in the train. The relationship between the areas of the two peaks will determine which of these is the case.

It is also possible to determine the amount of dispersive wave generated by the pulse if the signal-to-noise ratio is high enough to examine the pulses on a logarithmic scale [45].

7.4.2 Optical Spectrum Analyzer

An optical spectrum analyzer serves the purpose of determining the laser wavelength, the optical bandwidth, and to some extent the mode of operation of the laser. Resolutions of 0.1 nm are usually sufficient for this purpose. The typical spectrum of a modelocked fiber laser is shown in Fig. 6. The clear solitonic sidebands are observed on top of the hyperbolic secant spectrum of the pulse itself. Other modes of operation that can be observed readily are: CW oscillation (characterized by sharp peaks in the spectrum), noise generation, an extremely wide bandwidth, a single pulse per round trip, a strong coherence spike in the autocorrelation, and a noisy looking pulse.

7.4.3 RF Spectrum Analyzer

The RF spectrum analyzer is to assist in: analyzing the cavity prior to startup, adjusting the drive repetition rate in an actively modelocked laser, and estimating the phase and amplitude noise on the laser output. In all cases the spectrum analyzer looks at the output of a high-speed detector. The output will generally consist of beat notes at the cavity round-trip time. If there is feedback into the cavity, or a subcavity formed by a parasitic reflection, the envelope of the beat notes will exhibit a periodic modulation. This modulation corresponds to the period of the subcavity length. The strength of this modulation then will be inversely proportional to the self-starting threshold of the laser [29, 46].

In adjusting the repetition rate of the cavity for an actively modelocked laser, the modulator drive will produce an additional peak on the RF spectrum. The drive frequency should be adjusted until this driven peak corresponds to one of the free running peaks of the laser cavity. As coincidence is approached, the laser will go into driven relaxation oscillations characterized by a broadening of the RF beat note into a comb of peaks, and then it will collapse into a narrow peak with possibly a noise peak on each side corresponding to residual relaxation oscillations. This progression assumes that the laser is operating at the same frequency the entire time and that the modulator bias is properly adjusted.

The third area in which the RF spectrum analyzer is helpful is in analyzing the fidelity of the pulse train produced by the laser. By integrating the noise spectrum around the RF beat note, the phase and amplitude noise can be estimated. The details of that procedure can be taken from references [47, 48].

7.4.4 Fast Photodiode and Sampling Oscilloscope

A fast photodiode and sampling scope can give another handle on the noise present in the pulse train. In a harmonically modelocked laser, some designs have a tendency to produce multiple pulses in one time slot, or missing pulses in the cavity [20]. This can be easily seen with a fast scope as double-height pulses or a baseline underneath the pulse. The quality of the pulse train will also say a lot about the phase and amplitude stability of the pulses. In the case of a passively modelocked laser, if multiple pulses are generated in the cavity, the sampling scope can give the time sequence of the pulses in the cavity. Many pulsing conditions have been observed in passively modelocked fiber lasers that are not immediately apparent from either the autocorrelation or the spectrum [33].

7.4.5 Real-Time Oscilloscope

It can also be useful to have a real-time oscilloscope, although this is less often used, to determine the stability of the pulse train on a longer time scale. It is possible to have a reasonable autocorrelation and a nice trace on the sampling scope, and to have the laser exhibit strong relaxation oscillations at the same time. Such oscillations are easily verified on a real-time oscilloscope.

As a final word on diagnostics, it will be noticed that many of the reasons for using different diagnostics are application-dependent. In fact, most of these diagnostics are only necessary while the source is being developed and characterized and can be dispensed with once the operating regimes are established. In fact, it is often necessary to only have one or two of the diagnostics on-line to ensure that the laser is operating in the desired mode. For some lasers, operation is turnkey enough that no diagnostic is necessary after the initial setup sequence [49].

References

1. Shimizu, M., Tamada, M., Horiguchi, M., Takeshita, T., and Oyasu, M. (1990). *Electron. Lett.* **26**, 1641.
2. Lefevre, H. C. (1980). *Electron. Lett.* **16**, 778.
3. Heismann, F., Divino, M. D., and Buhl, L. L. (1990). *Appl. Phys. Lett.* **57**, 855.
4. Pistoni, N. C., and Martinelli, M. (1991). *Opt. Lett.* **27**, 711.
5. Duling III, I. N., and Esman, R. D. (1992). *Electron. Lett.* **28**, 1126.
6. Jiang, M., Rahman, L., Barnett, B. C., Andersen, J. K., Islam, M. N., and Reddy, K. V. (1996). In *Conference on Lasers and Electro-Optics*, 1996 Technical Digest Series 9, p. 525, Optical Society of America, Washington, DC.
7. Kafka, J. D., Baer, T., and Hall, D. W. (1989). *Opt. Lett.* **14**, 1269.
8. Optical fiber will exhibit elastic stretching up to about 1% of its overall length. Therefore, 10 m of fiber can be stretched up to 10 cm! In general, however, PZTs will not provide 1% stretch.
9. Shan, X., Cleland, D., and Ellis, A. (1992). *Electron. Lett.* **27**, 182.
10. Takara, H., Kawanishi, S., and Saruwatari, M. (1995). *Electron. Lett.* **31**, 292.
11. Siegman, A. E., and Kuizenga, D. J. (1974). *Optoelectronics* **6**, 43.
12. Duling, I. N., Goldberg, L., and Weller, J. F. (1988). *Electron. Lett.* **24**, 1333.
13. Pfeiffer, Th., and Veith, G. (1993). *Electron. Lett.* **29**, 1849.
14. Doerr, C. R., Wong, W. S., Haus, H. A., and Ippen, E. P. (1992). *Opt. Lett.* **19**, 1747
15. Siegman, A. E., and Kuizenga, D. J. (1974). *Optoelectronics* **6**, 43.
16. Mollenauer, L. F., Gordon, J. P., and Evangelides, S. G. (1991). *Laser Focus World* **27**, 159.
17. Romagnoli, M., Wabnitz, S., and Franco, P. (1995). *J. Opt. Soc. Am. B* **12**, 72.
18. Doerr, C. R., Haus, H. A., and Ippen, E. P. (1995). *Opt. Lett.* **19**, 1958.
19. Haus, H. A., Tamura, K., Nelson, L. E., and Ippen, E. P. (1995). *IEEE J. Quantum Electronics* **31**, 591.
20. Carruthers, T. F., and Duling III, I. N. (1997). *Opt. Lett.* **21**, 1927.

21. Barnett, B. C., Rahman, L., Islam, M. N., Chen, Y. C., Bhattacharya, P., Riha, W., Reddy, K. V., Howe, A. T., Stair, K. A., Iwamura, H., Friberg, S. R., and Mukai, T. (1995). *Opt. Lett.* **20**, 471.
22. Ober, M. H., Hofer, M., Chiu, T. H., and Keller, U. (1993). *Opt. Lett.* **18**, 1532.
23. Tsuda, S., Knox, W. H., Zyskind, J. L., Cunningham, J. E., Jan, W. Y., and Pathak, R. (1996). In *Conference on Lasers and Electro-Optics*, 1996 Technical Digest Series 9, p. 494, Optical Society of America, Washington, DC.
24. De Souza, E. A., Islam, M. N., Soccolick, C. E., Pleibel, W., Stolen, R. H., Simpsona, J. R., and Giovanni, D. J. (1993). *Electron. Lett.* **29**, 447.
25. Duling III, I. N. (1991). *Opt. Lett.* **16**, 539–541.
26. Doran, N. J., and Wood, D. (1988). *Opt. Lett.* **13**, 56.
27. Duling III, I. N., Chen, C. J., Wai, A. K., and Menyuk, C. R. (1994). *IEEE J. Quantum Electronics* **30**, 194–199.
28. Duling III, I. N. (1995). In *Compact Sources of Ultrashort Pulses*, I. N. Duling III (ed.), Cambridge Studies in Modern Optics, pp. 148, Cambridge Univ. Press, Cambridge.
29. Kraus, F., Brabec, T., and Spielmann, C. (1991). *Opt. Lett.* **16**, 235.
30. Haus, H. A., Ippen, E. P., and Tamura, K. (1994). *IEEE J. Quantum Electronics* **30**, 200.
31. Harvey, G. T., and Mollenauer, L. F. (1993). *Opt. Lett.* **18**, 107; Yoshida, E., Kimure, Y., and Nadazawa, M. (1992). *Appl. Phys. Lett.* **60**, 932.
32. Dennis, M. L., and Duling III, I. N. (1992). *Electron. Lett.* **28**, 1894.
33. Grudinin, A. B., Richardson, D. J., and Payne, D. N. (1992). *Electron. Lett.* **28**, 67–68.
34. Fermann, M. E. (1995). In *Compact Sources of Ultrashort Pulses*, I. N. Duling III (ed.), Cambridge Studies in Modern Optics, pp. 179–207, Cambridge Univ. Press, Cambridge.
35. Tamura, K., Doerr, C. R., Nelson, L. E., Haus, H. A., and Ippen, E. P. (1994). *Opt. Lett.* **19**, 46.
36. Tamura, K., Ippen, E. P., Haus, H. A., and Nelson, L. E. (1993). *Opt. Lett.* **18**, 1080.
37. Sala, K. L., Kenney-Wallace, A., and Hall, G. L. (1980). *IEEE J. Quantum Electronics* **16**, 990.
38. Dennis, M. L., and Duling III, I. N. (1993). *Appl. Phys. Lett.* **62**, 2911–2913.
39. Kelly, S. M. J. (1992). *Electron. Lett.* **28**, 806–807; Smith, N. J., Blow, K. J., and Andonovic, I. (1992). *J. Lightwave Tech.* **10**, 1329–1333; Noske, D. U., Pandit, N., and Taylor, J. R. (1992). *Opt. Lett.* **17**, 1515.
40. Dennis, M. L., and Duling III, I. N. (1994). *IEEE J. Quantum Electronics* **30**, 1469.
41. Gordon, J. P. (1992). *J. Opt. Soc. Am. B* **9**, 91.
42. Grudinin, A. B., Richardson D. J., and Payne, D. N. (1993). *Electron. Lett.* **29**, 1860.
43. Duling III, I. N. (1995). In *Compact Sources of Ultrashort Pulses*, I. N. Duling III (ed.), Cambridge Studies in Modern Optics, p. 165, Cambridge Univ. Press, Cambridge.

44. Tamura, K., Ippen, E. P., and Haus, H. A. (1994). *Phot. Tech. Lett.* **6**, 1433.
45. Tamura, K., Nelson, L. E., Haus, H. A., and Ippen, E. P. (1994). *Appl. Phys. Lett.* **64**, 149.
46. Tamura, K., Jacobson, J., Ippen, E. P., Haus, H. A., and Fujimoto, J. G. (1993). *Opt. Lett.* **18**, 220.
47. von der Linde, D. (1986). *Appl. Phys. B* **39**, 201.
48. Keller, U., Li, K. D., Rodwell, M., and Bloom, D. M. (1989). *IEEE J. Quantum Electronics* **25**, 280.
49. A laser that can be used with little or no monitoring would apply mostly to the semiconductor modelocked fiber lasers, as they tend to be relatively turnkey. In addition, well-engineered versions of the alternating dispersion laser described in Section 7.3.5 would also qualify.

8. CHARACTERIZATION OF SHORT LASER PULSES

T. Feurer and R. Sauerbrey

Institut für Optik und Quantenelektronik
Friedrich-Schiller-Universität
Jena, Germany

8.1 Introduction

8.1.1 Overview

In describing the characterization of short laser pulses we first have to define the terms "characterization" and "short." By "short" we mean laser pulses that are typically shorter than approximately 10 psec. This time interval is, of course, only approximate and marks the border where for longer times electronic or all optical methods for the characterization of pulses are commercially available. With "characterization" we mean that the temporal, spatial, and spectral properties of the laser pulse are measured by appropriate methods. The discussion of these methods is the subject of this chapter, where we put the emphasis on temporal characterization. It is not intended, however, to review this field but rather to provide an introduction into the most common methods for ultrashort pulse characterization.

The necessity to measure ultrashort time intervals has been driven by the needs of fast optics, chemistry, and electronics. In particular, experiments aimed at phenomena occurring on subpicosecond time scales require methods to measure optical pulses of femtosecond temporal duration. Historically the measurement of short time intervals has not always been important. Figure 1 shows the development of mankind in its ability to measure ultrashort time intervals. We see that for a long time, starting in the Renaissance until the mid-nineteenth century, progress was not very fast. Streak photography was invented around 1840, which increased our capability to measure short time intervals considerably into the microsecond regime. Considerable further progress was made after 1960, when the laser and appropriate modelocking techniques had been developed. Today we are able to measure times of several femtosecond duration, and the measurement of attosecond time intervals is under discussion.

The principle that has always led to the greatest progress in the measurement of short time intervals is the unique mapping of a short time interval into a spatial interval that can then simply be measured by a ruler or an analogous device. To set the scale we calculate the spatial extension l of an optical pulse of temporal length $\tau = 1$ psec. We find $l = c\tau = 300$ μm. For a 10-fsec pulse we would obtain

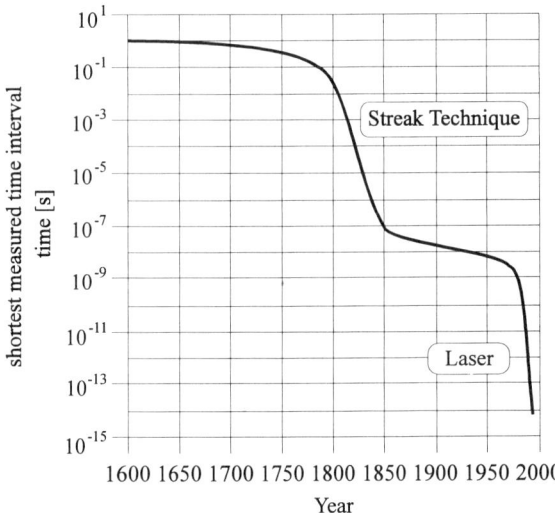

FIG. 1. The history of efforts measuring shorter and shorter time intervals.

a length of 3 μm. For much shorter pulses the optical period gets to the same order as the pulse duration. Such pulses will not be considered in this chapter. Both the spatial and spectral characterization of ultrashort laser pulses follow the rules that are also known for longer laser pulses or CW lasers. Conventional beam profilers or spectrometers can be used. With the exception of the last section, we will focus on temporal pulselength measurements applicable from the near-ultraviolet (~ 250 nm) to the infrared (~20 μm).

8.1.2 Definitions

Solving the wave equation in a vacuum for a scalar wave propagating in one spatial dimension z leads to a solution that is harmonic and makes use of the superposition principle:

$$E(t,z) = a_1 f_1(z - ct) + a_2 f_2(z + ct), \quad (1)$$

where a_1 and a_2 are constants, c is the speed of light, and f_1 and f_2 are largely arbitrary sufficiently well-behaved functions. The most important solution is the harmonic wave:

$$E(t,z) = A \sin(kz - \omega_0 t). \quad (2)$$

This can be written using Euler's theorem as a sum of exponentials:

$$E(t,z) = E_0 \exp[i(kz - \omega_0 t)] - c.c., \text{ with } E_0 = A/2i, \quad (3)$$

INTRODUCTION

where c.c. denotes the complex conjugate of the first term. Since we are considering light pulses, E_0 becomes a function of time and determines the temporal envelope of the laser pulse. Compared to the oscillations of the electric field ($1/\omega_0 \approx 1$ fsec), E_0 is a slowly varying function. When only linear operations are performed, it is sufficient to work with the complex notation only. For a laser pulse, not only the envelope is a function of time but also the phase can vary in time. The time dependence of an arbitrary pulse at a certain position z may be written as

$$E(z,t) = E_0\left(t - \frac{z}{c}\right) \exp\left\{i\left[\Phi\left(t - \frac{z}{c}\right)\right]\right\}. \tag{4}$$

If we are only interested in the time evolution of the pulse at a fixed location $z = 0$ (e.g., the position of the detector), we obtain

$$E(t) := E(0,t) = E_0(t) \exp[i\Phi(t)]. \tag{5}$$

We assume in the following that the envelope function is sufficiently well behaved and has its temporal maximum at $t = 0$. This maximum has by far the highest amplitude value compared to other extrema that may occur at other times ($t_{m1}, t_{m2}, \ldots$). In other words, $|E(0)| > \max_i |E(t_{mi})|$.

Another possibility to represent an arbitrary pulse is to express it by its Fourier components in the appropriate way:

$$E(t) = \frac{1}{\sqrt{2\pi}} \int_{-\infty}^{\infty} d\omega \, \widetilde{E}(\omega) \exp(i\omega t). \tag{6}$$

The full width half-maximum (FWHM) $\Delta\tau_E$ is defined by

$$\Delta\tau_E = 2t_{1/2} \quad \text{with} \quad E(t_{1/2}) = \tfrac{1}{2} E(0). \tag{7}$$

Usually the phase $\Phi(t)$ is expanded in terms of a Taylor series around the pulse center:

$$\Phi(t) = \sum_{\nu=0}^{\infty} \frac{1}{\nu!} \Phi^{(\nu)}(0) t^\nu = \Phi_0 + \omega_0 t + \tfrac{1}{2} b t^2 + \tfrac{1}{6} c t^3 + \ldots. \tag{8}$$

Now we consider the instantaneous frequency, which is defined by

$$\omega_{\text{inst}}(t) = \frac{d\Phi}{dt} = \sum_{\nu=0}^{\infty} \frac{1}{\nu!} \Phi^{(\nu+1)}(0) t^\nu = \omega_0 + bt + \tfrac{1}{2} c t^2 \ldots. \tag{9}$$

The constant b is called the linear chirp parameter, c the quadratic chirp, and so on. From definition (9), it is clear that the linear chirp describes a linear increase or decrease in the instantaneous frequency depending on the sign of parameter b.

Since neither the electric field nor its Fourier transform but the temporal and spectral intensity are measured directly, we need to define how these quantities can be obtained from the electric field. The instantaneous intensity of a light pulse is

$$I(t) = \varepsilon_0 c |E(t)|^2. \tag{10}$$

The spectral intensity of such a laser pulse is proportional to the square of its Fourier transform:

$$I(\omega) \approx |\tilde{E}(\omega)|^2 = |\mathbf{F}\{E(t)\}|^2 = \left| \frac{1}{\sqrt{2\pi}} \int_{-\infty}^{\infty} dt\, E(t) \exp(-i\omega t) \right|^2. \tag{11}$$

It is obvious that no phase information can be retrieved from the measured temporal intensity. The FWHM of the temporal and spectral intensity is narrower than that of the corresponding electric fields by a factor that is determined by the pulseshape.

Another important quantity is the time bandwidth product $\Delta\tau\Delta\omega$, which is just the product of the FWHM of the temporal intensity $\Delta\tau$ and the FWHM of the spectral intensity $\Delta\omega$. If the time bandwidth product is equal to the minimum value that is possible for a certain pulseshape, the pulse is called bandwidth-limited or transform-limited. This immediately implies that there is no phase modulation on the pulse. On the other hand, with the information that a Gaussian pulse is bandwidth-limited, the pulsewidth can be obtained by simply measuring the spectral intensity. Table I summarizes the minimum time bandwidth product for a variety of pulses [1].

8.2 Spatial Characterization and Focusing

For most experiments where the highest intensity possible is required at given laser parameters, it is necessary to focus the laser beam very tightly. For up to about 10^{15} W/cm^2 this is usually achieved by lenses. For even higher intensities, spherical mirrors or off-axis paraboloids have to be used. To reach these high intensities it may be necessary to eliminate many possible error sources.

The quality of the focus is determined by the spatial intensity profile of the laser pulse and the divergence distribution. If the spatial intensity distribution exhibits rotation symmetry with respect to the axis of propagation, no asymmetry in the focus is found. For a pulse that shows astigmatism, the position of the beam waist is different for two perpendicular axes (the x and y axes). Finally, the beam waist can have different widths in different directions (asymmetric beam waist). In order to obtain the best focal spot, these errors have to be corrected by the appropriate optics.

TABLE I. Temporal FWHM and Minimum Time Bandwidth Product for a Variety of Analytical Pulseshapes

	Intensity $I(t)$	FWHM (T)	Time bandwidth product $\Delta\tau\Delta\omega$
Square 1	$\|t\| \leq T/2$	1	5.5663
Parabolic $1 - (t/T)^2$	$\|t\| \leq T$	$\sqrt{2}$	4.5716
Gaussian $\exp[-(t/T)^2]$		$2\sqrt{\ln 2}$	2.772
Lorentzian $1/[1 + (t/T)^2]$		2	1.3861
Hyperbolic secant $\mathrm{sech}^2(t/T)$		1.7627	1.9779
Asymmetric exponential $\exp[t/t_1]$ $\exp[-t/t_2]$ $T = t_1 + t_2$ $r = t_1/T$	$t < 0$ $t \geq 0$	$\ln 2$	$(\ln 2)/2\pi \sqrt{(-b + \sqrt{b^2 + 4a}/2a}$ $a = r^2(1-r)^2$ $b = r^2 + (1-r)^2$

A measure of the focal spot size is the beam radius w (intensity equal to $1/e^2$ of the maximum intensity). If w is equal to the minimum beam radius w_0, the focus is called "diffraction-limited" (2θ is the divergence of the beam):

$$w_0 = \frac{\lambda}{\pi\theta}. \tag{12}$$

An effect that is a consequence of the fact that real lenses are not "thin lenses" is spherical aberration (on-axis focusing with a single lens). The minimum focal spot size can be estimated using

$$w = \frac{1}{2} k(n) \frac{a}{(f^\#)^2} + 1.22\lambda f^\#, \tag{13}$$

where $k(n)$ is a function depending on the shape of the lens (to first order, $k(n) \approx 1$), a is the beam radius, $f^\# = f/a$ is the f-number of the lens (f is the focal length), and n is the index of refraction.

The above-mentioned errors may occur for continuous wave lasers as well as for pulsed lasers and can be determined by standard techniques (Mode Master © Coherent). Short pulse lasers, however, as a consequence of their special characteristics, do show some new error sources. Due to the wide bandwidth, chromatic aberration does prevent optimal (diffraction-limited) focusing. Parts of the pulse traversing the lens at different radii reach the focus at different times, leading to a delay between pulse and phase fronts [2]. The delay of the outermost beam, t_a, determines to first order the pulse broadening in the focus and is given by

$$t_a = \frac{a^2 k_0}{2f(\omega_0)[n(\omega_0) - 1]} \frac{dn}{d\omega}\bigg|_{\omega = \omega_0}, \qquad (14)$$

where k_0 is the wave vector, f is the focal length of the lens, and $n(\omega_0)$ is its index of refraction [3]. This increase in pulsewidth can lower the intensity quite drastically. In order to avoid this effect, achromats must be used. A second effect that may lead to pulse lengthening is the group velocity dispersion (GVD) of the lens material. This is only important for very short pulses.

The spatial intensity profile of a laser pulse can have a dramatic effect on the laser beam propagation in air. At high intensities self-focusing and phase front distortion may occur. Laser pulses exceeding a power of about 1 TW have to be propagated in vacuum in order to maintain proper focusability.

8.3 Conventional Detectors for nsec to psec Pulses

The temporal characterization of ultrashort laser pulses relies mostly on either streak cameras or on auto- and cross-correlation techniques. For the spatial and spectral characterization conventional methods can be used. Since the main scope of the current chapter is the discussion of techniques special to ultrashort laser pulses, we will only give a brief overview of conventional radiation detectors. More detailed descriptions can be found in [4]. Table II summarizes the most important conventional photodetectors, their spectral range, their spatial resolution, and their temporal resolution. Furthermore, their typical applications for short pulse measurements are given. It is immediately apparent from the entry in the column "temporal resolution" that all these detectors can provide only temporally integrated information on ultrashort laser pulses. The entries in the columns "spectral range" and "spatial resolution" are typical values for commercially available instruments. For example, the spectral range into the VUV can usually be extended below 105 nm when the window is removed from the detector, requiring operation in a vacuum.

TABLE II. Conventional Detectors for nsec to psec Pulses

Detector	Spectral range	Spatial resolution	Temporal resolution	Typical application for short pulse measurements
Photomultiplier	105 nm – 1.1 µm	1 cm	1 nsec	Detection of low-photon fluxes, autocorrelators
Vacuum photodiode	105 nm – 1.1 µm	1 cm	50 psec	Relative pulse energy measurement, temporal alignment of delay lines
Semiconductor photodiode	105 nm – 16 µm	1 mm	20 psec	Relative pulse energy, alignment of delay lines
Microchannel plate (MCP)	1 nm – 140 nm	10 µm	100 psec	Energy of VUV and soft x-ray pulses, spatial resolution, streak camera
CCD array	105 nm – 1.1 µm	10 µm	1 msec	Measurement of beam profiles, spectra, FROG, streak camera
Quadrant diode	105 nm – 16 µm	<1 µm	1 nsec	Laser beam alignment
Energy meter	10 nm – 1 mm	1 cm	10 msec	Absolute pulse energy

8.4 Streak Camera

8.4.1 Principle of Operation

The name "streak camera" (though it is not a camera in the usual sense) comes from the days when light was reflected onto a film from a high-speed rotating drum. Light was "streaked" onto the film. Since the readout of the streak camera is two-dimensional, it is able to record the time evolution of one spatial coordinate, such as a line focus or a spectrum. Therefore, it can be used to perform temporally and spatially resolved measurements or time-resolved spectroscopy [5].

Modern streak cameras cover the spectral range from the x-ray region [6] to the near-infrared and have a maximum temporal resolution of about 0.5 psec [7]. One full-time scan can vary between about 50 psec and 10 msec. They are able to record single events or repetitive phenomena up to the GHz range. The sensitivity can be optimized to the point where every single photoelectron is detected.

The operating principle of a streak camera is shown in Fig. 2. Light passes through a slit and is imaged onto the photocathode of the tube. The number of electrons produced by the photoelectric effect is proportional to the number of incident photons and therefore to the intensity. They are accelerated towards the other end of the streak tube by an appropriate potential. Before impinging onto the

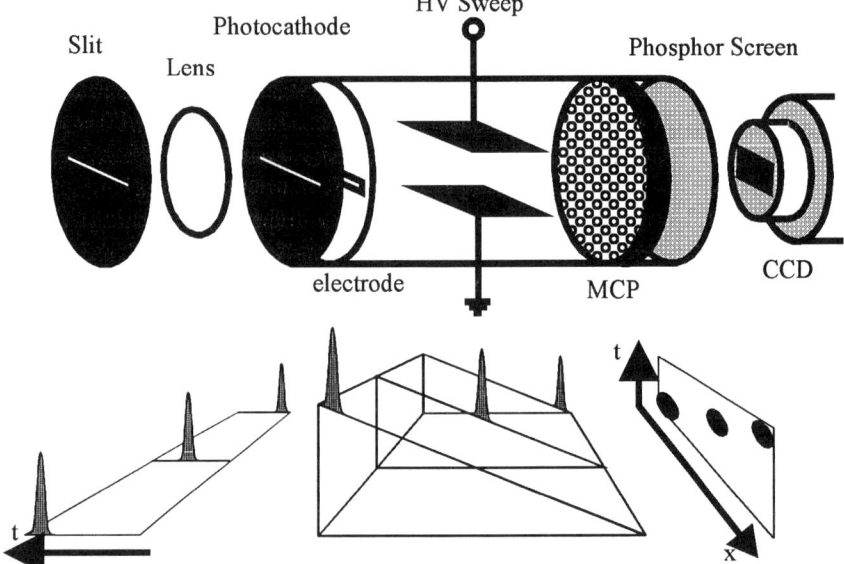

FIG. 2. Schematics of a streak camera. By applying a high voltage, sweep electrons at different times leave the pair of deflection plates under different angles. The time axis is therefore transferred into a spatial coordinate.

phosphor layer, they pass a pair of deflection electrodes and are multiplied by a microchannel plate (MCP) (gain = 10^3–10^4).

The sweep of the high voltage on the deflection plates is synchronized to the incoming light pulse. Hence, electrons at later times leave the pair of deflection plates under a larger angle than those electrons at earlier times. Therefore, they enter the MCP at different positions. The purpose of the MCP is to multiply the two-dimensional electron distribution so that the electrons bombarding the phosphor screen will produce enough phosphorescence that subsequently is recorded by a CCD camera. The intensity of the phosphorescence is directly proportional to the number of impinging electrons and therefore to the number of photons initially impinging on the photocathode.

8.4.2 Streak Camera Technology

8.4.2.1 Photocathode and Spectral Characteristics
A photocathode usually consists of several metallic films evaporated on the window material facing the interior of the camera. The spectral response characteristics of these layered photocathodes determines how many photons at a certain wavelength are necessary to produce one photoelectron. These characteristics can be given in

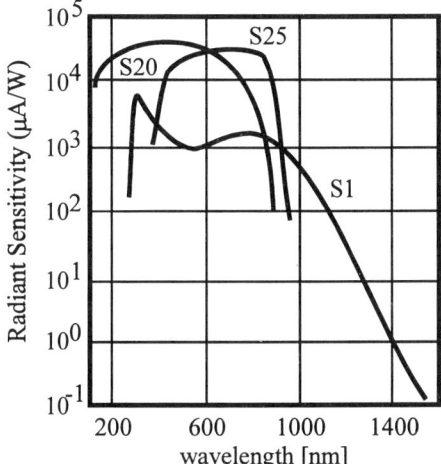

FIG. 3. The radiant sensitivity of the standard coatings (S1, S20, and S25) as a function of wavelength.

terms of radiant sensitivity (A/W), which is the photoelectric current as a function of the incident light. It can also be given in terms of the quantum efficiency (%) that is one over the number of photons necessary to produce one photoelectron.

The fact that the photocathode transforms the photon pulse into a pulse of electrons means that it influences not only the spectral characteristics and the sensitivity of the streak camera but also the time resolution of the system. The influence on the time resolution increases with decreasing wavelength because at shorter wavelengths more and more secondary electrons are produced. Their energy distribution is about one order of magnitude broader than the one of the primary electrons. In order to obtain a good time resolution, a cathode material with a very narrow energy distribution of the secondary electrons has to be chosen. For the visible spectral wavelength range, standard coatings such as S1, S5, or S20 are commonly used (see Fig. 3).

Close to the photocathode is the extraction mesh. It accelerates the photoelectrons preferably in a way that the velocity spread of the accelerated electrons is as narrow as possible (see Section 8.4.2.4).

8.4.2.2 Streak Tube The streak tube has to fulfill two requirements. First, it has to have electron optics that image the slit of the photocathode onto the phosphor screen at the back of the streak tube. The electrostatic lens system can be such that the image is magnified or demagnified. It also determines the spatial resolution of the system that is typically on the order of 100 μm. Second, it has to provide a pair of properly designed deflection electrodes that are perpendicular to the cathode slit. The rise times of the applied voltage can be as high as 50 V/psec.

The voltage sweep also has to be as linear as possible. In order to maximize the time resolution of the streak camera, the number of electrons between the deflection plates has to be as small as possible in order to avoid space charge shielding. This makes it necessary to multiply the electrons before they impinge on the phosphor screen. In most cases, MCPs are used for this purpose. In order to increase the signal-to-noise ratio, the high voltage that determines the gain of the MCP can be gated so that mostly photoelectrons from the light pulse will cause a phosphorescence signal.

8.4.2.3 Dynamic Range The dynamic range is a measure of the light intensity range that leads to a linear response of the streak camera. The lower limit is determined by the noise level. A high upper limit implies that the electron density between the deflection plates is high and space charge effects subsequently lower the temporal resolution. Therefore, there has to be a tradeoff between a large dynamic range and a high temporal resolution depending on the emphasis of the experiment.

8.4.2.4 Temporal Resolution There are several factors that limit the temporal resolution of a streak camera:

- First, there is the technical time resolution $\Delta \tau_t$ that is mainly determined by two parameters. The sweep velocity v_t of the electrons over the phosphor screen and the spatial resolution l_t of the detection system, which is usually a CCD camera:

$$\Delta \tau_t = \frac{l_t}{v_t}. \qquad (15)$$

- Another limiting factor is the time dispersion of the photoelectrons $\Delta \tau_d$. This effect is sometimes called the chromatic time dispersion and is due to the velocity distribution of the photoelectrons:

$$\Delta \tau_d \approx \frac{m \Delta v}{eE}, \qquad (16)$$

  where E is the extraction field and Δv is the width of the velocity distribution. A high extraction field can therefore reduce the time dispersion quite drastically. Using field strengths as high as 10 kV/cm can decrease $\Delta \tau_d$ to subpicosecond values.

- At high electron currents space charge effects tend to lengthen the electron bunch, leading to an increased pulsewidth ($\Delta \tau_s$) [8].

If all sources of error are independent of each other, their squares can be added in order to obtain the temporal response of the streak camera to an incoming pulse of length $\Delta \tau_L$:

$$\Delta\tau_{streak} = \sqrt{\Delta\tau_t^2 + \Delta\tau_d^2 + \Delta\tau_s^2 + \Delta\tau_L^2}. \tag{17}$$

8.4.2.5 Calibration Calibration of the time axis can be performed in various ways. The most convenient approach is to pass a short pulse through an etalon (consisting of two parallel flat mirrors each having a high-reflectivity coating on the inside surface and an antireflection coating on the outside surface). The train of pulses that emerges from such an arrangement has a temporal spacing between two individual pulses that is equal to the round-trip time of the etalon. By recording this train of pulses, the time axis can be calibrated and the linearity of the deflection unit (time axis) be tested. The fact that the intensity of the subsequent pulses decreases in a way that is determined by the product of the reflectivities of the two mirrors can be used to test the linearity of the amplification system.

8.4.2.6 Synchronization Synchronization becomes important when the streak camera is operated at high scan speeds, especially when signal averaging is applied in order to increase the signal-to-noise ratio. Typically, the pulse is split into two parts: one used to trigger the voltage ramp, the other delayed before falling onto the entrance slit. The delay is such that the beginning of the fluorescence signal at the slit exactly coincides with the starting of the voltage ramp at the deflection electrodes. In order to minimize the time jitter between the two events, a special switch has to be used. Most commonly this is a Krytron, a spark gap, or a stack of avalanche transistors. With these devices the jitter is 20–100 psec, which is still quite large compared to the time scales under consideration.

8.5 Autocorrelation and Cross-Correlation Techniques

8.5.1 Principles

Laser pulses considerably shorter than about 1 psec cannot be measured accurately using streak cameras. Autocorrelation and cross-correlation methods, however, provide a very convenient tool to determine the pulsewidth of such ultrashort pulses. Specially designed experimental arrangements can be used to focus different aspects, that is, temporal shape, phase information, and high contrast ratio.

Measuring short pulses with autocorrelation and cross-correlation techniques usually produces graphs that do not resemble the incoming light pulses directly but are related to those by some mathematical transformation. In order to obtain the original pulse the inverse operation has to be performed.

First, a schematic setup for a typical autocorrelator will be discussed. Using this scheme we subsequently explain differences between multishot and single-shot autocorrelators, intensity and interferometric autocorrelators, and autocorrelators and cross-correlators. The appropriate mathematical description will be developed in the text. As an example, all general formulas will be applied to a Gaussian pulse with linear chirp b, corresponding to an instantaneous time-dependent frequency $\omega_{inst} = \omega_0 + bt$:

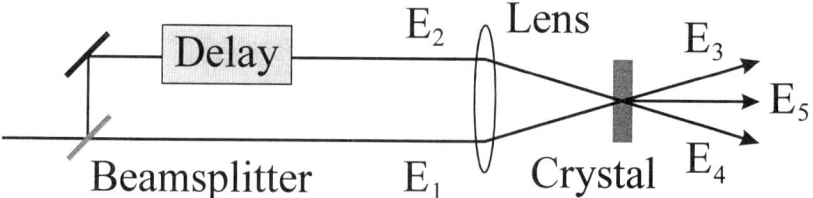

FIG. 4. Schematics of an autocorrelation measurement.

$$E(t) = \exp[-a^2(1 - iA)t^2] \exp(i\omega_0 t) \quad \text{with} \quad b = 2a^2 A. \tag{18}$$

Figure 4 shows a typical experimental setup of an autocorrelator. Two replicas of the incoming beam (E_1 and E_2) are crossed in a nonlinear medium. The main requirement for the nonlinear effect is that it has to be instantaneous (compared to the temporal width of the pulse to be measured). One of the pulses (E_2) can be shifted in time by varying the length of the beam path. The nonlinear medium must produce a measurable higher-order response for the incoming pulses. The order of an autocorrelation is determined by the order of the nonlinear optical process used to generate the response. Even for second-order effects it is usually necessary to focus both beams in order to exceed the intensities necessary to obtain a detectable signal (see Section 8.2). The two pulses can traverse the nonlinear medium either collinearly (the wavevectors of both pulses are parallel) or under a certain angle. The collinear case leads to an interferometric autocorrelation (if the spatial resolution of the detector is sufficiently high), and the latter setup produces an intensity autocorrelation. More strictly speaking, in Fig. 4 the interferometric autocorrelation is a consequence of the fact that all three second-harmonic fields, E_3, E_4, and E_5, interfere at the plane of detection, whereas in the case of an intensity autocorrelation only the second-harmonic field, E_5, that exists if both pulses overlap is measured. In both cases, however, a signal is recorded that depends on the temporal delay between the two replicas. In a multishot autocorrelator this delay is constantly increased or decreased between successive shots and the higher-order response is measured.

If one of the two replicas is changed in amplitude or phase before entering the nonlinear medium, the technique is called cross-correlation.

We now focus on a mathematical description of the autocorrelation measurement. The normalized nth-order autocorrelation function or the interferometric autocorrelation function with background is defined as

$$g_n^{(B)}(\tau_1, \tau_2, \ldots, \tau_{n-1}) = \frac{\int_{-\infty}^{\infty} dt \, |E(t) + E(t - \tau_1) + \ldots + E(t - \tau_{n-1})|^{2n}}{n \int_{-\infty}^{\infty} dt \, |E|^{2n}(t)}, \tag{19}$$

where $g_n^{(B)}$ is normalized such that the background level, $g_n^{(B)}(\infty,\ldots,\infty)$, is equal to 1. For those experiments that produce no background:

$$g_n^{(0)}(\tau_1,\tau_2,\ldots,\tau_{n-1}) = \frac{\int_{-\infty}^{\infty} |\,dt\, E(t)E(t-\tau_1)\ldots E(t-\tau_{n-1})\,|^2}{\int_{-\infty}^{\infty} dt\, |E|^{2n}(t)}. \qquad (20)$$

For a second-order autocorrelation, Eqs. (19) and (20) yield

$$g_2^{(B)}(\tau) = \frac{\int_{-\infty}^{\infty} dt\, |\,E(t)+E(t-\tau)\,|^4}{2\int_{-\infty}^{\infty} dt\, |E|^4(t)}, \qquad (21)$$

$$g_2^{(0)}(\tau) = \frac{\int_{-\infty}^{\infty} dt\, |\,E(t)+E(t-\tau)\,|^2}{\int_{-\infty}^{\infty} dt\, |E|^4(t)}. \qquad (22)$$

The detected signal is time-integrated because a slow (compared to the time scales under consideration) detector is used to measure the autocorrelation trace. Since the Fourier transform of the first-order autocorrelation $g_1^{(B)}$ is equal to $I(\omega) + I(-\omega)$, it does not provide more information than the spectrum of the pulse; hence the necessity for higher-order autocorrelators or nonlinear detectors.

The fringe-averaged portion of the nth-order normalized autocorrelation function is the intensity autocorrelation function

$$G_n^{(B)}(\tau_1, \tau_2, \ldots, \tau_{n-1}) = \langle g_n^{(B)}(\tau_1, \tau_2, \ldots, \tau_{n-1})\rangle_{\tau,n}, \qquad (23)$$

$$G_n^{(0)}(\tau_1, \tau_2, \ldots, \tau_{n-1}) = \langle g_n^{(0)}(\tau_1, \tau_2, \ldots, \tau_{n-1})\rangle_{\tau,n}. \qquad (24)$$

The multidimensional "optical time average" denoted in Eq. (23) is defined as

$$\langle [\,\ldots\,]\rangle_{r,n} := \frac{1}{T^{n-1}} \int_{\tau_1-T/2}^{\tau_1+T/2} \int_{\tau_2-T/2}^{\tau_2+T/2} \int_{\tau_{n-1}-T/2}^{\tau_{n-1}+T/2} d\tau_1\, d\tau_2 \ldots d\tau_{n-1}\, [\,\ldots\,]$$

$$\text{with}\quad \omega_0^{-1} \ll T \ll \Delta\tau. \qquad (25)$$

For second-order autocorrelation the average yields

$$G_2^{(0)}(\tau) = \frac{\int_{-\infty}^{\infty} dt\, E_0^2(t)\, E_0^2(t-\tau)}{\int_{-\infty}^{\infty} dt\, E_0^4(t)} \quad \text{and}\quad G_2^{(B)}(\tau) = 1 + 2G_2^{(0)}(\tau). \qquad (26)$$

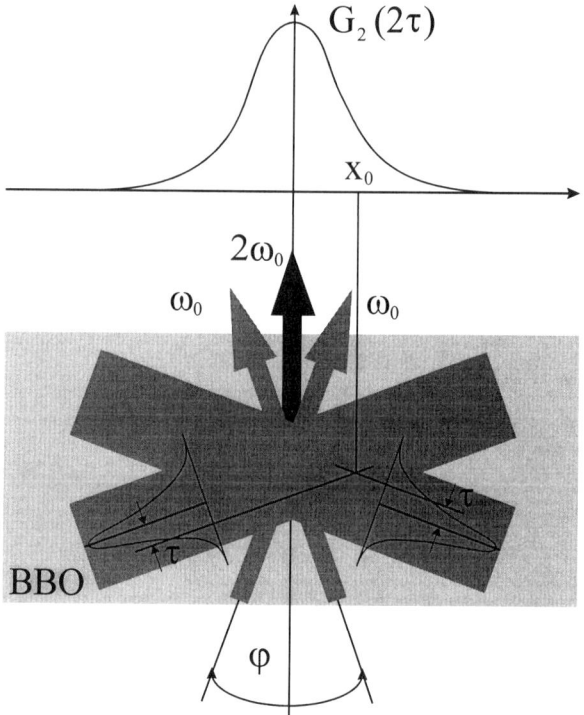

FIG. 5. Tilted pulse fronts in the case of a noncollinear single-shot autocorrelation measurement.

Second- or third-order correlation measurements are usually practical since the intensities required for higher-order correlations are only available for very few laser systems.

For low-repetition-rate laser systems or systems that exhibit relatively high shot-to-shot fluctuations, multishot autocorrelation becomes both cumbersome and subject to severe errors. Then single-shot autocorrelators are used. Now not only two circular foci but two line foci overlap in such a way that the temporal delay varies along the line focus. In the collinear case this can be achieved by tilting the pulse fronts of both pulses in the opposite direction (e.g., by wedged windows). In the noncollinear case the tilted pulse fronts are a consequence of the experimental setup (see Fig. 5).

The basic idea, therefore, is to map the temporal delay between the two pulses into a spatial profile that is recorded by a diode array. A very crucial point of this procedure is that both pulses are homogeneous in intensity along the line focus. An arbitrary point x_0 along the line focus is related to the temporal delay τ by

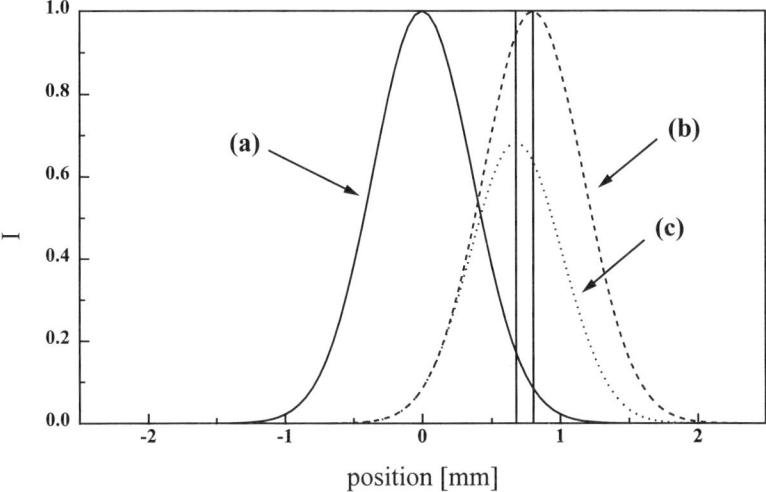

FIG. 6. Consequences of a nonhomogeneous spatial beam profile on the time calibration of a single-shot autocorrelator: (a) represents the autocorrelation trace for no external delay, (b,c) for a delay of 200 μm. (a) and (b) are obtained for a homogeneous intensity profile, whereas (c) is the result of a nonhomogeneous profile. The maximum of (c) is shifted and leads to an incorrect time calibration.

$$\tau = \frac{nx_0 \sin(\varphi/2)}{c}, \qquad (27)$$

where n is the index of refraction of the nonlinear medium, φ is the angle between the two incident pulses, and c is the speed of light in a vacuum. At a fixed resolution of the detector, the angle φ determines the temporal resolution of the autocorrelator. The smaller the angle, the higher the temporal resolution.

The external delay can be used to calibrate the system in a very simple way. Introducing a delay Δt_0 in one of the two beams shifts the autocorrelation trace along the x axis. The shift Δx_0 is related to the delay by

$$\Delta x_0 = \frac{c\Delta t_0}{2n \sin(\varphi/2)}. \qquad (28)$$

Note that the measured second-order autocorrelation trace is not exactly the second-order intensity autocorrelation $G_2(\tau)$ but is related to the function $G_2(2\tau)$ [9] (see also Fig. 5). Figure 6 shows the effect of a nonhomogeneous intensity distribution along the line focus. For this calculation we assume a Gaussian intensity distribution with an FWHM that is equal to the length of the line focus (4 mm). Note that the shift is smaller than the expected value and can therefore lead to a quite substantial error in determination of the pulsewidth.

TABLE III. Second-Order Autocorrelation Functions for the Pulseshapes Listed in Table I

Pulseshape	Autocorrelation function $G_2^{(0)}$		FWHM$_G$ [T]	FWHM$_G$/FWHM$_P$
Square	$1 - |\tau/T|$ 0	$|\tau| \leq T$ $|\tau| > T$	1	1
Parabolic	$1 - 5/4(\tau/T)^2 + 5/8$ $|\tau/T|^3 - 1/32 |\tau/T|^5$ 0	$|\tau| \leq 2T$ $|\tau| > 2T$	1.6226	1.1473
Gaussian	$\exp[-1/2(\tau/T)^2]$		$2\sqrt{2\ln 2}$	1.4142
Lorentzian	$[1 + (\tau/2T)^2]^{-1}$		4	2
Hyperbolic secant	$3[(\tau/T)\coth(\tau/T) - 1]$ $\times \sinh^{-2}(\tau/T)$		2.7196	1.5427
Asymmetric exponential	$[t_2/(t_2 - t_1)] \exp(-|\tau|/t_2) -$ $[t_1/(t_2 - t_1)] \exp(-|\tau|/t_1)$		$2\ln 2 +$ $0.6857r(1-r) +$ $1.93r^2(1-r)^2$	$(1/\ln 2)[2\ln 2 +$ $0.6857r(1-r) +$ $1.93r^2(1-r)^2]$

The last column shows the ratio of the FWHM of the autocorrelation function and the FWHM of the temporal intensity.

Another possibility to measure the single-shot autocorrelation function was demonstrated by Raksi and coworkers [11]. The incoming pulse is divided into two identical replicas. Both traverse through a Fabry–Perot interferometer with a slightly different spacing. The trains of pulses released are crossed in a nonlinear medium. Each pair of pulses corresponds to a different delay. The task is now to measure the envelope of the signal on a time scale that is about the spacing between two subsequent pulses.

A major problem that is true for almost all autocorrelation techniques is how to obtain the original laser pulse from the measured trace. Even more complications arise if, aside from the temporal intensity, the phase of the pulse has to be retrieved, that is, the complete complex electric field.

Theoretical models can usually be used to calculate the temporal intensity for a given laser system. Then it is only necessary to multiply the FWHM of the autocorrelation trace by a constant factor in order to obtain the width of the laser pulse. Table III lists the correction factors for a variety of pulseshapes (see Table I) along with the second-order autocorrelation function.

A variety of procedures have been proposed that are capable of providing enough information to fully retrieve the complex field of the pulse (see, e.g., [10]). But it is beyond the scope of this chapter to review this field in detail. An example for an iterative procedure is given in Section 8.5.3.2.

8.5.2 Intensity Autocorrelators

Intensity autocorrelation measurements can be the result of two different arrangements. First, in a noncollinear setup the intensity autocorrelation $G_n^{(0)}$ is measured. The signal does not contain portions that are produced by one of the two single beams, and therefore no interference patterns are observed. This measurement is inherently background-free. On the other hand, if all response fields superimpose at the plane of detection, we observe interference fringes with a modulated contrast. This is only true if the detector has the necessary spatial resolution; otherwise, the fringes are averaged and again the intensity autocorrelation is obtained. In this case, there is a background, and the signal-to-background ratio is

$$n!(n-1)! \sum_{i_1=0}^{n} \sum_{i_2=0}^{i_1} \sum_{i_3=0}^{i_2} \cdots \sum_{i_{n-1}=0}^{i_{n-2}}$$

$$\times \left[\frac{1}{(n-i_1)!\,(i_1-i_2)!\,(i_2-i_3)!\,\ldots\,(i_{n-2}-i_{n-1})!\,(i_{n-1})!} \right]^2 : 1 \quad (29)$$

for the nth-order autocorrelation [1]. For example, a second-order autocorrelation yields a ratio of 3:1 and a third-order autocorrelation 31:1.

We now discuss second-order autocorrelators. Especially in connection with second harmonic generation (SHG), they are widely used throughout the visible and infrared (IR) spectrum. Then we will briefly treat higher-order-intensity autocorrelators that mostly rely on higher-harmonic generation, multiphoton absorption, and multiphoton ionization in gases.

8.5.2.1 Second-Order Autocorrelators Second-order autocorrelators and especially those that use SHG are the "classic" autocorrelators. The intensities required to produce a detectable signal are not too high, and nonlinear crystals are available over a wide spectral range covering the visible and infrared regions. They have been realized in various experimental arrangements as multishot and single-shot autocorrelators [12, 13]. However, they exhibit a twofold ambiguity regarding the direction of the time axis. In particular, like all even autocorrelation functions, the second-order autocorrelation trace is always symmetric and therefore does not reveal any information on pulse asymmetries. For SHG there has been a successful attempt to remove this ambiguity by introducing a glass block in one of the beam paths and recording the autocorrelation traces with and without the glass block [14].

For the Gaussian pulse with a linear chirp the intensity autocorrelation yields

$$G_2^{(0)} = \exp(-a^2\tau^2) \quad \text{with background} \quad G_2^{(B)} = 1 + 2G_2^{(0)}. \quad (30)$$

A wide variety of crystals for SHG is available, covering the spectral range from about 0.2 to 20 μm. Table IV shows a list of crystals commonly used as nonlinear

TABLE IV. Most Commonly Used Nonlinear Crystals for Higher-Harmonic Generation

Crystal		Transparency range	Type
KH_2PO_4	KDP	0.177–1.7 μm	neg. uniaxial
KD_2PO_4	DKDP	0.2–2.0 μm	neg. uniaxial
$NH_4H_2PO_4$	ADP	0.184–1.5 μm	neg. uniaxial
LiB_3O_5	LBO	0.16–2.6 μm	neg. biaxial
$LiIO_3$	Lithium iodate	0.3–6.0 μm	neg. uniaxial
$LiNbO_3$	Lithium niobate	0.33–5.5 μm	neg. uniaxial
$\beta\text{-}BaB_2O_4$	BBO	0.198–2.6 μm	neg. uniaxial
Ag_3AsS_3	Proustite	0.6–13 μm	neg. uniaxial
Ag_3SbS_3	Pyrargyrite	0.7–14 μm	neg. uniaxial
$AgGaSe_2$	Silver gallium selenide	0.71–18 μm	neg. uniaxial
$CdGeAs_2$	Cadmium germanium arsenide	2.4–18 μm	pos. uniaxial
CdSe	Cadmium selenide	0.75–20 μm	pos. uniaxial

If the index of refraction for the ordinary beam is higher than that of the extraordinary beam, the crystal is negative; otherwise, it is positive.

media [15]. The lower limit for their application in SHG is in many cases approximately two times the lower bound of the transparency range.

For short pulsewidth measurements, phase matching is important for two reasons. First, it has to be fulfilled in order to get a conveniently measurable second-harmonic signal. Since the converted intensity scales with the length of the crystal, a longer crystal yields a higher signal. But short pulses cover a considerable spectral range, and phase matching must therefore be fulfilled for the whole spectral bandwidth of the laser pulse. This is only possible if the crystal is thin. Figure 7 shows the phase matching in the case of $\beta\text{-}BaB_2O_4$ (BBO). The fundamental wavelength is 497 nm and the cut angle 52.9°. It can be seen that a 100-μ thick crystal supports a bandwidth of about 5.2 nm, which for a bandwidth-limited Gaussian pulse leads to a pulsewidth of about 70 fsec, whereas a 300-μm thick crystal only allows for detection of a pulse that is longer than 220 fsec.

Table V lists the possible types of phase matching in uniaxial crystals. If both incident waves have the same polarization, the second harmonic is polarized in the perpendicular direction. This is called type I phase matching. When the polarizations of the incoming waves are perpendicular to each other, then the polarization of the second harmonic is in an extraordinary direction for negative crystals and

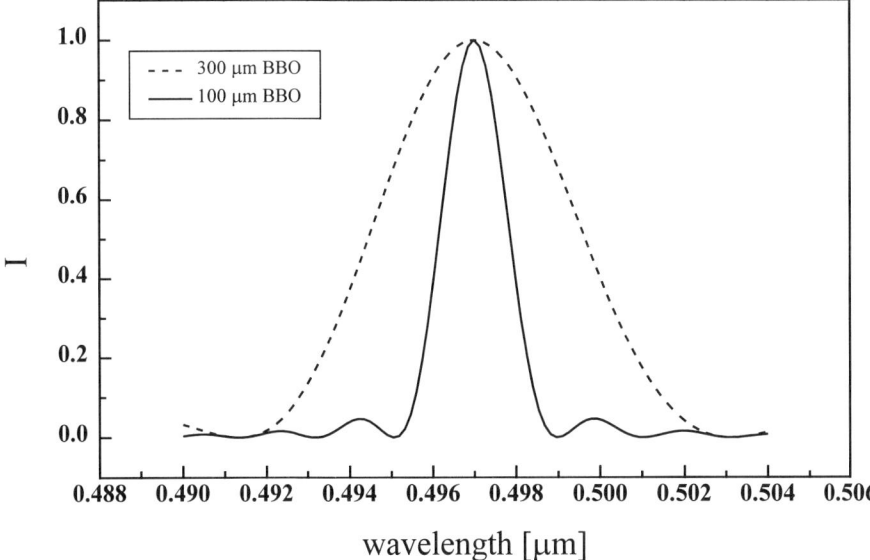

FIG. 7. Influence of the phase matching condition on the spectrum of the up-converted pulse.

TABLE V. Phase Matching Conditions in Uniaxial Crystals

Phase matching	Crystal type	Incoming beams	Second harmonic	Label
Type I	Negative	o–o	e	"ooe"
	Positive	e–e	o	"eeo"
Type II	Negative	o–e	e	"oee"
	Negative	e–o	e	"eoe"
	Positive	o–e	o	"oeo"
	Positive	e–o	o	"eoo"

"o" represents polarization in the ordinary direction and "e" polarization in an extraordinary direction.

in the ordinary direction for positive crystals. In this case, type II phase matching applies.

We now consider a typical autocorrelator based on SHG in BBO. The crystal has a thickness of 300 μm. Knowing the index of refraction for the ordinary and extraordinary beams, the angle can be calculated. For a fundamental of 497 nm

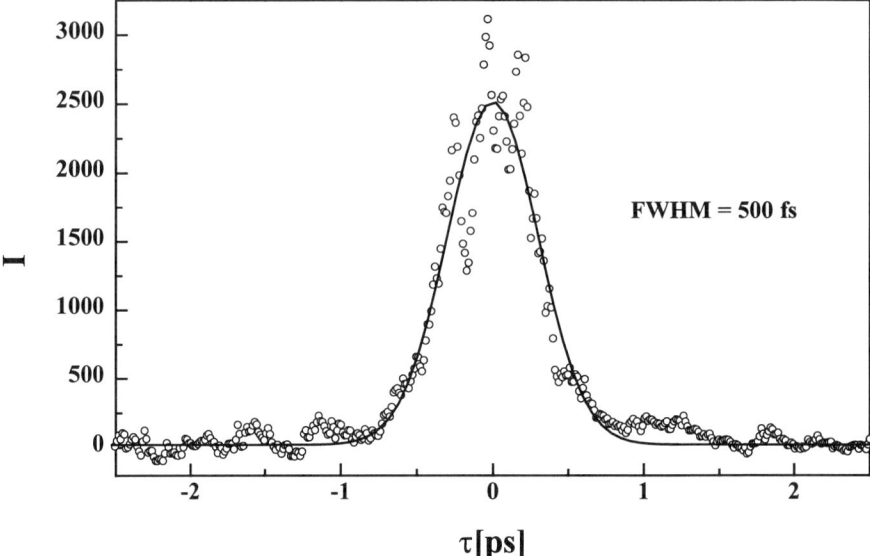

FIG. 8. Example of an intensity autocorrelation using SHG. Assuming a Gaussian pulseshape, the temporal FWHM of the pulse is 500 fsec.

and a second harmonic of 248 nm, we obtain a cut angle of 52.9°. The beam splitter produces two replicas of the incoming beam that cross each other in the BBO crystal. The image of the second harmonic is recorded using a CCD array. Since we are considering the case where the two pulses are not collinear, a second-harmonic signal is produced only if both pulses coincide spatially and temporally. The temporal overlap can be obtained by adjusting the delay in one arm. Figure 8 shows an example of a single-shot measurement. The pulse stems from a dye laser system that delivers pulses at 497 nm.

Another second-order effect is two-photon fluorescence. This method has mainly been used in order to measure short pulses in the UV regime since there are no suitable crystals with transmission in the VUV (see Table VI). The nonlinear medium can be an organic dye, a gas, or a solid and has to exhibit two-photon absorption at the wavelength of interest. Mapping the spatial profile of the consecutive fluorescence in the transverse direction onto a CCD array leads to the autocorrelation trace. It is well known that rare gases fluoresce on excimer emission bands and that multiphoton absorption can populate molecular states that lead to this fluorescence.

Similar to two-photon fluorescence is two-photon ionization of gases. This method is very sensitive, and its unique property is that the signal to be detected is an electric charge. With this technique, autocorrelation measurements have been

TABLE VI. Two-Photon Effects, the Center Wavelength of the Excitation (or Upper Bound), and the Emission Line or Band, Respectively

Medium	Process	Excitation wavelength [nm]	Emission wavelength [nm]	Reference
Xe_2	TPA	248.3	172	[16]
Diamond	TPA	<440	–	[17]
Cd	TPI	<276	508	[18]
NO	TPI	<268	–	[19]
BaF_2	TPEF	<273	310	[20, 21]
Si, GaAsP, CdS	TPC	$2h\nu > E_g > h\nu$	–	[22]
Water	TPP	–	–	[23]

TPA = two-photon absorption; TPI = two-photon ionization; TPEF = two-photon excited exciton fluorescence; TPC = two-photon conductivity; E_g = energy gap; TPP = two-photon induced photoacoustic signal.

performed at 193, 248, 276, 308, and 351 nm using triethylamine, DABCO, and NO gases [19]. Efforts to build single-shot autocorrelators based on two-photon ionization are very rare since measuring a spatial charge distribution in a gas with sufficient spatial resolution is very difficult.

Measuring the two-photon induced photoacoustic signal in water, Nishioka and coworkers were able to successfully demonstrate a single-shot autocorrelator [23]. This method was first applied by Schmid et al. for a train of 1.06-µm pulses [24].

Near-surface second-harmonic generation (SSHG) can also be used to realize autocorrelation measurements in a wavelength range from <200 nm to 3.3 µm [25]. This was first shown by Gierulski et al. [26]. The main advantage of SSHG for short-pulsewidth measurement is that the phase matching condition does not apply for this process, since the penetration depth of the fundamental and the second harmonic are on the order of 5 to 10 nm. In other words, there is no bandwidth limitation for the up-conversion. SSHG on Si(111) has been used to measure femtosecond UV pulses [27] and on GaAs(111) and Pd(111) to measure pulses from 200 nm to 3.3 µm.

8.5.2.2 Higher-Order Autocorrelators Higher-order autocorrelators generally rely on higher harmonic generation, multiphoton absorption, multiphoton ionization, and the optical Kerr effect (third-order effect).

It has been shown that the fluorescence intensity of the C–A transition in XeF (480 nm) is proportional to the third power of the laser intensity. Such a transition can be used to develop a third-order autocorrelator in the wavelength range from 204 to 306 nm [28]. Another third-order effect that has been successfully applied

TABLE VII. Mechanisms that Change the Static Index of Refraction and Approximate Relaxation Times

Effect	Relaxation time [sec]
Nonlinear electronic polarization	$10^{-14} \ldots 10^{-15}$
Molecular orientation in liquids	$10^{-11} \ldots 10^{-12}$
Electronic state distortion	$10^{-8} \ldots 10^{-9}$
Thermal effects	$1 \ldots 10^{-1}$

by Schulz *et al.* to measure 120-fsec pulses at 308 nm is degenerate four-wave mixing [29].

A widely used third-order technique is the optical Kerr shutter. A short laser pulse induces a birefringence in a Kerr medium, which is mounted between a crossed polarizer–analyzer pair. The transmission through this setup is related to a change in the index of refraction. Such deviations from the static index of refraction $n = n_0 + \Delta n(t)$ can be the result of a variety of physical processes, some of which are listed in Table VII.

If the relaxation time is small compared to the pulsewidth, then the transient birefringence follows exactly the intensity envelope of the pump pulse. In the opposite case, the time evolution of the birefringence is determined by dynamics of the Kerr medium. The first case can be used to realize an autocorrelator that is background-free. A less intense pulse (usually pump:probe = 10:1) that is linearly polarized with a certain angle (usually 45°) with respect to the pump pulse is used to probe the induced birefringence. Hence, the temporal shape of the probe pulse resembles transmission through the Kerr shutter. If we consider a medium with a short relaxation time, the transmitted probe pulse is proportional to the square of the pump pulse. If the maximum of the induced birefringence is sufficiently small, that is, $(n_\parallel - n_\perp < \lambda/8L$, where $\parallel$ and $\perp$ are the parallel and perpendicular directions of the polarization of the pump pulse, respectively, λ is the wavelength, and L is the length of the crystal, then the time-integrated transmitted probe signal is proportional to the third-order autocorrelation function [30]. The Kerr shutter technique can be used both as multishot and single-shot correlator. The dynamic range of the autocorrelation measurement is determined by the leakage of the probe beam through the polarizer analyzer pair. Figure 9 shows a single-shot autocorrelation trace of a femtosecond UV pulse (KrF: 248 nm) recorded with the Kerr shutter technique. Assuming a Gaussian pulseshape, the FWHM of the instantaneous intensity is 400 fsec.

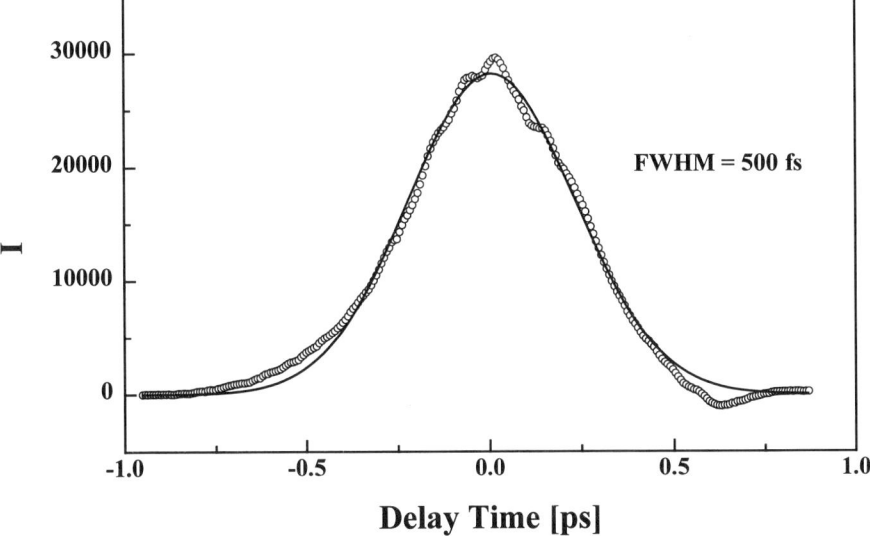

FIG. 9. Example of an intensity autocorrelation using the optical Kerr effect. Assuming a Gaussian pulseshape, the pulsewidth (FWHM) is 500 fsec.

Not only the induced birefringence but also the induced beam deflection as a result of the temporal variation of the index of refraction can be used to realize an autocorrelator. The pump beam induces a transient prism in the Kerr medium that deflects the probe beam. The magnitude of the deflection is a direct measure of the intensity of the pump pulse [31].

8.5.3 Phase-Sensitive Autocorrelators

In order to produce the shortest pulses possible in a laser system linear, quadratic and even cubic chirp introduced by intracavity material has to be compensated. To achieve this in a controlled manner requires autocorrelation techniques that are able to reveal the phase structure of a laser pulse. Here we will focus on two methods: interferometric autocorrelators, and a technique that has been developed called Frequency Resolved Optical Gating (FROG). The latter combined with a third-order nonlinear effect provides a tool that yields very intuitive time–frequency distributions of a laser pulse.

8.5.3.1 Interferometric Autocorrelators   Autocorrelation measurements that are performed with interferometric accuracy yield autocorrelation traces that exhibit interference fringes equally spaced by one half period of the carrier frequency. At zero delay, the electric fields of both arms are in phase and lead to constructive interference, whereas at a delay of one half optical cycle both fields have the opposite phase and interfere destructively. The lower and upper envelopes of the interference fringes merge into the intensity autocorrelation trace for delay times larger than the coherence time of the pulse. The contrast ratio of the nth-order interferometric autocorrelation is $n^{(2n-1)}$:1.

The interferometric autocorrelation can provide information on the phase modulation of the pulse, since phase modulations change the autocorrelation function in a specific way. Rather large phase modulations have the effect that the interference between the pulse front and the pulse tail is lost, which shows in a decreasing contrast of the interference fringes. Again the most commonly used nonlinear effect is SHG [32], and the interferometric autocorrelation for a Gaussian pulse with linear chirp is

$$g_2^{(B)}(\tau) = 1 + 2\exp(-a^2\tau^2) + 4\exp\left(-a^2\tau^2 \frac{3+A^2}{4}\right)$$
$$\times \cos\frac{Aa^2\tau^2}{2}\cos\omega\tau + \exp[-a^2(1+A^2)\tau^2]\cos 2\omega\tau. \quad (31)$$

The lower and upper envelopes can be found by setting $\omega\tau = 2\pi$ or π, respectively. The second-order autocorrelation with no background yields

$$g_2^{(0)}(\tau) = \frac{2}{3}\exp(-a^2\tau^2). \quad (32)$$

Figure 10 shows four calculated second-order autocorrelation traces, $g_2^{(B)}$, for different values of linear chirp A. Figure 11 shows an example of a single-shot measurement. The pulse stems from a 0.8-µm self-modelocked Ti:sapphire laser oscillator equipped with chirped dielectric mirrors for dispersion control [33].

In the ultraviolet, two-photon fluorescence has been used successfully for the operation of a phase-sensitive autocorrelator [20].

8.5.3.2 Frequency-Resolved Optical Gating Frequency-resolved optical gating (FROG) is a technique that combines a nonlinear effect with a dispersive element. Most commonly used is SHG or the optical Kerr effect, the latter being a third-order effect that provides more information but requires higher intensity. It is also possible to use self-diffraction, higher-harmonic generation, or the cascaded $\chi^{(2)}$ process. The measured autocorrelation trace (FROG trace) is related to the spectrogram [34] of the incoming pulse as

$$S(\tau,\omega) = \left|\frac{1}{\sqrt{2\pi}}\int_{-\infty}^{\infty} dt\, e^{-i\omega t}E(t)\,h(t-\tau)\right|^2. \quad (33)$$

The window function $h(t)$ selects a small portion of the signal centered around t, calculates its energy spectrum, and repeats this procedure for each time (short-term Fourier transform). In the case of an SHG–FROG, the window function is

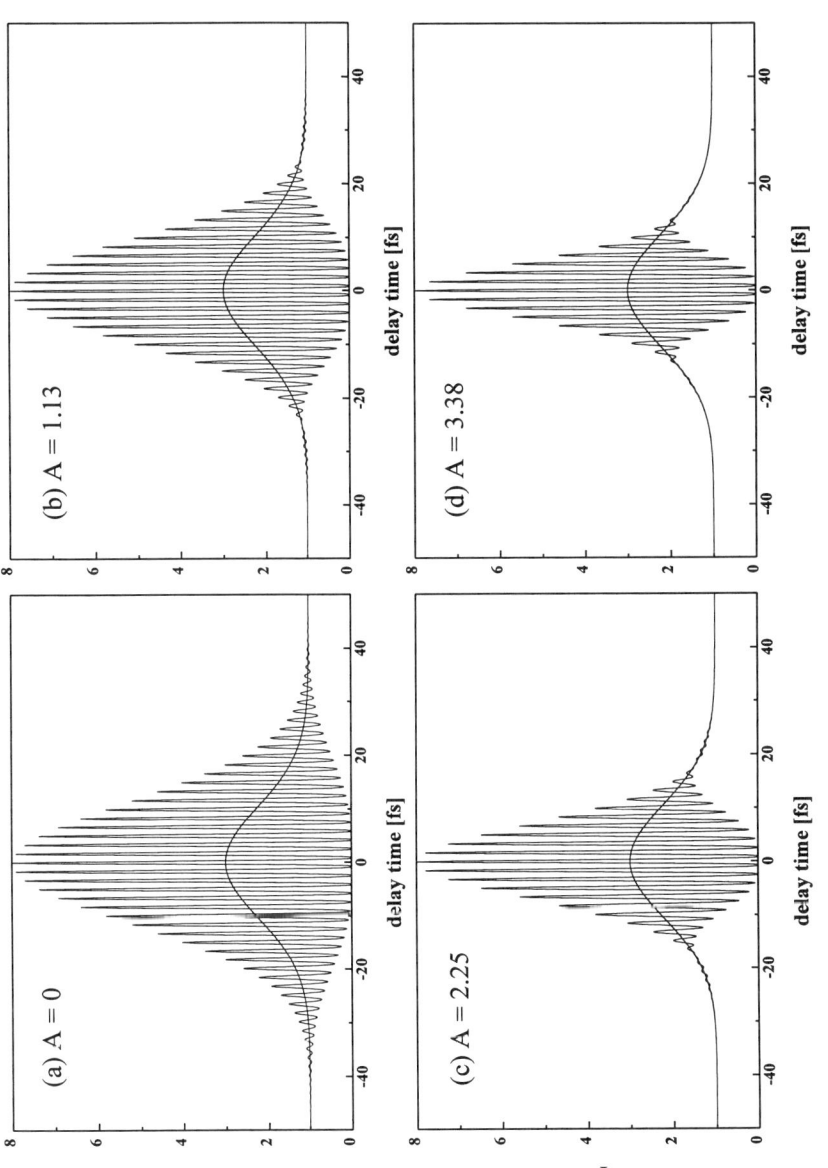

FIG. 10. Interferometric autocorrelation traces for four Gaussian pulses (FWHM = 25 fsec, λ = 497 nm). (a) $A = 0$, (b) $A = 1.13$, (c) $A = 2.25$, (d) $A = 3.38$. Included in (a)–(d) is the intensity autocorrelation trace.

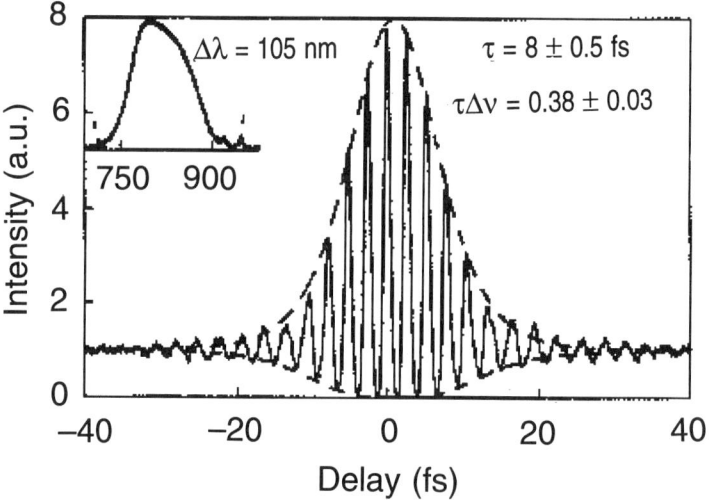

FIG. 11. Trace of a single-shot interferometric autocorrelation using SHG. Assuming a sech² pulse envelope, the FWHM of the temporal intensity is 8 fsec. The time bandwidth product $\Delta\tau\Delta\omega$ is 2.39 and therefore higher than the minimum value of 1.98 for a sech² pulse (see Table I).

$h(t) = E(t)$ in the case of the Kerr–FROG $h(t) = |E(t)|^2$. Using an iterative algorithm, the complex electric field can be calculated without any assumptions on the actual amplitude or phase of the original pulse [35].

The first stage of a FROG autocorrelator resembles a single-shot intensity autocorrelator using two line foci in a noncollinear setup (see Fig. 12). In the second stage, this line focus (response of the nonlinear medium) is imaged on the entrance slit of a high-resolution spectrometer. At the exit plane of the spectrometer, a two-dimensional image with one axis being the delay and the other the wavelength is obtained. Since the SHG has a twofold ambiguity in time, the FROG traces will always be symmetric with respect to the delay.

Figure 13 shows a measured Kerr–FROG trace for a UV pulse at 248 nm. The pulselength is about 400 fsec and the bandwidth in the spectral range about 1 nm.

It can be seen that the wavelength of the pulse decreases with time, meaning that the pulse has a positive almost linear chirp. Integration over the wavelength axis yields a quantity that is related to the pulsewidth and integration over the time axis a quantity that is related to the spectrum. For the Gaussian pulse with the linear chirp the Kerr-FROG signal can be calculated as

$$S_{\text{Kerr}}(\tau,\omega) \approx \exp\left[-\frac{1}{3+(A^2/12)}\left[4a^2\tau^2\left(1+\frac{A^2}{12}\right)\right.\right.$$
$$\left.\left.+\frac{1}{2a^2}(\omega-\omega_0)^2 - \frac{2}{3}A\tau(\omega-\omega_0)\right]\right]. \quad (34)$$

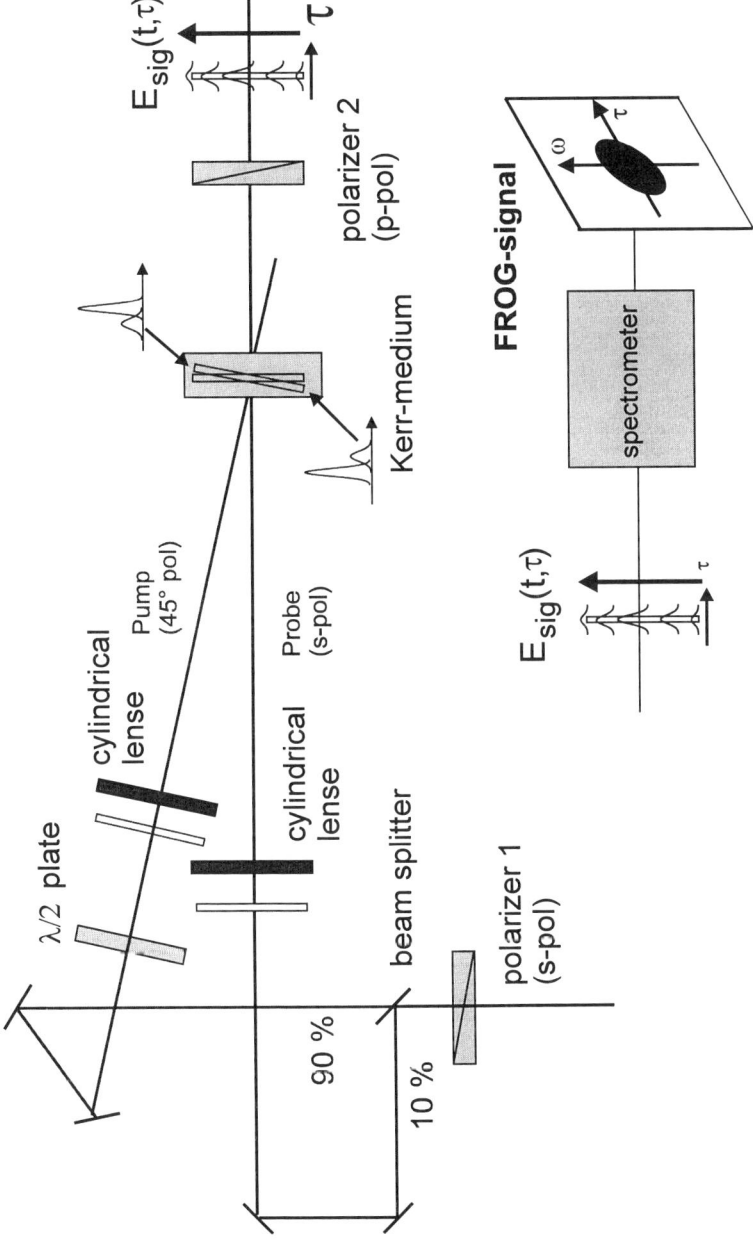

FIG. 12. Schematic setup of a typical FROG measurement.

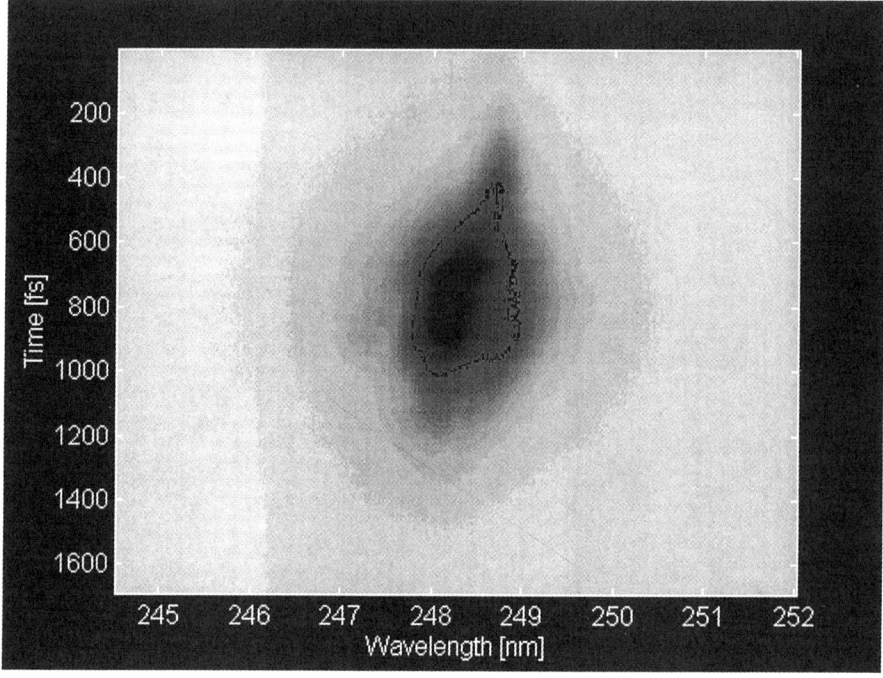

FIG. 13. Kerr–FROG trace of a 400-fsec UV pulse (248 nm).

8.5.4 Cross-Correlation Techniques

Cross-correlation techniques require that one of the pulses has been modified in some respect before entering the nonlinear medium. This has been realized in various ways using different effects. The main goal of cross-correlation is usually to use a part of the incoming beam to probe the transient evolution of the effect that is produced by the other part. It also can be used to obtain additional information on the incoming beam itself that is hard to measure otherwise. Pulses can be stretched by changing the pulse properties in the frequency domain and subsequently being probed by the unstretched pulse. In the following we give a few examples of cross-correlation measurements.

Down-conversion of UV pulses has been used to measure the cross-correlation function of a UV and a visible laser pulse in order to determine the pulsewidth of the UV pulse [36]. Chilla and Martinez demonstrated a method that provides the frequency dependence of the phase by performing phase measurements in the frequency domain [37]. A chirp-sensitive cross-correlation technique that does not

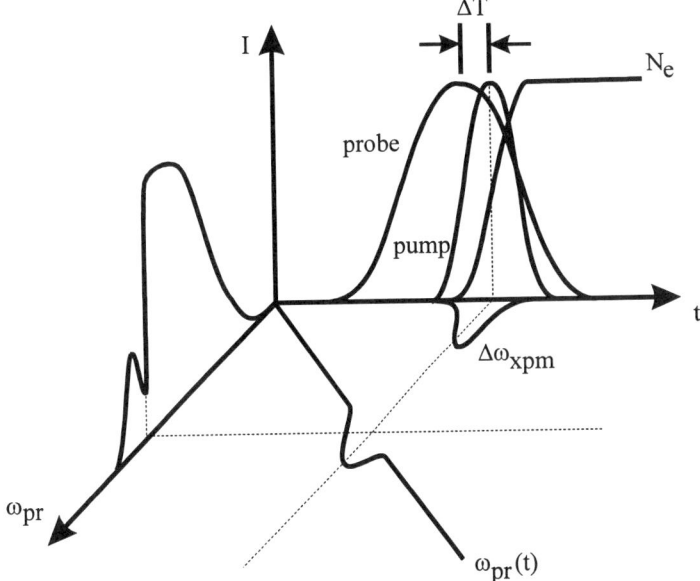

FIG. 14. Qualitative sketch of the technique of chirp measurement using plasma-induced cross-phase modulation.

require a nonlinear medium and is applicable over a wide wavelength range from the visible to the VUV is plasma-induced cross-phase modulation [38]. The basic idea is shown in Fig. 14.

The intense pump pulse ionizes a gaseous target (which represents the nonlinear process) and causes a step-like increase in the electron density. The changing electron density causes a small temporal segment of the probe pulse (delay pump probe: τ) to be blue-shifted by $\Delta\omega_{pr}$ [39]. If the probe pulse has a phase modulation, then the original instantaneous frequency curve of the probe pulse acquires a blue shift at the time τ. In the frequency domain, the blue shift moves a small segment of the original probe spectrum to a higher frequency and leaves a spectral hole. Hence, a different spectral segment will be blue-shifted as τ changes. A plot of these shifted spectral segments as a function of the delay τ yields the frequency sweep of the original pulse. Experimental results were obtained with a dye laser pulse that was phase-modulated by the rapidly changing electron density in a gas ionized with a temporally correlated excimer laser. The chirp of the dye laser pulse was manipulated by the combination of an optical fiber and a compressor and could be positive as well as negative. Results are shown in Fig. 15. It is apparent that negative as well as positive chirp of arbitrary order has been measured.

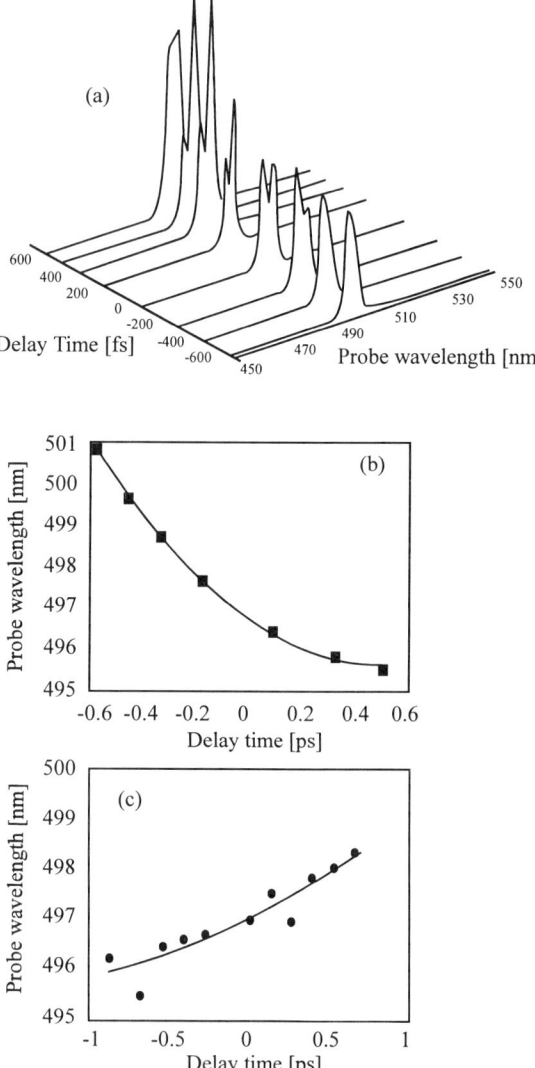

FIG. 15. (a) Experimentally measured probe spectra as a function of the delay time between a chirped 1-psec probe pulse at 497 nm and a-400 fsec pump pulse at 248 nm (5×10^{14} W/cm^2). (b) Position of the spectral hole in the probe spectrum plotted as a function of the delay time. The plot shows that the dye laser pulses have a positive chirp. (c) When the pulses from the fiber pass through a prism compressor that is too long for optimal pulse compression, the chirp of the probe pulse is dominated by the negative group velocity dispersion of the prism pair.

8.6 Special Techniques for the VUV and X-Ray Regions

8.6.1 Generation and Detection of Short-Wavelength Ultrashort Pulses

It has become possible to also generate ultrashort vacuum ultraviolet [41, 43, 44] and x-ray pulses [6]. In principle, the methods to characterize such pulses temporally are similar to those for ultrashort pulses in the optical regime. Streak cameras as well as cross-correlation methods are used. In practice, however, there are important differences that require a separate treatment of the techniques to measure ultrashort pulses of short-wavelength light. Most importantly, no optical materials are available that transmit below 105 nm. Consequently, soft x-ray streak cameras have to be in a vacuum, since no entrance window material exists. Nonlinear techniques used for autocorrelators in an optical regime such as SHG are in principle limited to wavelengths longer than 210 nm. In practice, the shortest wavelength where autocorrelators based on SHG are used is about 400 nm, since no crystals that simultaneously transmit below 200 nm have suitable nonlinear coefficients and are phase-matchable are available at the present time. In the ultraviolet regime, 200 nm < λ < 400 nm, however, other nonlinear techniques such as surface harmonic generation [27] or two-photon excited exciton fluorescence [20] can be conveniently used for the construction of autocorrelators. Both streak cameras as well as cross-correlation techniques for the wavelength range below 200 nm will be described briefly.

8.6.2 X-Ray Streak Camera

In principle, x-ray streak cameras operate like streak cameras in the optical regime. A high-temporal-resolution x-ray streak camera is described in reference [6]. Such x-ray streak cameras are now commercially available. The entrance window has to be extremely thin material with a low absorption coefficient for the energy range of the x-rays under consideration. Pressure differences should be held below ≈1 mbar. In commercial instruments, windows of Lexan (100 nm), aluminum (25 nm), and KBr (150 nm) are used. The main difference between optical streak cameras and x-ray streak cameras lies in the nature of the photoelectrons produced at the photocathode. In the optical regime, the emitted electrons are primaries with a narrow energy distribution (≈ 0.1 eV). Transit time dispersion is therefore $\tau_d \leq 200$ fsec. In the VUV and soft x-ray regimes, however, photon energies are high enough (>10 eV) to also produce secondary electrons with a much broader energy distribution depending on the material and the photon energy. Transit time dispersion is consequently larger in the x-ray regime than in the optical regime and is a function of photon energy. KBr is used as a photoemitter because of its narrow secondary electron distribution in the soft x-ray regime.

Commercial instruments with a response of 2 psec (FWHM) and a timing precision of 0.2 psec are now available. In order to reach such a performance, the appropriate photocathode material has to be used, space-charge broadening has to be avoided by using sufficiently low intensities limiting the dynamic range to about 50, and a high extraction field, a fast well-calibrated sweep speed, and dynamic focusing have to be used. Improvements to the instrumental time resolution will require new photocathode materials with narrower electron distributions or novel low dispersive electron focusing geometries.

8.6.3 Cross-Correlation Techniques

Short-pulse VUV and x-ray radiations are in all cases generated by high-intensity laser pulses in the optical regime. The intensity of the VUV and soft x-ray pulses, however, is moderate. Autocorrelation techniques based on a nonlinear interaction are therefore difficult. Since the short-wavelength radiation is generated by a nonlinear process (high harmonics, parametric processes) or in a laser-produced plasma (incoherent x-rays), the short-wavelength pulses are temporally correlated with an intense short-pulse laser. Consequently, cross-correlation techniques employing the intense optical laser are possible. The temporal resolution is then limited by the pulsewidth of the high-intensity laser. This restriction has up to now prevented a direct measurement of the temporal duration of high harmonics in the VUV and XUV. Estimates have been obtained based on the blue shift of the harmonics in the ionizing gas [40]. The temporal duration of VUV radiation generated by nonlinear mixing processes and soft x-rays from laser-produced plasmas, however, has been measured successfully by cross-correlation techniques. Two examples are described.

The first technique relies on a fast plasma shutter driven by the practically instantaneous process of optical field ionization [40, 42]. The experimental arrangement is shown in Fig. 16.

A femtosecond pulse is split into two replicas of the original pulse. One is used to generate a VUV pulse by a four-wave difference-frequency mixing process in Xe [43, 44]. The second pulse is used to create a plasma along the VUV propagation path. Ionization-induced defocusing [42] at aperture 2 causes the transmission through aperture 3 to decrease. The inset shows the VUV spectrum generated in xenon. To demonstrate pulsewidth measurements based on plasma defocusing, the transmission of the 147 nm pulse as a function of the delay time between the KrF pulse (10^{14} W/cm^2) to create the plasma and the pulse to generate the VUV light is measured.

As shown in Fig. 17, the transmission of the VUV pulse decreases once the VUV pulse becomes defocused. Field ionization causes the electron density to increase to about 10^{17} cm^{-3} in a time interval equal to half the pump pulsewidth. For the described setup, this change in electron density causes the probe pulse

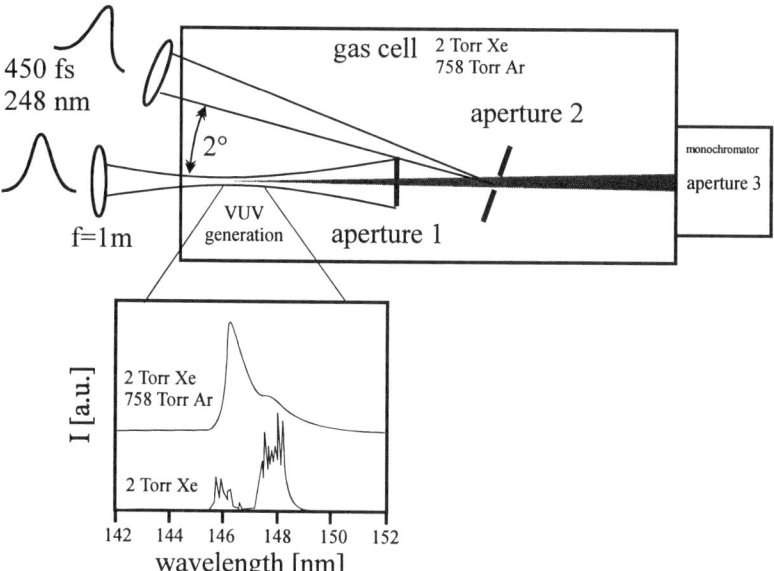

FIG. 16. Experimental setup for generating and measuring femtosecond VUV pulses.

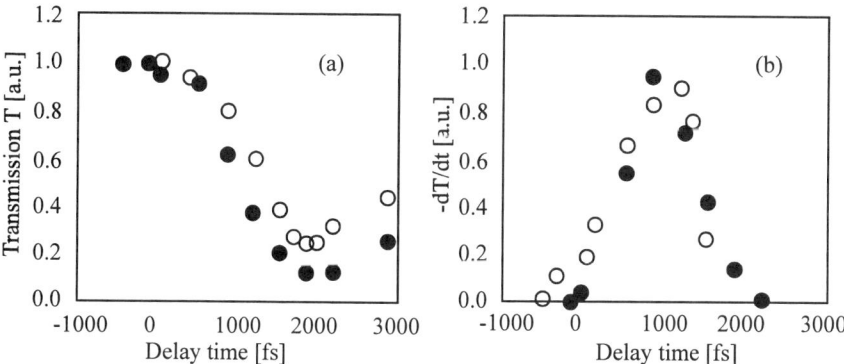

FIG. 17. VUV pulsewidth measurement by plasma defocusing. (a) Transmission of the 147-nm pulse as a function of the delay between the VUV pulse and the plasma creating pulse. (b) Cross-correlation pulseshape and Gaussian fit.

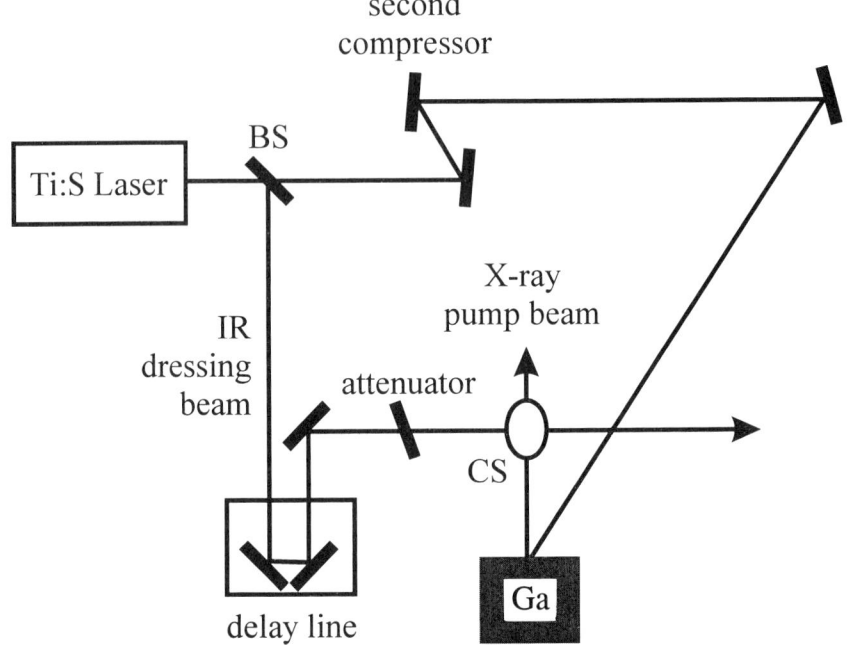

FIG. 18. Schematic of the experimental setup.

divergence to change by ~3 mrad. We can obtain a cross-correlation measurement of the 147 nm pulsewidth by taking the derivative of the transmission curve. Assuming that only group-velocity dispersion broadens the pulse, then a temporal width of about 400 fsec can be obtained from Fig. 17.

An elegant technique to measure the temporal duration of soft x-rays emitted from a laser-produced plasma driven by a several hundred femtosecond titanium–sapphire laser based on laser-assisted Auger decay in argon has been demonstrated [45]. The experimental setup is shown in Fig. 18.

An amplified Ti:sapphire laser delivers pulses that are divided by a beam splitter. The x-ray pump beam passes through a grating pair compressor before it is focused on the gallium target. The plasma radiates x-ray pulses that traverse the argon jet in the interaction region. The second beam, after passing a delay line, is focused into the interaction region of an electron spectrometer. The Auger electrons produced by the soft x-rays (~ 200 eV) are detected. The interaction of the "dressing beam" with the argon atoms causes sidebands to appear in the Auger electron spectrum, since the Auger electrons originate from dressed states in argon. Observing the appearance of these sidebands as a function of the delay between the "dressing beam" and the plasma-generating beam yields a cross-correlation

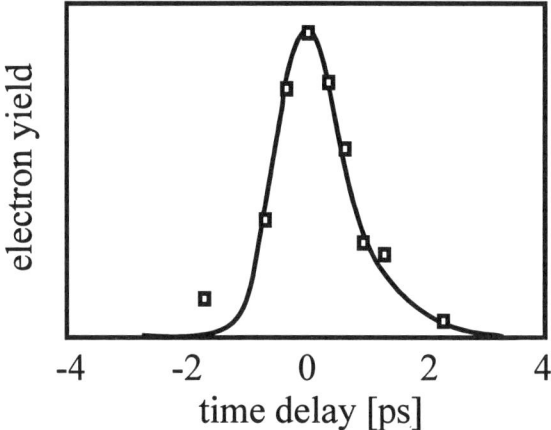

FIG. 19. Cross-correlation of the x-ray and optical pulses.

curve between the x-ray pulse and the "dressing beam." An experimental result is shown in Fig. 19. From the data, an x-ray pulse duration of 0.7 psec is inferred.

Further experiments [46, 47] have measured the cross-correlation between an intense laser pulse and soft x-rays by modifying the absorption of the x-rays in a narrow absorption line by the intense laser pulse,

References

1. Sala, K. L., Kenney-Wallace, G. A., and Hall, G. E. (1980). *IEEE J. Quantum Electronics* **16**, 990.
2. Bor, Z. (1988). *J. Mod. Opt.* **35**, 1907.
3. Kempe, M., and Rudolph, W. (1993). *Opt. Lett.* **18**, 137.
4. *Handbook of Optics*, Vol 1 (1995). Chapters 15–23, McGraw-Hill, New York.
5. Campillo, A. J., and Shapiro, S. L. (1983). *IEEE J. Quantum Electronics* **19**, 585.
6. Murnane, M. M., Kapteyn, H. C., and Falcone, R. W. (1990). *Appl. Phys. Lett.* **56**, 1948.
7. Kinoshita, K., Ito, M., and Suzuki, Y. (1987). *Rev. Sci. Instrum.* **58**, 932.
8. Niu, H., and Sibbett, W. (1981). *Rev. Sci. Instrum.* **52**, 1830.
9. Jansky, J., Conadi, G., and Gyuzalian, R. N. (1977). *Opt. Comm.* **23**, 293.
10. Chilla, J. L. A., and Martinez, O. E. (1991). *Opt. Lett.* **16**, 39.
11. Raksi, F., Heuer, W., and Zacharias, H. (1991). *Opt. Comm.* **86**, 1.
12. Salin, F., Georges, P., Roger, G., and Brun, A. (1987). *Appl. Opt.* **26**, 4528.
13. Brun, A., Georges, P., LeSaux, G., and Salin, F. (1991). *J. Phys. D* **24**, 1225.

14. Naganuma, K., Mogi, K., and Yamada, H. (1989). *Appl. Phys. Lett.* **54**, 1201.
15. Dmitriev, V. G., Gurzadyan, G. G., and Nikogosyan, D. N. (1991). *Handbook of Nonlinear Optical Crystals*, Springer, Berlin–Heidelberg.
16. Hutchinson, M. H. R., McIntyre, I. A., Gibson, G. N., and Rhodes, C. K. (1987). *Opt. Lett.* **12**, 102.
17. Dadap, J. I., Focht, G. B., Reitze, D. H., and Downer, M. C. (1991). *Opt. Lett.* **16**, 499.
18. Tünnermann, A., Eichmann, H., Henking, R., Mossavi, K., and Wellegehausen, B. (1991). *Opt. Lett.* **16**, 402.
19. Szatmari, S., Schäfer, F. P., and Jethwa, J. (1990). *Rev. Sci. Instrum.* **61**, 998, and the references therein.
20. LeBlanc, S. P., Szabo, G., and Sauerbrey, R. (1991). *Opt. Lett.* **16**, 1508.
21. Osvay, K., Ross, I. N., Hooker, C. J., and Lister, J. M. D. (1994). *Appl. Phys. B* **59**, 361.
22. Takagi, Y., Kobayashi, T., and Yoshihara, K. (1992). *Opt. Lett.* **17**, 658.
23. Nishioka, H., Ishiguro, M., Kawasumi, T., Ueda, K., and Takuma, H. (1993). *Opt. Lett.* **18**, 45.
24. Schmid, A., Horn, P., and Braulich, P. (1993). *Appl. Phys. Lett.* **43**, 151.
25. Plass, W., Rottke, H., Heuer, W., Eichhorn, G., and Zacharias, H. (1992). *Appl. Phys. B* **54**, 199.
26. Gierulski, A., Marowsky, G., Nikolaus, and B., Vorob'ev, N. (1985). *Appl. Phys. B* **36**, 133.
27. Canto-Said, E. J., Simon, P., Jordan, C., and Marowsky, G. (1993). *Opt. Lett.* **18**, 2038.
28. Sarukura, N., Watanabe, M., Endoh, A., and Watanabe, S. (1988). *Opt. Lett.* **13**, 996.
29. Schulz, H., Schüler, H., Engers, T., von der Linde, D. (1989). *IEEE J. Quantum Electronics* **25**, 2580.
30. St. Albrecht, H., Heist, P., Kleinschmidt, J., van Lap, D., and Schröder, T. (1992). *Appl. Phys. B* **55**, 362.
31. Heist, P., and Kleinschmidt, J. (1994). *Opt. Lett.* **19**, 1961.
32. Szabo, G., Bor, Z., and Müller, A. (1988). *Opt. Lett.* **13**, 746.
33. Stingl, A., Lenzner, M., Spielmann, Ch., Krausz, F., and Szipöcs, R. (1995). *Opt. Lett.* **20**, 602.
34. Cohen, L. (1989). *Proc. IEEE* **77**, 941.
35. DeLong, K. W., Trebino, R., and Kane, D. J. (1994). *J. Opt. Soc. Am. B* **11**, 1595.
36. Noordam, L. D., ten Wolde, A., and van Linden van den Heuvell, H. B. (1989). *Rev. Sci. Instrum.* **60**, 835.
37. Chilla, J. L. A., and Martinez, O. E. (1991). *IEEE J. Quantum Electronics* **27**, 1228.
38. LeBlanc, S. P., and Sauerbrey, R. (1994). *Opt. Comm.* **111**, 297.
39. LeBlanc, S. P., Sauerbrey, R., Rae, S. C., and Burnett, K. (1993). *J. Opt. Soc. Am. B* **10**, 1801.

40. LeBlanc, S. P., Qui, Z., and Sauerbrey, R. (1995). *Opt. Lett.* **20**, 312.
41. Macklin, J. J., Kmetec, J. D., and Gordon III, C. L. (1993). *Phys. Rev. Lett.* **70**, 766.
42. LeBlanc, S. P., and Sauerbrey, R. (1996). *J. Opt. Soc. Am. B* **13**, 72.
43. LeBlanc, S. P., Qui, Z., and Sauerbrey, R. (1995). *Appl. Phys. B* **61**, 439.
44. Tünnermann, A., Momma, C., Mossavi, K., Windolph, C., and Wellegehausen, B. (1993). *IEEE J. Quantum Electronics* **29**, 1233.
45. Schins, J. M., Breger, P., Agostini, P., Constantinescu, R. C., Muller, H. G., Grillon, G., Antonetti, A., and Mysyrowicz, A. (1994). *Phys. Rev. Lett.* **73**, 2180.
46. Sher, M. H., Mohideen, U., Tom, H. K. W., Wood II, O. R., Aumiller, G. D., and Freeman, R. R. (1993). *Opt. Lett.* **18**, 646.
47. Barty, C. P. J., Raksi, F., Rose-Petruck, C., Schafer, K. J., Wilson, K. R., Yakovlev, V. V., Yamakawa, K., Jiang, Z., Ikhlef, A., Cote, C. Y., and Kieffer, J. C. (1995). *SPIE Proc.* **2521**, 246.

9. NONLINEAR OPTICAL FREQUENCY CONVERSION TECHNIQUES

U. Simon and F. K. Tittel

Department of Electrical and Computer Engineering
Rice University
Houston, Texas

9.1 Introduction

Using classical light sources, transparent optical materials are essentially passive, unaffected by the light wave traveling through them. For very intense light fields such as laser light, however, the presence of light can indeed affect the properties of the medium. These changes can act back on the light itself in a nonlinear way. The nonlinear response of the medium can convert the laser light into new spectral components, for example, harmonics of the optical frequency and, when more than one frequency is present in the input wave, sum- and difference-frequencies.

The nonlinear response of transparent dielectric media can be described by expanding the dipole moment per unit volume P in a Taylor series in terms of the incident oscillating electrical field [1]:

$$P = \varepsilon_0(\chi_1 E + \chi_2 EE + \chi_3 EEE + \ldots), \tag{1}$$

where χ is the weakly dispersive dielectric susceptibility; ε_0 is the vacuum permeability; and E is the incident light field. Classically, the dielectric susceptibility can be viewed as the response of an electron driven by an electromagnetic field in an anharmonic potential well resulting from the interatomic electric field E_A in the solid. The interatomic field is on the order of 10^8 V/cm. For driving optical fields much weaker than E_A, the polarization response is essentially linear. This low-optical-intensity regime is covered by the χ_1 term, which is responsible for ordinary optical phenomena like reflection and absorption. For optical fields intense enough to drive the electron beyond the quadratic minimum of the interatomic potential, the response becomes increasingly nonlinear, as described by the higher-order susceptibility terms. The second-order susceptibility χ_2 is a third-rank tensor and must therefore vanish if the medium is symmetric under inversion through the center of symmetry. This term is responsible for second-harmonic sum- and difference-frequency generation. The χ_3 term is responsible, for example, for the

Raman effect, the Kerr effect, and third-harmonic generation. A comprehensive introduction to the basics of these phenomena can be found in references [1, 2].

The generation of new frequencies (via χ_2 and χ_3) is of extreme practical importance because, although there have been a large number of lasers demonstrated, each type of laser typically generates only one or a few optical frequencies, and only a few lasers have proved practical and commercially viable. The necessity of achieving new wavelengths and to develop practical tunable coherent light sources has led to the exploration of nonlinear optics. The essential goal of practical nonlinear frequency conversion is the efficient generation of new spectral components from the input frequencies. At microwave frequencies, the availability of diode rectifiers and other strongly nonlinear elements permit straightforward design of efficient "lumped" mixers, smaller than a wavelength in size. At optical frequencies, by contrast, nonlinear responses are quite weak, so that efficient mixing requires "distributed" devices many wavelengths long. Many of the practical problems associated with optical frequency mixing result from the distributed nature of the mixing process. In particular, the difference in phase velocities of the interacting waves of different frequencies in a nonlinear medium produces a phase difference that accumulates along the length of the device and can significantly limit the efficiency of the mixing process. Thus, special steps have to be taken in order to "phase-match" nonlinear processes (see Section 9.2.1).

Some 30 years after the first demonstration of frequency conversion from the red to the blue [3], nonlinear optical devices have widespread applications in fields as diverse as laser fusion, biomedical instrumentation, femtosecond spectroscopy, and precision metrology. Though nonlinear optical technology is now well into its fourth decade, a renaissance is under way, driven by improvements in solid-state laser and nonlinear optical material technology.

The availability of high-power semiconductor diode lasers [4] has opened the door to rapid advances in diode-pumped solid-state laser sources such as Nd:YAG, Nd:YLF, Er:YAG, Tm–Ho:YAG, and Yb:YAG. These lasers operate at various discrete near-infrared wavelengths between 946 and 2010 nm in CW, Q-switched, or modelocked operation. Highly stable single-frequency diode-pumped monolithic YAG lasers [5] provide the frequency and amplitude stability as well as the power level required for highly efficient nonlinear frequency conversion. Single-mode GaAlAs diode lasers followed by high-power broad-area or tapered semiconductor amplifiers [6, 7] also serve as intense pump laser sources for nonlinear frequency conversion in the visible and near-infrared wavelength range. The large gain bandwidth of Ti:Al_2O_3 lasers, whose output can be tuned from 0.7 to 1.06 µm, makes it possible to generate modelocked pulses as short as 10 femtoseconds [8] to pump nonlinear optical devices for applications in ultrafast time-resolved spectroscopy.

New and improved crystalline nonlinear optical materials, such as BBO, LBO, KTP, $KNbO_3$, MgO:$LiNbO_3$, $AgGaS_2$, and $AgGaSe_2$, which all meet the key

FIG. 1. Simple single-pass frequency doubling scheme. Part of the incident fundamental radiation P_ω is converted to the second harmonic $P_{2\omega}$.

requirements for efficient nonlinear optical frequency conversion, can be used to generate virtually any wavelength from the ultraviolet (UV) to the far-infrared (FIR). By using quasi-phase-matching (Section 9.2.5), virtually any nonlinear process can be phase-matched throughout the entire transparency range of the nonlinear optical crystal. The nonlinear frequency conversion efficiency can be significantly enhanced by introducing the nonlinear medium into the pump laser cavity (9.2.2) or a passive optical buildup cavity (9.2.3) that resonates the interacting light waves or by using guided-wave nonlinear optical devices (9.2.5).

A summary of progress in second-harmonic generation is given in Section 9.2. Section 9.3 describes sum- and difference-frequency mixing sources that can provide tunable radiation covering the entire wavelength region from the UV to the FIR. An overview of progress in third-harmonic generation is given in Section 9.4, followed by a description of optical parametric oscillators (OPOs) in Section 9.5. OPOs are unique in that they convert fixed-frequency lasers into coherent light sources tunable over extended wavelength regions. Stimulated Raman scattering (SRS) can be used to upshift and downshift the wavelength of tunable and fixed-frequency lasers (Section 9.6). Finally, up-conversion lasers (Section 9.7) pumped by near-infrared sources emit coherent radiation at shorter wavelengths (red to UV).

9.2 Second-Harmonic Generation

9.2.1 The Second-Harmonic Generation Process

In second-harmonic generation (SHG), the electric field of an incident light wave at frequency ω interacts with a nonlinear optical medium, giving rise to an output at twice the frequency of the incident radiation. Figure 1 illustrates the basic geometry of the SHG process. For plane waves and negligible pump power depletion, the generated harmonic intensity is given by [9]

$$P_{2\omega} = [KL^2 P_\omega^2/A]\, \text{sinc}^2(\Delta kL/2), \tag{2}$$

where $K = 2z^3\omega^2\varepsilon_0^2 d_{\text{eff}}^2$; $z = \sqrt{\mu_0/\varepsilon_0\varepsilon} = 377\Omega/n_0$ is the plane-wave impedance; n_0 is the refractive index of the nonlinear material; d_{eff} is the effective nonlinear coefficient; ε_0 is the vacuum permeability; P_ω is the incident fundamental power; A is the cross-section of the fundamental beam; L is the length of the nonlinear medium in the beam-propagation direction; and Δk is the wave–vector mismatch between the fundamental and harmonic wave.

The wave–vector mismatch Δk occurs because of the natural dispersion in the refractive index that is present in all materials. A plane wave at the fundamental frequency ω propagates with the phase velocity $c/n(\omega)$ through a nonlinear medium of length L, where $n(\omega)$ is the frequency-dependent refractive index. A polarization wave proportional to $\chi_2 E(\omega)E(\omega)$ is thus generated with twice the temporal and spatial frequency of the fundamental. This polarization wave serves as an oscillatory current density that generates the electromagnetic wave at frequency 2ω, propagating with phase velocity $c/n(2\omega)$. Quantum-mechanically, SHG involves the destruction of two photons at the fundamental frequency ω and generation of one photon at the second harmonic 2ω. The wave–vector mismatch Δk is thus given by:

$$\Delta k = k(2\omega) - 2k(\omega) = 4\pi(n_{2\omega} - n_\omega)/\lambda. \tag{3}$$

For the wave–vector mismatch $\Delta k \neq 0$, the generated harmonic wave gradually gets out of phase with the driving polarization as it propagates through the nonlinear medium (Fig. 2a). The relative phase of the interacting waves varies along the medium, and the direction of power flow oscillates, initially generating the second harmonic but then converting the second harmonic back into fundamental radiation after the relative phase reaches π. In this case, maximum conversion to the harmonic is obtained for a nonlinear medium whose length L is an odd multiple of the coherence length $L_c = \pi/\Delta k$. If $\Delta k = 0$, for example, $n_\omega = n_{2\omega}$, the phase velocities of the fundamental and second-harmonic waves are equal and the relative phase of the driving polarization and generated field remains constant throughout the nonlinear medium. The interaction is then said to be phase-matched and the harmonic power grows as the square of the length L (Fig. 2b).

Equation (2) indicates that in an interaction that is not phase-matched the efficiency is reduced by the factor $\text{sinc}^2(\Delta kL/2)$. For a typical coherence length of 5 μm, which corresponds to a reduction of $\sim 10^{-6}$ in a 1-cm nonlinear medium. For minimal reduction of the conversion efficiency, L_c must be larger than L. This requires matching the refractive indices to a part in 10^5 for L on the order of 1 cm. For typical media the indices differ by $\sim 10\%$ between the fundamental and the second harmonic. Thus, phase-matching does not occur without special steps.

A systematic approach to compensate for dispersion of the refractive index is quasi-phase-matching, which is discussed in Section 9.2.5. Most commonly, how-

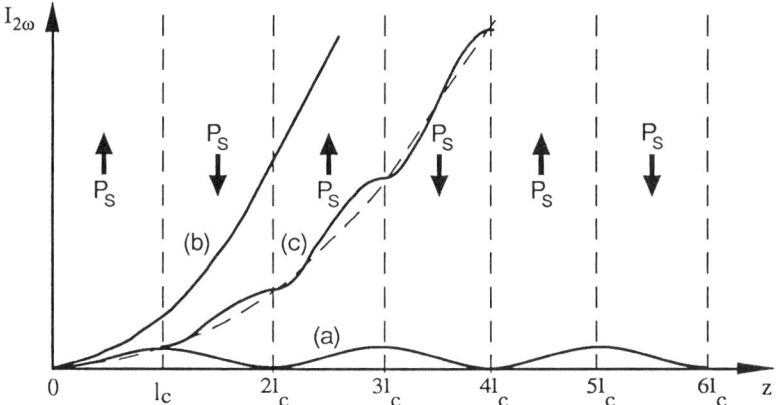

FIG. 2. The effect of phase-matching on the growth of second-harmonic intensity with distance in a nonlinear crystal; (a) nonphase-matched interaction; (b) perfect phase-matching in a uniformly poled crystal; (c) first-order QPM by flipping the orientation of the ferroelectric spontaneous polarization P_s (linked to the sign of the nonlinear coefficient) every coherence length for interaction curve (a) [37]. Reproduced with permission from the IEEE.

ever, SHG is phase-matched by exploiting the birefringence of nonlinear optical crystals. In uniaxial birefringent crystals, the two orthogonal polarization eigenmodes, known as ordinary and extraordinary waves, in general have different phase velocities: c/n_0 and c/n_e. Only the velocity of the extraordinary wave depends on the propagation direction in the crystal. In crystals with an appropriate balance of birefringence and dispersion, one can achieve phase-matching by choosing the propagation direction of the light such that the phase velocity of one polarization mode at the second-harmonic frequency is equal to that of the other polarization mode at the fundamental frequency (Fig. 3). This configuration is referred to as type I phase-matching. Thus, birefringent phase-matching makes use of those elements of the nonlinear susceptibility tensor χ_2 that couple orthogonally polarized waves. The value of Δk cannot only be adjusted by varying the angle Θ between the direction of propagation of the waves and the crystal optical axis but also by varying the temperature of the crystal for a fixed direction of propagation. The angle Θ at which $\Delta k = 0$ is called the phase-matching angle, and the temperature at which $\Delta k = 0$ and $\Theta = 90°$ is called the phase-matching temperature.

The usefulness of a particular nonlinear material for SHG is determined by the magnitude of its effective nonlinear coefficient and its ability to transmit and phase-match the harmonic and fundamental frequencies. The short wavelength limit of SHG (~205 nm by using BBO [10]) is currently set by an inability to phase-match the harmonic process at short wavelengths, while the infrared limit

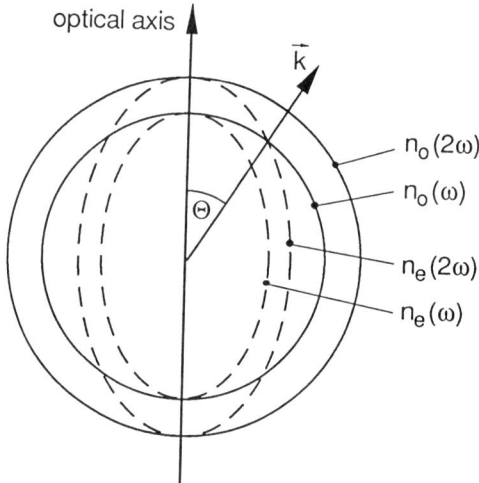

FIG. 3. Refractive index matching for SHG in a uniaxial positive birefringent crystal ($n_e < n_o$). At the phase-matching angle Θ between the beam propagation direction and the crystal optical axis the refractive indices of fundamental and harmonic match if they propagate as an *o*-ray and an *e*-ray, respectively.

is determined by increasing absorption at the fundamental. Some important properties of the most commonly used nonlinear optical crystals in SHG are summarized in Table I.

For phase-matched SHG with high fundamental intensities, the SHG intensity can grow so large that the depletion of the fundamental intensity must be accounted for. In this case, the SHG power is given by [9]

$$P_{2\omega} = P_\omega \tanh^2 [\sqrt{KL^2 P_\omega /A} \ \mathrm{sinc}(\Delta kL/2)] . \tag{4}$$

The inverse dependence of the conversion efficiency on the beam cross-section requires that the beam in practice be focused into the nonlinear crystal. A theoretical treatment of SHG for focused Gaussian beams was worked out by Boyd and Kleinman [11].

The SHG conversion efficiency is determined by parameters related to the pump source, such as power density, beam divergence, and spectral linewidth, and to the harmonic generator, such as nonlinear coefficient, crystal length, angular and thermal deviation from the optimum operating point, absorption, and inhomogeneities in the crystal. As indicated by Eq. (2), the conversion efficiency of the SHG process, $\eta = P_{2\omega}/P_\omega$, scales as the incident pump power if pump wave depletion is negligible. High-power continuous-wave (CW) or pulsed lasers are therefore easily converted with high efficiencies using a simple single-pass scheme

TABLE I. Properties of Nonlinear Optical Materials

Material	Transparency [μm]	Effective nonlinear coefficient [pm/V]	Refractive index @ 1064 nm	Pulsed damage threshold for 1-nsec pulses [GW/cm^2]	Absorption [cm^{-1}] @ 1064 nm
KDP	0.18–1.8	0.44	1.49/1.46	5	0.07
KTiOPO$_4$ (KTP)	0.35–4.5	3.2	1.74/1.75/1.83	9–20	0.006
LiB$_3$O$_5$ (LBO)	0.16–2.3	0.85	1.56/1.59/1.61	20	0.005
β-BaB$_2$O$_4$ (BBO)	0.19–2.6	2	1.66/1.54	14	0.005
KNbO$_3$ (KN)	0.4–5.5	1.3	2.12/2.14/2.20	7	0.005
LiNbO$_3$	0.35–5	4.7	2.24/2.15	10	0.002
LiNbO$_3$	0.31–5	1.8	1.86/1.71	2	0.002
AgGaS$_2$	0.5–13	10.4	2.35/2.29	0.02 (10 ns)	0.09 @ 10.6 μm
AgGaSe$_2$	0.7–18	28	2.59/2.56	0.03 (50 nsec/2 μm)	0.05 @ 10.6 μm
ZnGeP$_2$	0.7–12	70	3.07/3.11	0.05 (25 nsec/2 μm)	0.9 @ 10.6 μm

[12], as shown in Fig. 1. However, in the case of low-power CW or diode-pumped CW modelocked lasers, the low single-pass doubling efficiency can be increased by two to three orders of magnitude either by resonantly enhancing the fundamental in the bulk nonlinear sample using a high-finesse resonator in either a passive or active configuration or by improving the optical confinement in the nonlinear medium with a single-pass waveguide structure.

9.2.2 Intracavity Frequency Doubling

For a laser resonator with an output coupler of transmission T, the intracavity circulating power is approximately a factor $1/T$ larger than the output power. Therefore, highly efficient harmonic conversion of low-power laser radiation can be achieved by placing the nonlinear crystal in an auxiliary beam waist of the laser cavity itself. In this case, the laser output coupler is replaced by a mirror that is highly reflective at the fundamental and highly transmissive at the harmonic. The doubler crystal inserted into the laser cavity acts as an output coupler of transmission T_{eff} in an analogous manner to the transmitting mirror of the normal laser, but couples out power at twice the laser frequency, $P_{2\omega} = T_{\text{eff}} P_\omega = \gamma P_\omega^2$, where $T_{\text{eff}} = \gamma P_\omega$. If the coupling parameter γ is chosen so that the conversion efficiency is equal to the optimum mirror transmission T of the fundamental laser, the available output at the fundamental will completely be converted to the harmonic [13]. For example, for a CW-pumped Nd:YAG laser with an optimum output coupling of $T = 0.1$, an intracavity conversion efficiency of only 10% will produce an external conversion efficiency of 100%, so that the total 532-nm power generated in both directions by the doubler is equal to the maximum 1.064-μm power that can be extracted from the cavity without the nonlinear crystal.

In order to avoid degradation of the pump laser performance, high optical quality nonlinear crystals are required for intracavity frequency doubling. Moreover, spatial hole burning present in a standing-wave laser cavity allows the laser to operate in multiaxial mode. Since the losses of these axial modes are coupled through sum-frequency generation (Section 9.3) in the crystal, this gives rise to intrinsic large-amplitude fluctuations of the doubled output [14]. These instabilities can be eliminated by forcing the laser to oscillate in single-axial mode through the elimination of spatial hole burning or by using intracavity mode-selecting elements. The elimination of spatial hole burning can be accomplished by using either a unidirectional ring laser cavity [15, 16] or a twisted-mode linear resonator [17]. An example for stable intracavity frequency doubling of an Nd:YAG laser with an intracavity Brewster plate as a mode-selecting element [18] is shown in Fig. 4a. The fundamental is linearly p-polarized by the Brewster plate and is split by the KTP crystal into ordinary and extraordinary ray components, as shown in Fig. 4b. After traversing the KTP crystal, reflecting off the high reflector and passing through the crystal again, the wave returns to the Brewster plate with a phase difference $\delta = 4\pi \Delta n L_{KTP}/\lambda$ between the ordinary and extraordinary ray components, where L_{KTP} is the length of the KTP crystal, λ is the fundamental wavelength, and Δn is the birefringence of KTP at the fundamental. Only when δ is an integral multiple of π is the round-trip fundamental wave still p-polarized and transmitted by the Brewster plate without loss. With this setup, ~3 mW of stable green light at 532 nm was generated from 250 mW of incident diode laser pump power at 809 nm.

The number of intracavity components can be reduced by using a gain medium that also functions as the frequency-doubling nonlinear optical material (self-doubling NYAB; $Nd_xY_{1-x}Al_3(BO_3)_4$). In this case, the use of a monolithic laser cavity is particularly attractive for construction of a very compact diode laser-pumped CW laser source emitting at 531 nm [19]. However, since NYAB has an appreciable absorption at 531.5 nm that limits the useful crystal length, and since the thermally sensitive phase-matching process takes place in the presence of the pump-related thermal gradients, most reported NYAB laser outputs are low compared to the outputs of CW intracavity-doubled Nd:YAG lasers pumped at the same power level. Hemmati [20] reported 50 mW of CW 531-nm output power with an optical-to-optical conversion efficiency of 4% for a diode-pumped NYAB laser. Using the same pump laser, 140 mW of CW 532 nm was delivered from an Nd:YAG laser with an intracavity KTP doubler corresponding to 7% optical-to-optical conversion efficiency.

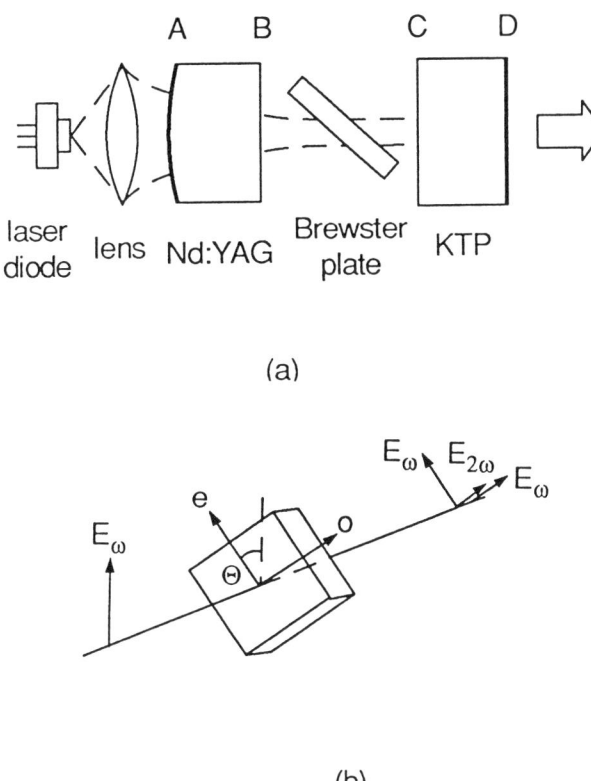

FIG. 4. (a) Design of an intracavity-doubled Nd:YAG laser with the Brewster plate and the KTP crystal serving as a frequency-selective birefringent filter [18]. Reproduced with permission from the IEEE. Facets A and D are HR-coated, and B and C are AR-coated for 1064 nm; A is AR-coated and B is HR-coated for the pump light at 809 nm; C is HR-coated and D is AR-coated for 532 nm. (b) Directions of the wave polarization and the birefringent crystal axes. The angle $\Theta = 45°$ is formed between the p-polarized light wave and the extraordinary ray in type II phase-matched KTP ($o \to e + o$).

9.2.3 External Enhancement Cavity

While intracavity frequency doubling has the advantage of being relatively easy to implement, it does not allow for independent optimization of the laser and doubling processes, a feature which is especially important in low-gain laser systems [21]. This degree of freedom is provided by the use of an external optical cavity to enhance the fundamental intensity inside the nonlinear crystal, a technique that was first demonstrated in 1966 by Ashkin *et al.* [22]. A well-designed cavity can increase the conversion efficiency by as much as three orders of magnitude, making possible efficient devices with pump powers as small as 50 mW [23]. To couple the pump laser beam efficiently into the external resonator, the resonance of the enhancement cavity needs to be locked to the pump laser frequency [24, 25]. Thus, a frequency-stable single-axial-mode pump source and either an active servo cavity-length control scheme [26] or an optical feedback laser locking scheme [27] are required to maintain coincidence of the cavity resonance with the pump laser frequency. Moreover, only very low-loss doubling crystals and careful external cavity designs allow large fundamental enhancement factors to be achieved.

The theoretical treatment of external singly resonant SHG given in reference [22] was extended by Kozlovsky *et al.* [23] taking into account the effects of pump laser depletion. The harmonic power $P_{2\omega}$ generated by the circulating fundamental power P_c is given by

$$P_{2\omega} = \kappa \gamma_{2\omega} P_c^2, \tag{5}$$

where the parameter $\gamma_{2\omega}$ is the nonlinear conversion parameter, which can be derived using the formalism of Boyd and Kleinman [11] for focused Gaussian beams; κ is equal to 1 for a ring resonator and 2 for a standing-wave cavity. If r_m represents the fraction of resonated fundamental left after one round trip inside the cavity and $r_\omega = 1 - t_\omega$ is the fundamental input transmission of the cavity, the ratio of circulating power to input power is given by [23]

$$P_c/P_\omega = t_\omega / (1 - \sqrt{r_\omega r_m})^2. \tag{6}$$

For a given geometry and input power, the amount of in-coupling t_ω must be chosen such that $r_m = r_\omega = 1 - t_\omega$, for example, the input transmission is equal to the total round-trip cavity loss including the depletion loss due to SHG [23]. This condition, for which the buildup factor is $P_c/P_\omega = 1/t_\omega$, corresponds to zero reflection for the fundamental and represents the optimum for second-harmonic conversion. The cavity is then said to be "impedance-matched", with all of the pump power driving the load and none being reflected. Proper impedance matching is critical for highly efficient SHG.

External Mirror Cavity Efficient resonant SHG requires that the total cavity round-trip loss of the resonated fundamental is dominated by harmonic conversion. It is the linear cavity losses (scatter and absorption) and not the magnitude of the nonlinear coupling constant that prevents achieving a conversion efficiency of 100%. A typical setup of an external cavity SHG experiment, using a bow-tie ring resonator [28], is shown in Fig. 5a. With a linear cavity round-trip loss of 1.7%, a fundamental power enhancement factor of 21 resulted in 36% SHG conversion efficiency at 18 W of CW 1.064-μm input pump power (Fig. 5b). A reduction in the linear cavity round-trip loss to 0.5% would increase the SHG

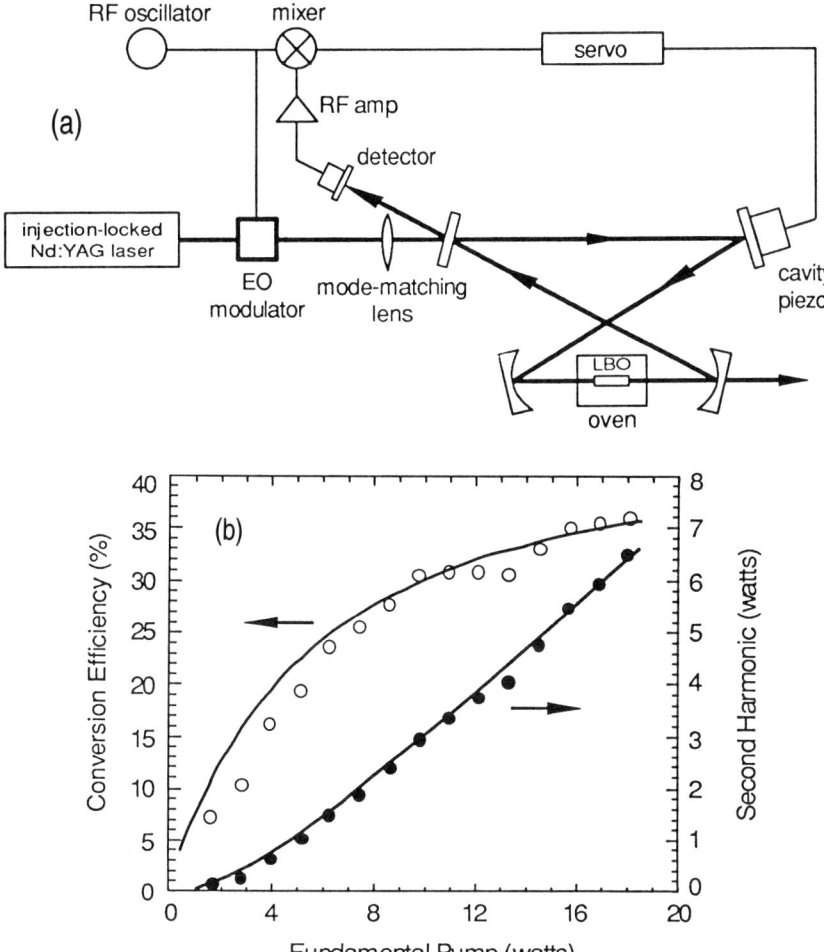

FIG. 5. (a) Experimental setup of a typical external SHG bow-tie cavity resonating an injection-locked Nd:YAG laser at 1.064 µm; (b) conversion efficiency and corresponding second-harmonic power as a function of the fundamental input power [28]. Reproduced with permission from the Optical Society of America.

conversion efficiency to 80% assuming perfect impedance and spatial mode matching into the external cavity. Using a very low-loss external ring cavity, SHG conversion efficiencies up to 85% have been achieved in 90° type II phase-matched KTP [29] for 700 mW of CW 1.08-µm input power [30]. The linear cavity losses of 0.32% were mainly due to the reflection losses at the surfaces of the crystal and resulted in a fundamental buildup factor of ~43.

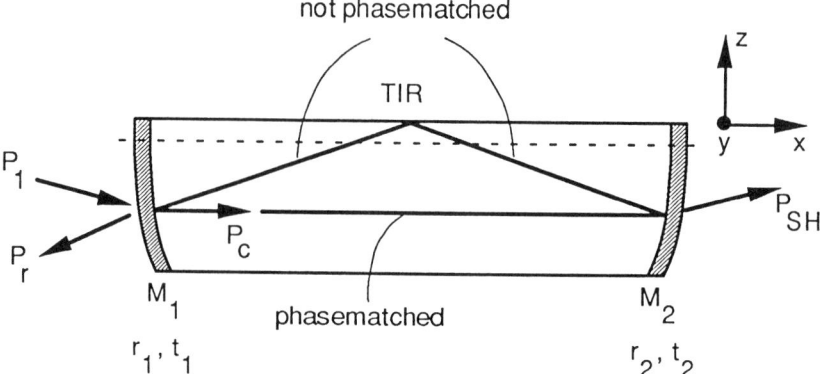

FIG. 6. Schematic diagram of a typical monolithic resonant frequency doubler. Mirrors are deposited directly onto the crystal, and a second harmonic is generated unidirectionally in the base leg of the regular beam path [31]. Reproduced with permission from the Society of Photo-Optical Instrumentation Engineers.

Monolithic Cavity In the case of a monolithic cavity design, the resonator is fabricated from a single piece of crystal with all radii, total internal reflection surfaces, and dielectric coatings directly formed on the crystal surfaces [31]. This technique minimizes losses and maximizes the rigidity of the resonator, yielding a highly stable and efficient narrow-linewidth device. The high passive frequency stability of these devices facilitates the locking of the laser frequency to the cavity. Either the pump laser frequency must be brought into coincidence with the cavity resonance or the intrinsic electrooptic and piezoelectric properties of the crystal, or a piezoelectric ceramic bonded to the monolithic resonator can be used to tune the cavity into resonance with the laser. Figure 6 shows a typical design of a monolithic ring resonator with spherical radii polished on each end of the crystal and a total internal reflection surface polished parallel to the mirror axis to define the ring resonator path. Only the base leg of the triangular beam path is phase-matched, and harmonic radiation is only generated in forward direction. The nonlinear crystal selected must be uniaxial with the resonated fundamental as the ordinary wave, or else the beam will bireflect at the mirrors whose reflection normals are not parallel or perpendicular to the c-axis of the nonlinear crystal. For the same reason, these resonators are not suitable for type II phase-matched SHG. A 12.5-nm $MgO:LiNbO_3$ monolithic ring cavity with a linear cavity round-trip loss of 0.42% and a fundamental power enhancement factor of 60 yielded 56% SHG conversion efficiency at 53 mW CW input power at 1.064 μm [23]. The same resonator device can be used as a standing-wave monolithic cavity by aligning the beam along the mirror axis. Both forward- and backward-propagating beams are then phase-matched, resulting in two harmonic outputs. An interesting variant of this technique makes use of a monolithic ring Fabry–Perot resonator and total internal reflection to produce an extremely low-loss cavity that yields SHG of 1.064-μm beams with input power of less than 1 mW [32].

Devices are now commercially available that can generate more than 100 mW of 532-nm light by intra- or extracavity resonant SHG with diode-pumped Nd:YAG lasers as pump sources and either KTP or MgO:LiNbO$_3$ mixing crystals.

9.2.4 Synchronously Pumped Frequency Doubling

Resonant frequency-doubling schemes are also commonly used in the case of low-power CW modelocked lasers and have been theoretically described [33]. The picosecond-pulsed output of a modelocked laser described by Malcolm *et al.* [34] is impedance and spatially mode-matched into an external ring enhancement cavity that resonates the fundamental. The external ring cavity requires an active cavity-length stabilization scheme [26] with the round-trip time matched to the repetition rate of the modelocked laser so that adjacent pulses coincide in the external cavity and enhance the fundamental light. This is equivalent to matching the free spectral range (FSR) of the buildup cavity to the mode spacing of the modelocked laser. Two ring enhancement cavities in series using LBO and BBO as the doubling crystal, respectively, have been used to sequentially double (524 nm) and quadruple (262 nm) the frequency of a diode-pumped modelocked Nd:YLF laser [34]. With 1.4 W of CW modelocked Nd:YLF laser power (225-MHz repetition rate; 12-psec pulses) green output was generated with a 54% conversion efficiency, while the green-to-UV conversion efficiency was 11%.

9.2.5 Quasi-Phase-Matching

The essential difficulty in a nonphase-matched interaction is the difference in the phase velocities between the nonlinear polarization and the generated harmonic output, which produces a phase shift of π over every coherence length L_c, with a concomitant reversal of the energy flow between the waves (Fig. 2a). Birefringent phase-matching (Section 9.2.1) is the most commonly used technique to compensate for the phase velocity dispersion, but it limits the choice of useful SHG materials to those having an appropriate relationship between dispersion and birefringence and utilizes only those elements of the nonlinear susceptibility tensor that couple orthogonally polarized waves. Efficient birefringently phase-matched SHG has therefore been limited to a few nonlinear optical materials and a limited wavelength range. Phase-matching capabilities can be significantly extended by using quasi-phase-matching [35–37] (QPM). This technique involves repeated inversion of the relative phase between the driving polarization and the generated free harmonic by periodically resetting the phase of the polarization wave by π. This is accomplished by periodically varying the sign of the nonlinear coefficient along the beam propagation direction and in synchronism with the generated harmonic. Figure 2c illustrates the effect of first-order QPM on the growth of harmonic intensity with distance in the nonlinear material. Using this technique, the interacting waves propagate at different phase velocities $v_{ph} = \omega/k$, but accumulated phase-mismatch is prevented. Therefore, power keeps flowing continuously from the fundamental to the harmonic output. The sign of the nonlinear coefficient is reversed spatially with a grating period $\Lambda = 2mL_c$, where $m = 1, 3, 5, \ldots$, is the QPM order; L_c is the coherence length, given by $L_c = \lambda/4(n_{2\omega} - n_\omega)$; λ is the wavelength of the fundamental; and the subscripts ω and 2ω refer to the fundamental and second-harmonic frequency, respectively. The QPM condition is $\mathbf{k}_{2\omega} - 2\mathbf{k}_\omega - \mathbf{K} = 0$, with $\mathbf{K}$ the wave vector of the nonlinear coefficient grating, having the magnitude $\mathbf{K} = 2m\pi/\Lambda$. The effective nonlinear coefficient for QPM is given by $d_{eff} = (2/m\pi)d_{ij}$, which indicates that the conversion efficiency for mth-order QPM is reduced by a factor of $(2/m\pi)^2$ as compared to perfect phase-matching. A more detailed theoretical treatment of QPM has been given by Fejer et al. [37]. The advantages of QPM result from the decoupling of phase-matching from the birefringence of the nonlinear material. QPM can phase-match virtually any nonlinear process at any temperature within the entire transparency range of a crystal, even if the crystal is isotropic. High conversion efficiencies can be achieved by taking advantage of any of the components of the nonlinear susceptibility tensor, including those coupling waves of the same polarization that allow the use of single-mode waveguide devices that support only one polarization. In the case of $LiNbO_3$, interactions involving wavelengths between 350 nm and 5 µm are possible utilizing a-axis grown $LiNbO_3$ and $d_{33} = 34$ µm/V, which is seven times larger than the birefringently phase-matchable d_{31}.

QPM in Bulk Devices The difficulty in implementing the QPM scheme lies in creating a medium with the requisite sign reversal every coherence length L_c, which, for visible interactions, is on the order of several microns. Early work for infrared interactions involved stacks of polished crystal plates, each being one coherence length thick with adjacent plates rotated by 180°. This technique has been demonstrated in such materials as CdTe [38], GaAs [39], $LiNbO_3$ [40], and later in LBO [41], but this proved to be difficult to handle and excessively lossy. A more practical approach is generation of a QPM grating of periodically inverted domains [42] in ferroelectric materials such as $LiNbO_3$ (periodically poled $LiNbO_3$, PPLN). This poling technique involves periodic perturbation of the growth conditions during crystal growth [43, 44], thereby patterning the nonlinear susceptibility. PPLN has been used for quasi-phase-matched SHG [45, 46]. An external SHG conversion efficiency of up to 42% was achieved for first-order QPM in a 1.24-nm a-axis-grown bulk sample of $LiNbO_3$ (domain length 3.47 µm) [46]. The PPLN crystal was placed inside an external bow-tie buildup cavity and generated as much as 1.7 W of green, with 4.2 W of incident CW 1.064-µm Nd:YAG power. Up to an interaction length of 0.56 mm (160 domains), SHG conversion efficiency increased quadratically with crystal length, as expected for perfect domain periodicity.

QPM Using Surface Domain Gratings For a useful QPM device, the domain periodicity must be accurately controlled. If the spacing of the domains deviates from the ideal structure, a phase error accumulates between the interacting waves, resulting in a reduced nonlinear conversion efficiency [37]. Very precise lithographically controlled methods can be used to create QPM structures of alternating polarity at the surfaces of such ferroelectric substrates as $LiNbO_3$, $LiTaO_3$, or $KTiOPO_4$ (KTP). Using chemical gradients, periodic in-diffusion of dopants by means of a patterned mask induce domain reversals, thus achieving periodic poling on the wafer surface. Such surface domain gratings are used primarily in waveguide devices, which will be discussed in the next section. Moreover, the application of periodic fields with precise periodic electrodes is especially useful to pattern such centrosymmetric media as polymers [47] and liquids [48], in which second-order nonlinearity is induced by the applied electrical field.

QPM in Guided-Wave Devices The efficiency of single-pass bulk interactions is limited by the tradeoff between tight focusing for the sake of high intensities and loose focusing for the sake of large effective interaction lengths. Thus, for the frequency conversion of low-power lasers, often low-loss dielectric waveguides rather than bulk materials are used as the nonlinear media, as they allow strong modal confinement and thus high optical intensities over long interaction lengths. Compared to focused bulk interaction, the use of planar and channel waveguides improve the SHG conversion efficiency by factors of $(L/\lambda)^{1/2}$ and L/λ, respectively, where L is the length of the waveguide and λ the fundamental wavelength. This corresponds to improvement factors of 10^2 and 10^4, respectively, for SHG of 1-µm radiation in a 1-cm device. Thus, efficient nonresonant frequency conversion can be achieved with only milliwatts of pump power.

A channel waveguide nonlinear device is usually fabricated from a periodically poled planar nonlinear material wafer ($LiNbO_3$ [49, 50], $LiTaO_3$ [51–53], and KTP [54]). Waveguide fabrication requires creation of a core region of high refractive index surrounded by low-index cladding. In the oxide crystals widely used for frequency conversion, this as well can be accomplished by in-diffusing dopants into the wafer through a lithographic mask [49, 55]. Depending on the size and refractive index change of the core region, one or more modes are supported by the waveguide at a given wavelength [56]. Figure 7 depicts a typical channel waveguide device [52].

The advent of lithographic QPM techniques and compact diode laser pumps has stimulated much of the work on waveguide frequency conversion [37]. Con-

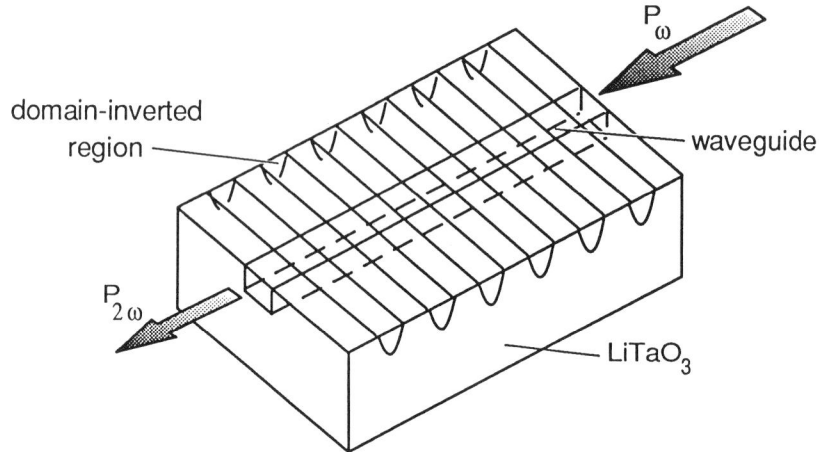

FIG. 7. Structure of a first-order quasi-phase-matched SHG LiTaO$_3$ waveguide [57]. Reproduced with permission from the Optical Society of America.

version efficiencies approaching 1% per mW have been demonstrated in 10-mm devices, and 10 to 20 mW of blue has been generated with pump powers below 100 mW by SHG in LiNbO$_3$, LiTaO$_3$, and KTP. Using a 10-nm first-order QPM LiTaO$_3$ SHG device, Yamamoto et al. [57] obtained 23 mW of blue light from 121 mW of Ti:Al$_2$O$_3$ laser input at 873 nm, corresponding to a normalized conversion efficiency ($\eta_{norm} = P_{2\omega}/(P_\omega L)^2$) of 157 %/W cm^2. Yamada et al. [58] reported the fabrication of a 3-mm periodic domain structure in a thin LiNbO$_3$ wafer by applying an external field at room temperature. After poling the substrate, a waveguide was fabricated in the first-order QPM SHG device using an annealed proton exchange process. As much as 20.7 mW of blue output was obtained for 196 mW of fundamental Ti:Al$_2$O$_3$ laser input power at 852 nm, corresponding to a normalized harmonic conversion efficiency of 600 %/W cm^2.

Waveguide QPM is not restricted to SHG. One can readily adapt this technology to a variety of interactions simply by changing the lithographic masks. It has already yielded SHG of wavelengths as short as 365 nm and difference-frequency generation (Section 9.3) of infrared wavelengths from 1.3 to 3 µm. Lim et al. [59] demonstrated quasi-phase-matched difference-frequency generation at 2 µm (1.8 µW) by mixing 1 mW from a 1.32-µm Nd:YAG laser with 160 mW from a tunable Ti:Al$_2$O$_3$ laser in a PPLN waveguide corresponding to a normalized conversion efficiency of 410 %/W cm^2. Laurell et al. [60] demonstrated efficient sum-frequency generation (Section 9.3) of two Ti:Al$_2$O$_3$ lasers in segmented KTP waveguides. For ~100 mW in each input beam, more than 2 mW of blue light was generated between 390 and 480 nm with ~3-nm tunability and a maximum conversion efficiency of 84 %/W cm^2.

QPM opens the search for better nonlinear media to new classes of materials, such as poled-polymer and fused-silica films, diffusion-bonded stacks of plates, laterally patterned semiconductors, and asymmetric quantum wells. It is likely that these techniques will lead to significant progress in the performance of existing device types, as well as new devices that would not be possible with conventional media. For the near future, waveguide QPM holds out the promise of high-gain parametric amplification and CW parametric oscillation (Section 9.5).

9.3 Sum- and Difference-Frequency Generation

The general features and operating principles of sum- and difference-frequency generation are very similar to those discussed in detail in the previous section on SHG. As a matter of fact, SHG is a special case of sum-frequency generation (SFG), with the two incident photons having the same fundamental frequency ω.

9.3.1 Sum-Frequency Generation

SFG is used to up-convert two input pump sources with frequencies ω_1 and ω_2, generating a new frequency ω_3 according to the relation $\omega_3 = \omega_1 + \omega_2$. At low conversion efficiencies the generated intensity $I(\omega_3)$ grows as the product of the incident pump intensities $I(\omega_1)I(\omega_2)$. Complete conversion of the total radiation in both pump waves with perfect phase-matching is in principle possible if they start with an equal number of photons. Sum-frequency mixing is performed in the same types of nonlinear materials as used for SHG (Table I) and can generate tunable and fixed-frequency radiation from the near-vacuum ultraviolet (VUV) to the infrared region.

Generally, as two different wavelengths are combined in the SFG process, a wider frequency range can be covered by SFG than is accessible by SHG; especially the generation of short wavelengths below 200 nm becomes generally critical due to unattainable phase-matching for SHG (the SHG short-wavelength limit in BBO is 205 nm [61] and increasing absorption in all known nonlinear crystals. The first generation of CW coherent radiation below 200 nm was demonstrated by SFG in a KB_5 crystal [62]. With the invention of BBO as a new nonlinear material with excellent optical properties, this material became the most important optical crystal for the generation of short wavelengths down to 189 nm [61]. By mixing a tunable Ti:Al$_2$O$_3$ laser with a frequency-doubled Ar$^+$ laser (257 nm) in BBO, CW radiation tunable down to 191 nm was generated [63]. Output powers of more than 10 µW were obtained by resonating the Ti:Al$_2$O$_3$ laser in an external buildup cavity, and using a walkoff compensated crystal configuration

[64]. Using a doubly resonant cavity, coherent 194-nm radiation has been obtained with a CW output power of 31 µW [65].

Mückenheim et al. [66] demonstrated continuous coverage of the wavelength range from 189 to 197 nm by SFG in BBO using a tunable infrared dye laser (780 to 950 nm) and a frequency-doubled dye laser fixed to a wavelength of 245 nm. Peak energies of typically 10 to 100 µJ were obtained throughout this spectral range. More recently, by using a similar SFG-scheme Heitmann et al. [67] demonstrated generation of 100 µJ (20 kW) at 196 nm from 67-mJ Nd:YAG laser power used to pump two dye lasers. High output energies and conversion efficiencies could be achieved compared to competing methods for the generation of near-VUV radiation such as stimulated Raman scattering ($\sim$10 kW, $\eta \approx 10^{-2}\%$) [68] or four-wave difference-frequency generation and third-harmonic generation in rare gases ($\sim$20 to 60 W; $\eta \approx 10^{-4}\%$) [69].

The technique of SFG is also being used in the development of compact blue CW light sources based on near-infrared diode lasers. Resonantly enhanced SFG of several mW of blue light near 460 nm by mixing a single-mode diode laser with a diode-pumped 1064-nm Nd:YAG laser was demonstrated by using an intracavity SFG scheme [70] and by resonating both infrared pump lasers in an external monolithic KTP resonator [71].

Another application of SFG is the generation of the third and fourth harmonics of certain laser frequencies through two and three subsequent second-order parametric processes, respectively. Under certain conditions, these multistep processes are more efficient than direct third- or fourth-harmonic generation (Section 9.4). Nebel and Beigang [72] generated mW of VUV radiation tunable from 192 to 210 nm by SFG of the fundamental and third harmonic of a CW modelocked Ti:Al$_2$O$_3$ laser in BBO with a maximum conversion efficiency of $\eta_{4\omega} = P_{4\omega}/(P_\omega P_{3\omega})^{1/2}$ of $\sim$4%.

9.3.2 Difference-Frequency Generation (DFG)

Difference-frequency mixing is used to down-convert radiation from two incident waves at frequencies ω_1 (pump) and ω_2 (signal) to a third wave at frequency ω_3 (idler) according to the relation $\omega_3 = \omega_1 - \omega_2$. In the low-conversion regime, as for SFG, the generated DFG intensity grows as the product of the pump intensities $I(\omega_1)I(\omega_2)$. Because mixing efficiency is proportional to the intensities of both the pump and the signal, the signal wave grows exponentially at the expense of the pump wave (until pump depletion sets in), a phenomenon known as parametric gain. DFG is an important technique for the generation of coherent tunable infrared radiation from CW and pulsed pump laser sources in the visible or near-infrared.

Early tunable CW DFG sources mixed a CW dye laser with its Ar$^+$ pump laser in temperature-tuned 90° type I phase-matched LiNbO$_3$ [73]. They proved very

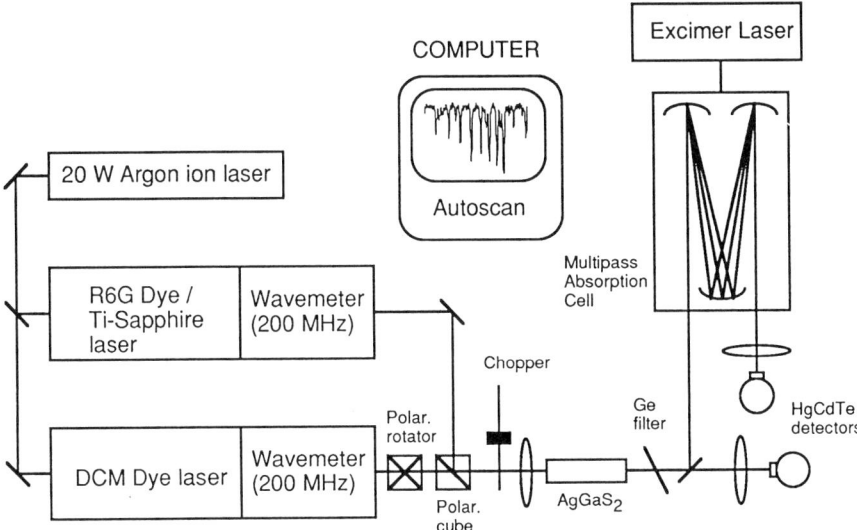

FIG. 8. Schematic of a DFG source used to generate continuously tunable narrow-band radiation from 3 to 9 μm utilizing type I noncritical phase-matching in AgGaS$_2$. The crystal is pumped by two dye and/or Ti:Al$_2$O$_3$ lasers [75]. Reproduced with permission from the Optical Society of America.

useful for high-resolution molecular spectroscopy but were limited to wavelengths shorter than 4 μm by the transmission characteristics of LiNbO$_3$. The use of LiIO$_3$ as the nonlinear medium extended the long-wavelength limit for CW DFG to ~5 μm [74]. Advances in nonlinear optical materials, such as AgGaS$_2$ and AgGaSe$_2$, now allow the generation of infrared radiation continuously tunable up to 18 μm by means of DFG, providing access to virtually all molecular fundamental vibrational modes. The narrow-band ($\delta\lambda \approx 1$ MHz) CW DFG source demonstrated by Canarelli et al. [75] was based on 90° type I phase-matching in AgGaS$_2$ and pumped by single-mode dye and/or Ti:Al$_2$O$_3$ lasers. It covered the infrared wavelengths from 4 to 9 μm with CW output powers of several 10 μW, exceeding the noise equivalent power (NEP) of liquid N$_2$-cooled infrared detectors by four to five orders of magnitude. The infrared source, shown schematically in Fig. 8, has been used to detect high-resolution kinetic spectra of such various free radical species as HOCO, DOCO, and HCCN [76].

Significant progress in the field of CW III–V semiconductor diode lasers [77] makes these sources attractive as pump lasers in compact and robust infrared CW DFG spectrometers for applications in chemical analysis, environmental sensing, industrial process monitoring, and medical technology. The nonlinear conversion efficiency of early all-diode laser CW DFG [78] has been limited by the low output

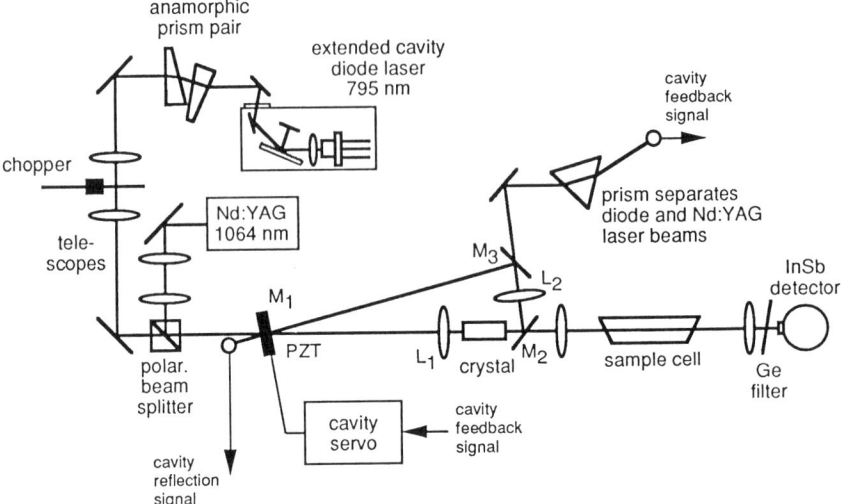

FIG. 9. Schematic of a DFG source used to generate infrared light near 3.2 μm by mixing an external-cavity diode laser (795 nm) with a compact diode-pumped Nd:YAG laser (1064 nm) in $AgGaS_2$. The 1064-nm radiation is enhanced by using a three-mirror buildup cavity. As much as 2 μW of IR has been generated and has been used to detect methane near 3.2 μm [80]. Reproduced with permission from Optical Society of America.

power of single-mode III–V diode lasers. The infrared output can be increased, however, by using optical semiconductor amplifiers to boost the output power of the single-mode diode lasers [79], or by placing the nonlinear crystal in an external buildup cavity [80] or in the laser cavity [81] of one of the pump lasers (intracavity DFG). Simon et al. [79] obtained as much as 50 μW of CW infrared radiation tunable near 4.3 μm by mixing a diode laser injection-seeded GaAlAs tapered traveling wave optical amplifier with a $Ti:Al_2O_3$ laser in 90° type I phase-matched $AgGaS_2$. The same authors demonstrated an external cavity DFG source [80] (Fig. 9) by mixing a resonantly enhanced compact diode-pumped 1064-nm Nd:YAG laser with 15 mW from a 795-nm single-mode GaAlAs diode laser in a 5-mm 90°-cut $AgGaS_2$ crystal. With 2 μW of CW infrared output, angle-tunable from 3.16 to 3.42 μm, the source was applied to the detection of the v_3-asymmetric stretch motion of methane.

DFG devices have also been used for the generation of ultrashort pulses in the femtosecond (fsec) time domain to study fast reaction dynamics in areas such as optoelectronics and spectroscopy. Color center lasers and OPOs (Section 9.5) have provided fsec pulse trains up to 2.5 μm, whereas DFG allows the generation of tunable fsec pulses at even longer wavelengths. Many pump probe experiments also require independently tunable excitation and probe pulses that are precisely synchronized on an ultrafast time scale. By difference-frequency mixing the outputs of a 100-fsec colliding pulse modelocked (CPM) dye laser/amplifier and a spectrally broad 200-fsec traveling-wave dye laser (TWDL) in two 0.3-mm $LiIO_3$ crystals, Ludwig et al. [82] generated two 10-nJ infrared pulses in the 2.5 to 5 μm wavelength range with pulse durations of 300 fsec ($\delta\nu \sim 250$ cm^{-1}). For fixed crystal orientations, the acceptance bandwidth of the nonlinear processes was less than the spectral bandwidth of the TWDL output, that is, only part of the TWDL output was converted to the mid-infrared. Consequently, the infrared wavelength produced in the two crystals can be tuned independently by changing the respective phase-matching angle, and the IR tuning range is determined by the spectral width of the TWDL. As an application of this source to pump probe experiments, the bleaching of the interband absorption in a PbSe direct narrow-bandgap semiconductor was observed. This technique can be easily extended to other infrared wavelengths by using different pump sources or nonlinear mixing crystals. By difference-frequency mixing the outputs of a 120-fsec Ti:Al_2O_3 laser/regenerative amplifier and a 200-fsec TWDL/amplifier in $AgGaS_2$, Hamm et al. [83] generated 10-nJ infrared pulses in the 4.5 to 11.5 μm region with a pulse duration of 400 fsec at a repetition rate of 1 kHz. The spectral bandwidth of the two times bandwidth-limited pulses was 60 cm^{-1}. The generated IR power was limited by strong two-photon absorption of the intense Ti:Al_2O_3 laser light in the mixing crystal.

For spectroscopy of molecular rotational transitions, hyperfine transitions, and Rydberg transitions, tunable far-infrared (FIR) radiation from the GHz microwave range up to 6 THz is required. While these frequencies cannot be reached by microwave generators, DFG by mixing a fixed-frequency CO_2 laser with a tunable infrared spin-flip Raman laser in GaAs has been used to cover this spectral range [84]. Tunable FIR radiation has also been synthesized by mixing two CO_2 lasers on a metal–insulator–metal (MIM) diode [85] either by using a fixed-frequency CO_2 laser and a tunable waveguide laser (tunable over ±120 MHz) or two fixed-frequency CO_2 lasers plus tunable microwave sidebands [86]. Due to its nonlinear current–voltage characteristic, the MIM diode generates harmonics and mixing frequencies of the incident frequencies and radiates the FIR radiation in a long wire antenna pattern. Typical FIR powers of a few tens of μW have been obtained from 200 mW of CO_2 power. Using two pressure-broadened CO_2 lasers, line-tunable over several hundred lines from 9 to 10 μm (using different CO_2 isotopomers) and a microwave generator, the entire FIR region has been covered using this DFG technique [87]. These tunable FIR sources have been used primarily for highly accurate FIR measurements of stable species to serve as frequency calibration standards [88]; to measure frequencies of transient species including molecular ions for astronomical searches [89]; and to study line broadening and lineshape parameters [90] for atmospheric spectroscopy.

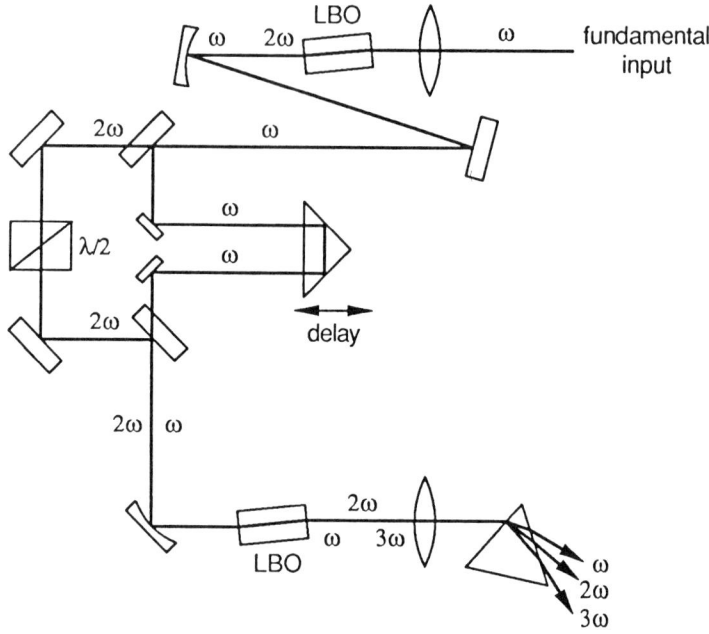

FIG. 10. THG based on two sequential second-order processes in LBO [93]. Reproduced with permission from the Optical Society of America.

9.4 Third-Harmonic Generation and Four-Wave Mixing

9.4.1 Third-Harmonic Generation as a Two-Step Process (Second-Order Parametric Interaction)

Third-harmonic generation (THG) can be performed either directly by a THG process or as a two-step process involving two sequential harmonic generation steps. In a first nonlinear crystal (doubler), a fraction of the fundamental is converted to the second harmonic. The unconverted fundamental and the second harmonic emerging from the first crystal are then coupled into the second crystal (tripler) for sum-frequency mixing. Efficient tripling requires the fundamental and the second-harmonic photons to emerge from the first crystal in a 1:1 ratio over a broad wavelength range that can be accomplished by means of an appropriate choice of polarization angle in the doubler crystal [91, 92]. In the case of short pulses, the group velocity dispersion of the fundamental and second-harmonic waves in the SFG process must be taken into account using a delay line. Figure 10 shows a typical experimental setup for THG of psec pulses [93].

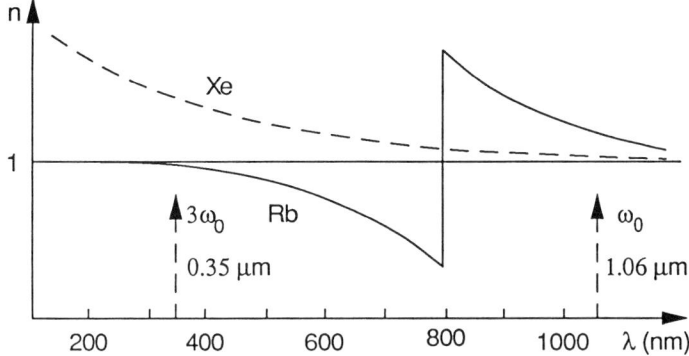

FIG. 11. Refractive indices of rubidium vapor and xenon, illustrating the THG phase-matching principle in centrosymmetric media [95]. Reproduced with permission from Springer-Verlag.

9.4.2 Four-Wave Mixing Interactions (Third-Order Parametric Processes)

Third-order parametric processes can be observed in materials with any symmetry. However, such centrosymmetric materials as gases, liquids, and some solids have been used most commonly as third-order processes and are the lowest-order nonzero nonlinearities present in these media. Generally, the lower conversion efficiencies of the higher-order processes require high pump laser intensities. As a result, third-order processes have been used primarily for generation of radiation in the extreme VUV at wavelengths too short to be reached with second-order interactions in nonlinear optical crystals.

The plane wave phase-matching condition for third-order parametric interactions is $\Delta k = 0$. The conversion efficiency is often increased by using tightly focused beams, in which case optimal performance can require either a positive or a negative value of Δk, depending on the interaction involved (focused THG requires $\Delta k < 0$) [94]. As the isotropic media used for third-order parametric processes are nonbirefringent, alternative phase-matching techniques must be used. In gases, negative dispersion occurs near allowed transitions. Therefore, phase-matching can be accomplished by using a homogeneous mixture of a rare gas and a metal vapor with different signs of dispersion (Fig. 11). Each component makes a contribution to the wave–vector mismatch in proportion to its concentration in the mixture. The appropriate phase-matching condition is met by adjusting the relative concentration of the two gases. However, in the case of THG using focused beams, only one gas with negative dispersion (e.g., Kr or Xe) is needed. The range that can be covered by THG in gas mixtures is determined by the extent of the negative dispersion region in the nonlinear materials and can be extended far into the VUV.

The conversion efficiency can be increased significantly if resonances are present between certain energy levels of the medium and the incident and generated frequencies. The effectiveness of single-photon resonances in enhancing nonlinear processes, however, is limited because of the absorption and dispersion that accompany them. Resonances in nonlinear effects can also occur when multiples of the incident frequencies match the frequency of certain types of transitions. The most commonly used resonance is a two-photon resonance involving two successive dipole transitions between levels whose spacing matches twice the value of an incident frequency. Near two-photon resonances, the nonlinear susceptibility can increase by as much as four to eight orders of magnitude, depending on the relative linewidth of the two transitions and the input radiation, resulting in a significant increase in the generated power as the input frequency is tuned through the two-photon resonance. Resonant enhanced THG has been proved very useful in allowing effective conversion of tunable radiation from dye lasers to the VUV to provide high-brightness narrow-band sources of radiation for high-resolution spectroscopy and other applications. THG from high-power pulsed lasers has been used to generate fixed-frequency and tunable radiation at various wavelengths from 3.5 μm to 57 nm.

Four-wave sum-frequency generation (SFG) and difference-frequency generation (DFG) of the form $\omega_4 = 2\omega_1 \pm \omega_3$ and $\omega_4 = \omega_1 + \omega_2 \pm \omega_3$ can also be used to produce radiation in wavelength ranges that are inaccessible by other means. These processes can be favored over possible simultaneous THG by the use of opposite circular polarization in the two pump waves, since THG with circularly polarized pump light is forbidden by symmetry. These interactions can also be used to generate tunable radiation in resonantly enhanced processes. In this case, the pump frequency at ω_1 or the sum combination $\omega_1 + \omega_2$ is adjusted to match a suitable two-photon resonance, while the remaining pump frequency at ω_3 is varied, producing tunable generated radiation.

For the DFG process phase-matching can be achieved with focused beams in media with either sign of dispersion [94]. Therefore, the usefulness of DFG is not restricted to narrow wavelength ranges above dispersive resonances, and it has been used to generate tunable radiation over broad spectral ranges in the VUV between 140 and 200 nm in rare gases Kr and Xe [96].

Four-wave DFG has also been used to generate tunable radiation in the infrared region by using pump radiation from visible and near-infrared sources. Because the gases used in these nonlinear interactions are not absorbing at FIR wavelengths, they allow more efficient generation of tunable FIR using pump sources in the visible and near-infrared region, than can be achieved in second-order nonlinear optical processes. To date, four-wave DFG has been used to produce coherent radiation at wavelengths out to 25 μm.

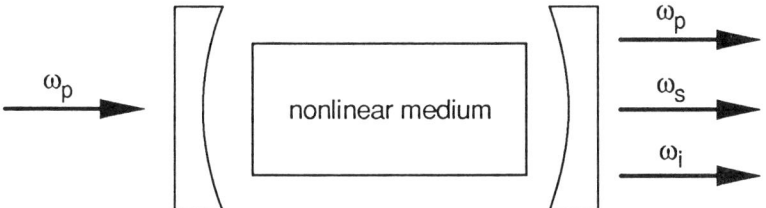

FIG. 12. Basic OPO scheme. The cavity resonates either the signal wave (singly resonant oscillator: SRO) or both signal and idler waves (doubly resonant oscillator: DRO).

9.5 Optical Parametric Amplifiers (OPAs) and Oscillators (OPOs)

9.5.1 The Optical Parametric Process

Ever since the invention of the laser, there has been a great deal of interest in the development of efficient continuously tunable coherent light sources for high-resolution spectroscopy. This field has attracted renewed interest in connection with a variety of applications in combustion diagnostics, process control, remote sensing, and environmental sensing. Picosecond and femtosecond pulses of tunable radiation are needed for time-resolved studies of chemical reactions or carrier dynamics in semiconductors. Replacement of the widely used dye and color-center lasers with more convenient sources based on Ti:sapphire and diode-pumped solid-state lasers is therefore of considerable practical utility.

OPOs [97–100] are powerful solid-state sources that provide laser radiation with potentially very large continuous-wavelength tuning ranges by parametric conversion of a fixed-frequency pump in a nonlinear optical material. Figure 12 illustrates the basic geometry of an OPO device. The nonlinear parametric interaction between the electrical field of the pump source and the nonlinear material converts pump photons at frequency ω_p in photons at frequencies ω_s (signal) and ω_i (idler) according to the conservation of energy, $\omega_p = \omega_s + \omega_i$, a process known as optical parametric amplification (OPA). For a given ω_p, there can be a continuous range of choices of ω_s and ω_i that gives rise to wavelength tunability of the parametric device. The specific pair of frequencies that will result in any given situation is determined by the conservation of momentum, or the phase-matching condition $\mathbf{k}_p = \mathbf{k}_s + \mathbf{k}_i$, with $\mathbf{k}$ the momentum vector of a wave [101]. Because the relationship between photon momentum and energy depend on the refractive index of the medium, both conditions can be simultaneously met by manipulating the refractive index of the nonlinear material, as in the case of SHG, SFG, DFG, and THG. This is usually done by using a birefringent crystalline nonlinear medium and setting the wave polarizations, crystal orientation, and temperature appropriately to establish the phase-matching condition. Wavelength tuning is then achieved by changing the angle, temperature, or electric field across the crystal to modulate the effective refractive indices.

For the degenerated case in OPA ($\omega_p/2 = \omega_s = \omega_i$), the small signal gain is numerically equal to the conversion efficiency of SHG. Thus, the gain can be as high as 1% per watt of input pumping power. Although such gain is too small to be of interest for signal amplification, it can nonetheless exceed the round-trip losses in a Fabry–Perot cavity containing the gain medium, allowing coherent output to build up from a noise input. The addition of an optical cavity transforms the parametric amplifier into an optical parametric oscillator. If the optical cavity provides optical feedback for the signal wave only, it is known as a singly resonant oscillators (SRO). In this case, the threshold condition is that the parametric gain exceed twice the signal losses. For losses of 1% in the cavity, typical threshold power is on the order of 5 W. If the optical cavity provides feedback at both the signal and the idler frequencies, the device is known as a doubly resonant oscillator (DRO) and the threshold condition requires only that the gain exceed the product of signal and idler losses. The threshold power is thus lowered to tens of milliwatts at the cost of complicating the tuning behavior of the oscillator.

An OPO has several potential advantages over a laser in generating widely tunable narrow-band coherent radiation. Tuning ranges of lasers are generally limited by relatively narrow-gain bandwidths and fixed-gain centers. OPOs offer a wide wavelength operation range (0.2–18 μm) by using a solid-state nonlinear crystal as the gain medium. Since the parametric gain does not depend on any form of atomic or molecular resonances, the gain center of an OPO is tunable by changing the phase-matching, yielding a tuning range that is limited only by the dispersion and transmittance bandwidth of the nonlinear crystal. Therefore, OPOs are particularly useful when they generate radiation in spectral regions where no tunable laser sources exist (e.g., ultraviolet [102, 103], and infrared [104, 105]).

After their invention in 1965 [106, 107], early OPO development was impeded by practical difficulties related to the nonlinear optical materials, such as low optical damage thresholds and poor optical quality, as well as inadequate frequency stability of the laser pump sources. Moreover, convenient frequency control has been difficult to obtain. The sudden resurgence of interest in OPO devices is largely attributable to significant advances in the growth and fabrication of new and improved nonlinear optical materials [108]. Simultaneously, the performance of solid-state pump lasers for driving OPOs has also improved [109]. OPOs now offer the potential reliability of an all-solid-state coherent laser source. They can provide high peak power on the order of tens of MW and high average power on the order of watts with diffraction-limited beam quality. An energy conversion efficiency as high as 62% has been demonstrated [110]. The spectral bandwidth control is similar to that required for pulsed dye lasers and linewidths as small as 0.001 cm^{-1} [111] have been obtained by using etalons or injection-seeding with

TABLE II. Characterstics and Phase-Matching Ranges for OPO Materials [171]

Material	Trans-mission range (μm)	Phase-matching range (μm)					Nonlinear figure of merit C^2 $(GW)^{-1}$	Optical damage threshold (GW/cm^2)
		0.266 μm pump	0.355 μm pump	0.532 μm pump	1.064 μm pump	2.05 μm pump		
BBO	0.19–2.56	0.3–2.5	0.415–2.5	0.67–2.5	–	–	40	~1.5
LBO	0.16–2.6	0.3–2.5	0.41–2.5	0.67–2.5	–	–	5.4	~2.0
KNbO$_3$	0.35–4.2	–	–	0.61–4.2	1.43–4.2	–	44	~1.2
KTP	0.35–4.0	–	–	0.61–4.0	1.45–4.0	–	45	~1.5
LiNbO$_3$	0.35–4.3	–	–	0.61–4.3	1.42–4.3	–	15	~0.20
AgGaS$_2$	0.8–9.0	–	–	1.2–9.0	2.6–9.0	75	~0.040	
AgGaSe$_2$	1.0–15	–	–	–	2.4–15	100	~0.040	
ZnGeP$_2$	2.0–8.0	–	–	–	~2.7–8	270	~0.040	

single-mode diode lasers. They have also proven ideal for OPAs since the parametric gain is polarized, forward-directed, and bandwidth-limited, thus avoiding the superfluorescence problems associated with dye laser amplifiers.

9.5.2 OPA/OPO Materials

As increasingly sophisticated frequency-conversion applications require operation at higher efficiencies over a broader range of peak and average power and over extended spectral ranges, the demands on nonlinear materials place increasing emphasis on parameters other than birefringence and nonlinear susceptibility [112]. Notable among these are low absorption and scatter losses, high surface damage threshold, high thermal conductivity, low thermooptic coefficients, and environmental stability. Another, often crucial, issue is ease of crystal growth and processing.

Early OPOs tended to operate in the infrared wavelength region with LiNbO$_3$ as the most commonly used nonlinear optical material. During the last decade, the range of viable OPO materials has increased, as discussed in connection with SHG, SFG, and DFG. The parametric tuning range in a given crystal is determined by the pump wavelength, the transparency of the crystal, and the range over which phase-matching can be achieved. Table II summarizes the properties of nonlinear optical materials for OPAs/OPOs. These materials collectively provide good coverage of the entire wavelength region from the infrared to the UV. Most of the materials listed have a significantly higher nonlinear coefficient than LiNbO$_3$; all have equal or higher optical damage thresholds. The OPO performance can be related to the figure of merit (C^2). The single-pass optical gain G through the

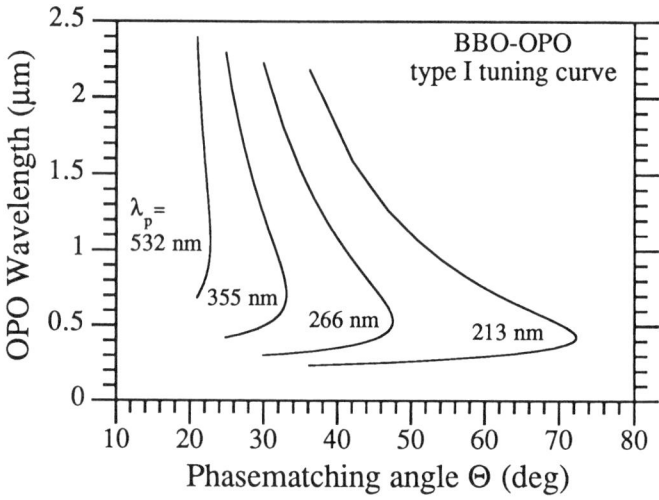

FIG. 13. Calculated wavelength tuning characteristics of a BBO OPO pumped by the second to fifth harmonic of an Nd:YAG laser. The Sellmeier equations for BBO are taken from reference [61]. Reproduced with permission from the IEEE.

crystal is given by $G = \sqrt{c^2 L^2 I_p}$, where L is the conversion length and I_p the intensity of the pump laser. As an example, Fig. 13 demonstrates the calculated wavelength tuning characteristics of a BBO OPO as a function of the phase-matching angle, pumped by the second to fifth harmonic of an Nd:YAG laser.

For the ultraviolet and visible regions, the borates BBO [102, 103, 113–119] and LBO [117–121] are the materials of choice for OPO applications. Although the smaller birefringence in LBO as compared to BBO tends to limit the phase-matching spectral range and leads to smaller tuning rates $\partial\lambda/\partial\Theta$ (requiring large-aperture crystals), it results in inherently narrower OPO linewidths and larger acceptance angles. Relatively high efficiencies can be maintained over the entire tuning range in the UV by temperature tuning the LBO OPO under noncritical phase-matching.

From the deep red to about 4.5 μm, the nonlinear crystal KTP [122–130] can be phase-matched and meets many of the key requirements for high-efficiency OPO applications, including applications in the femtosecond regime.

The large nonlinear coefficients and extended infrared transparency ranges make chalcopyrite crystals an important class for OPA/OPO work. However, the useful crystals in this family that are phase-matchable for nonlinear processes in the infrared region—$ZnGeP_2$ [131, 132], $AgGaS_2$ [133–135], and $AgGaSe_2$ [136–138]—suffer from lower optical damage intensities. However, a very attractive feature is the possibility of tuning the entire IR range from 2.3 to 18 μm using only a single AgGaSe$_2$ crystal [136] pumped by a Q-switched 2.05-μm Ho:YLF laser.

With suitable nonlinear optical crystals and pump sources, virtually any wavelength ranging from the UV to the infrared can now be reached with OPOs. The technology is better developed for the near-UV to 4.5 μm range, largely due to the availability of large high-quality BBO, LBO, MgO:LiNbO$_3$, and KTP crystals. As in the case of SHG, quasi-phase-matching (Section 9.2.5) promises to impact the development of periodically poled OPOs and the use of such materials as GaAs and ZnSe [139, 140] with very high figures of merit. Simultaneously, developments in organic nonlinear materials (single-crystal and polymeric media) with outstanding promise for extremely large nonlinear susceptibilities designed with molecular-engineering techniques suggest that practical applications will grow rapidly in the future [141].

9.5.3 Pump Laser Sources

Unlike conventional lasers, OPAs/OPOs require diffraction-limited pump radiation. High peak pump powers are required for the OPO to operate well above threshold, so that parametric oscillation can build up from noise during the pump pulse. As the OPO begins to oscillate, the pump power is converted to signal and idler power, thereby depleting the pump. Thus, an ideal OPO pump laser delivers good beam quality, high peak power and pulse energy, and a high pulse repetition rate to yield high average pump power. Also, narrow bandwidth is critical for pumping narrow-band OPOs.

The primary type of pump laser for OPOs remains the lamp-pumped Nd:YAG laser (using first and higher harmonics). Most existing OPOs use Nd:YAG pump lasers operating in the Q-switched (nsec), modelocked (psec), or femtosecond pulse regimes [142]. Dielectric graded reflectivity nonstable optical resonators provide large-size flat-topped TEM$_{00}$-output beams of high quality while simultaneously providing excellent near-field beam uniformity and high spatial extraction efficiency [143].

As the peak power stability of the pump laser leads to a corresponding stability in OPO output, advances in lasers with improved spatial and temporal coherence result in more stable OPO operation. In particular, the CW single-axial mode sub-kHz-linewidth output from diode laser-pumped monolithic nonplanar Nd:YAG laser ring oscillators [144–146] has proven ideal for pumping highly stable OPOs. To produce high peak powers while maintaining good spatial and temporal coherence for pumping OPOs, the nonplanar ring resonator has been long-pulse amplified [147] or used to injection seed an 18-W CW lamp-pumped Nd:YAG oscillator [148].

Excimer lasers can also serve as pump sources for OPOs as they provide a number of ultraviolet wavelengths at relatively high overall efficiencies and high average power levels. BBO [102, 149] and LBO [117, 150] OPOs pumped with narrow-band injection-seeded XeCl or KrF excimer lasers with narrow-linewidth and low-divergence-angle output have been demonstrated. The achievable optical-to-optical efficiencies are comparable to Nd:YAG laser-pumped OPOs for comparable pump beams. The linewidth of the output, however, is nearly an order of magnitude larger for excimer laser-pumped OPOs than for Nd:YAG laser-pumped OPOs.

A variety of other lasers—e.g., argon ion, dye, CO_2, copper vapor, color-center, He–Ne, diode, and Ti:sapphire—have achieved widespread use in research, industrial, and commercial applications. While almost all of these laser types have served as pumps for nonlinear optical devices, the Ti:sapphire lasers (particularly for generation of ultrashort pulses), diode lasers, and solid-state lasers pumped by diode lasers stand out as particularly important for contemporary developments.

9.5.4 Narrow Linewidth and Frequency Control

Many applications such as high-resolution spectroscopy, lidar, and remote-sensing require continuously tunable single-frequency light sources. In general, OPO spectral linewidths are set by the oscillator gain bandwidth (which in turn is determined by the birefringence, dispersion, and length of the crystal) and the spectral and angular spread of the pump laser beam [102, 151]. OPO linewidths are usually fairly broad, typically a few to many hundreds of wavenumbers, and often vary as a function of wavelength. Narrow-linewidth operation is achieved by providing active control for the signal and/or idler waves, either by inserting intracavity dispersive elements [152, 153] or injection seeding radiation from another narrow-band laser source [113, 154]. Type II phase-matching, known to lead to narrower linewidth than type I phase-matching [151], has also been used to generate narrow-band tunable radiational.

The use of intracavity dispersive elements in much the same way as in pulsed dye lasers led Bosenberg and Guyer [153] to a KTP OPO (Fig. 14) that produced single-longitudinal-mode single-frequency pulses ($\tau \approx 3.5$ nsec) over the range of 690 to 950 nm (signal) and 1.4 to 2.2 µm (idler) with a significantly reduced near-transform-limited bandwidth of 200 MHz. The SRO, which was pumped by the second harmonic of an injection-seeded single-mode 10-Hz Q-switched Nd:YAG laser, combined a grazing-incidence-grating resonator (Littman configuration) with an active-servo resonator-length control scheme and allowed continuous single-mode scans without mode hops of up to 100 cm^{-1}. It operated with up to 3 mJ of output, corresponding to an external photon conversion efficiency of 12%. Due to the lower efficiency of the grating at grazing incidence, the oscillator threshold was significantly higher than for an oscillator using the first-order Littrow configurational. The improvement in linewidth, however, makes the Littman design a good option for high-resolution applications.

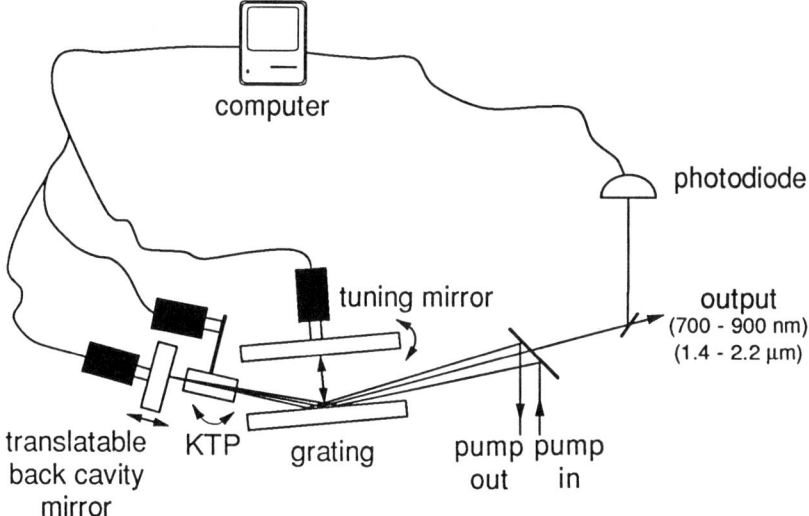

FIG. 14. OPO setup for generation of wavelength-tunable radiation with near-transform-limited bandwidth ($\Delta \nu \approx 100$ MHz) [153]. Reproduced with permission from the American Institute of Physics. A diffraction grating in Littman configuration serves as a frequency-selective element in the OPO cavity.

Haub *at el.* [154] demonstrated a BBO OPO with an effective infrared linewidth of 4 GHz and continuous tuning over ~30 cm^{-1}. The singly resonant OPO was pumped by a frequency-tripled Q-switched ($\tau = 5$ nsec) Nd:YAG laser and seeded at its visible signal wavelength with light from a tunable narrow-band (2 GHz) pulsed Rh6G dye laser. Less than 0.1 mJ dye laser energy per pulse was necessary to lock the OPO to the seed laser. For 100 mJ of 355-nm pump energy, typical pulse energies were ~20 and ~12 mJ for the idler and signal, respectively. The application of this source to rotationally resolved coherent anti-Stokes Raman spectroscopy (CARS) of N_2 in ambient air has been demonstrated [156].

Using a two-crystal walkoff compensated scheme [89], along with crystals cut for type II interaction, Bosenberg et al. [103] demonstrated a 355-nm pumped BBO OPO tunable from 480 to 630 nm and from 810 to 1360 nm with linewidths varying from 2 to 8 cm^{-1} over the entire tuning range without narrowing optical elements. Conversion efficiencies of 12% have been achieved with this design.

These examples show that OPOs can be efficient, broadly tunable, and narrow-band sources of coherent radiation.

9.5.5 Short-Pulse OPOs

Nanosecond pumping appears to provide the simplest method to reach the threshold of efficient parametric oscillation [102, 113, 152, 154, 157]. Nanosecond

OPOs that presently cover the wavelength region from 400 nm to 3.5 μm have been reported [158]. For example, the highly efficient nsec OPO reported by Wang et al. [115] was pumped by a 30-Hz Q-switched frequency-tripled Nd:YAG laser. The OPO used a 10-mm BBO crystal to generate continuously tunable radiation from 415 to 2400 nm. Using a double-pass pump configuration to return the undepleted portion of the pump beam back into the OPO cavity, quantum efficiencies as high as 57% and average output powers up to 507 mW at the signal wavelength 490 nm have been demonstrated from this compact continuously tunable oscillator. The exceptionally broad tuning range, ease of operation, high conversion efficiency, and potentially high output powers attainable from this OPO device make it a very practical and competitive alternative to tunable dye laser sources.

Nanosecond OPO systems have been most successful where materials with relatively high damage threshold resistance can be used [159]. For low-damage-threshold materials it is common to move to even shorter pump pulselengths. Modelocked synchronous pumping of OPOs (with the materials of choice being BBO and KTP) provides high peak power to bring the process above threshold and high average power to maximize the OPO output. For the same pump pulse energy, the peak power of the pulse can be very much higher, and thus more efficient pumping of the OPO can be achieved. Pump pulselengths are typically 5 to 100 psec, allowing peak pump powers well above the level of nsec systems without causing crystal damage. As the pulselength of a single modelocked pump pulse is too short to allow oscillation buildup, synchronous pumping of the nonlinear medium is required, for example, the total time for signal or idler waves to make a cavity round trip is equal to the modelocked pump pulse separation. When the OPO and pump laser cavities are synchronized, the circulating signal and idler pulses interact with multiple pump pulses during successive cavity round trips, allowing adequate buildup and thus high OPO efficiency. Figure 15 shows an example of a temperature-tuned LBO SRO synchronously pumped at 523 nm (Nd:YLF) [121]. The OPO was tunable from 652 to 2650 nm with four-color operation over an extensive portion of the wavelength range. An oscillation threshold of ~2.5 GW/cm^2 and an internal conversion efficiency of 20% were achieved. Guyer and Lowenthal [160] demonstrated an efficient synchronously pumped two-stage OPO driven by a modelocked and Q-switched Nd:YAG laser at 1.064 μm. KTP was used in the first OPO stage to degenerately down-convert the 1.064-μm pulses to 2.128 μm with ~55% conversion efficiency (signal plus idler). This first OPO stage was nearly nonresonant, with high parametric gain and ~95% output coupling, yielding extremely stable operation. Normally, a degenerate OPO has a large frequency bandwidth. However, in KTP the phase-matching must be type II, and this provides a narrow bandwidth of approximately 5 cm^{-1} at degeneracy. This fact is extremely important for driving subsequent nonlinear stages. The signal and idler 2.128-μm power was then used to drive the second

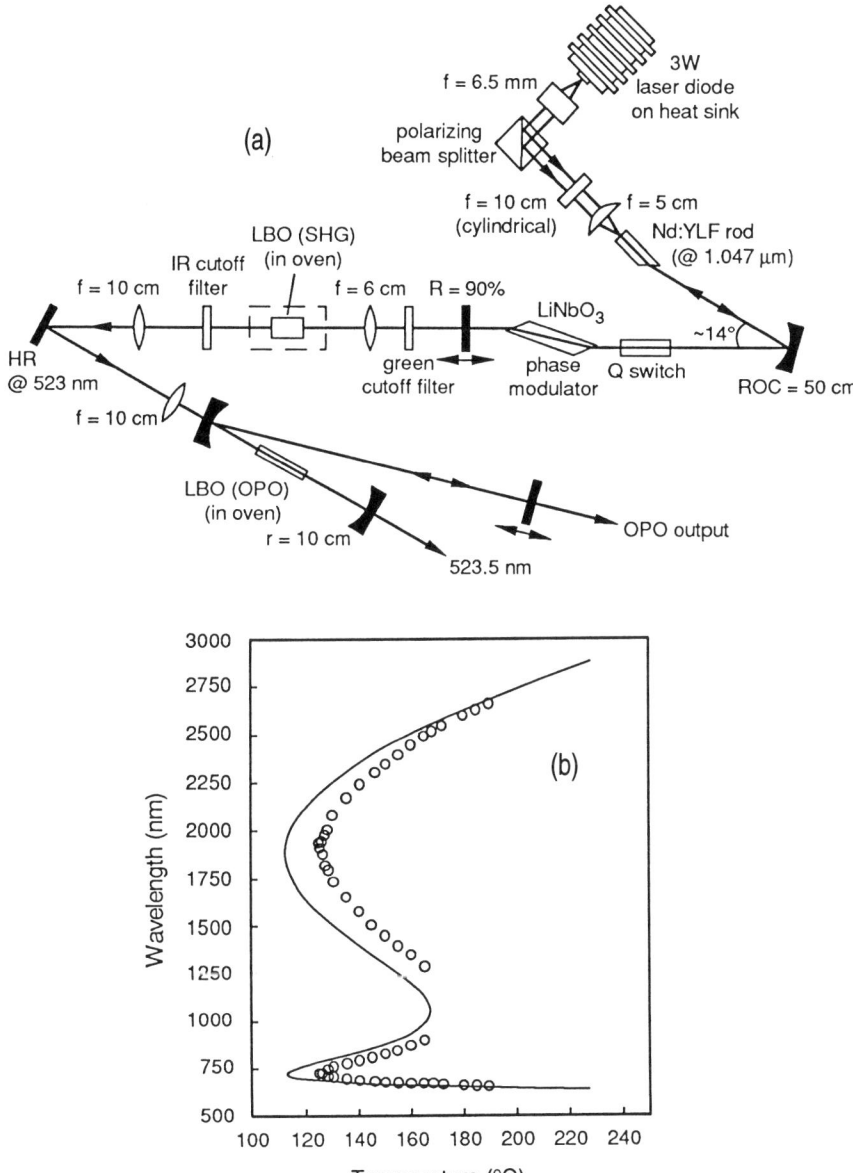

FIG. 15. (a) Setup of a synchronously pumped temperature-tuned LBO OPO [121]. Reproduced with permission from the Optical Society of America. (b) Dependence of the OPO wavelength as a function of the temperature for the LBO SRO with four-color operation over an extensive portion of the tuning range.

AgGaSe$_2$ mid-IR OPO stage. This stage achieved power conversion efficiencies of over 40% (2.128 μm converting to 3.8 and 4.8 μm) [105]. The temporal properties of a short turn-on time (2–3 nsec) and a low pulse-to-pulse temporal jitter (<500 psec) make OPOs particularly attractive devices for these multistage mixing experiments.

The synchronous pumping scheme has also been used for the generation of very short OPO output pulses required for studies of ultrafast processes in such diverse areas as physics, chemistry, electronics, and biology. Because of possible pulse broadening due to group velocity dispersion (GVD) in the nonlinear medium, the effective interaction length for femtosecond pulses is severely restricted to the order of several mm or less [161]. Thus, KTP is the prime candidate for such an application because of its relatively large effective nonlinear coefficient and small GVD. However, to reach the OPO threshold the KTP crystal must be synchronously pumped at an intracavity focus of the pump laser cavity. The intracavity fsec KTP OPO demonstrated by Edelstein *et al.* [162] provided 105-fsec short pulses at a 100-MHz repetition rate in the deep red at the mW level using a colliding-pulse modelocked (CPM) Rh6G dye laser as the pump source ($\tau = 170$ fsec). With one set of mirrors, the OPO output was tunable from 820 to 920 nm and from 1.92 to 2.54 μm. With three additional sets of mirrors, the entire range from 700 nm to 4.5 μm can be covered. Externally pumped by a self-modelocked 125-fsec Ti:Al$_2$O$_3$ laser, a high-power high-repetition-rate fsec KTP OPO yielded an ultrashort nearly transform-limited OPO output (57 fsec) at a power level of 115 mW, tunable in the infrared region where ultrafast dye or Ti:Al$_2$O$_3$ lasers are not available [163].

9.5.6 Continuous-Wave OPOs

For many applications such as high-resolution laser spectroscopy and optical frequency division applications, a continuously tunable source with single-mode CW output and narrow linewidth is desirable. Significant progress has been made to develop OPO devices with reduced threshold energies. Two of the most commonly used low-threshold schemes that finally resulted in CW OPO operation are DROs, as well as SROs, with double-passed pump beam. CW OPOs offer the potential for widely tunable output that preserves the coherence of the pump laser source.

A DRO resonates both the signal and idler waves. As the OPO threshold is proportional to the product of the signal and idler losses, this device has a lower pump threshold than an SRO. Doubly resonant CW OPOs have been operated at high efficiency and narrow linewidth [164, 165]. However, as DROs are over-constrained by the requirements of energy conservation, phase-matching, and simultaneous cavity resonances at both signal and idler waves, perturbations on the pump frequency or OPO cavity length can cause large power fluctuations and

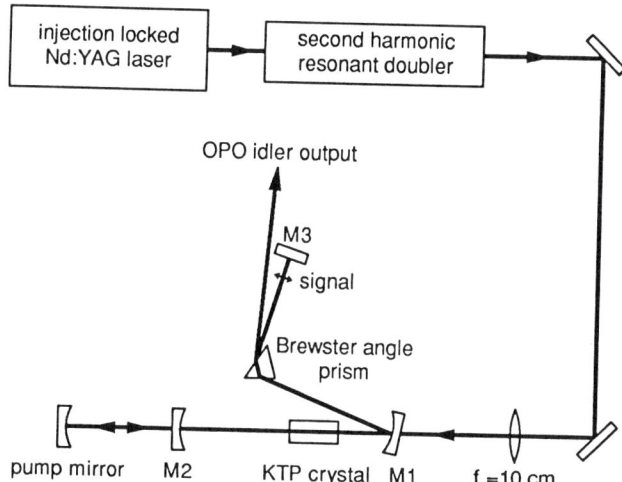

FIG. 16. Continuous-wave singly resonant KTP OPO [167]. Reproduced with permission from the Optical Society of America. Double-passing the pump wave reduces the threshold to a fourth of the calculated single-pass OPO threshold value.

mode hopping of the OPO output [166]. The monolithic ring cavity MgO:LiNbO$_3$ OPO developed by Nabors et al. [164] was operated near degeneracy and showed low cavity losses and good cavity length stability that are important for efficient and stable DRO operation. Pump laser stability was achieved by using the frequency-doubled output of a CW single-axial-mode diode laser-pumped monolithic Nd:YAG nonplanar ring laser. The threshold for CW OPO operation was 10 mW. With a total output power of ~8 mW, the DRO achieved 34% conversion efficiency from 532 nm pump light to signal and idler power. Wavelength tuning over 10 THz was achieved by varying the electric field applied to the nonlinear MgO:LiNbO$_3$ crystal. The OPO operated single frequency with a linewidth reproducing the 10-kHz linewidth of the pump laser source. At degeneracy, where the two output frequencies are at the subharmonic of the pump frequency, the OPO subharmonic output was found to phase lock to the pump phase to form an optical frequency divider [167]. The frequency tuning characteristics of the CW monolithic OPO have been studied theoretically in detail [142]. It appears that by tuning two variables simultaneously, such as the voltage and the pump frequency, the DRO can be continuously tuned over 10 GHz without axial mode jumps.

SROs resonate only the signal wave generated in the parametric process. Therefore, SRO designs show higher thresholds than DRO designs but result in wider and smoother OPO tuning ranges. A CW singly resonant KTP OPO (Fig. 16) demonstrated by Yang et al. [168] produced up to ~1 W of idler and 36 mW of signal output with 3.2 W of pump input from a frequency-doubled injection-locked Nd:YAG laser, representing a total OPO power conversion efficiency of ~35%. By double-passing the pump through the crystal, they achieved a threshold of 1.4 W, which corresponds to a fourth of the calculated single-pass OPO threshold [169]. This device is much less susceptible to cavity-length and pump-frequency fluctuations than a doubly resonant OPO.

A CW OPO using temperature-tuned noncritically phase-matched LBO and resonating all three waves was demonstrated by Colville et al. [170]. As the OPO threshold pump power scales as the third power of the pump wavelength (i.e., $P_p^{th} \sim \lambda^3$), pumping the OPO in the UV at 364 nm (Ar$^+$ laser) resulted in a low pump power threshold of 115 mW. Wavelength tunable from 502 to 494 nm and from 1.32 to 1.38 μm, the OPO provided a maximum of 103 mW total CW OPO output (signal and idler) for 1.1 W of incident pump power, corresponding to an external conversion efficiency of ~10%.

9.6 Raman Shifters

Raman shifters are based on the stimulated Raman scattering (SRS) process [172] and provide an attractive simple and efficient means of up- and down-shifting the frequency of intense laser sources by a multiple of a Raman-active molecular transition. In this nonlinear interaction, an incident pump wave at frequency v_p is converted to a scattered wave at a frequency $v_S = v_p - mv_{mol}$ (Stokes wave) or $v_{AS} = v_p + mv_{mol}$ (anti-Stokes wave), where m is a positive integer, and the difference in photon energy is being taken up or supplied by the nonlinear medium, which undergoes a transition with frequency v_{mol} between two internal molecular energy states. Most commonly used forms of Raman shifters involve interactions with molecular vibrations in high-pressure gases or in liquids (v_{mol} = 600–4155 cm^{-1}) or molecular rotations in high-pressure gases ($v_{mol} \approx$ 10–450 cm^{-1}) (Table III). The generated scattered wave has exponential growth, with the gain being proportional to the propagating distance and to the pump laser intensity. Phase-matching is generally not required for this stimulated processes, since the phase of the molecular excitation adjusts itself for maximum gain automatically. Figure 17 shows a typical Raman shifter setup. The output of a pulsed tunable dye laser is focused into a capillary waveguide cell (length ~50 cm) containing compressed hydrogen gas at ~40 atm [173], allowing the observation of several Stokes and anti-Stokes orders for a given incident pump wavelength. The waveguide extends the focal region, thereby increasing the gain length. Due to the large Raman shift of 4155 cm^{-1} in H$_2$, the use of several dyes in combination with a dye laser frequency doubler allowed the generation of various Stokes and anti-Stokes bands covering the entire wavelength range from 185 to 880 nm [174] and from 0.7 to 7 μm [175] without gaps. Mannik and Brown [173] observed an

TABLE III. Typical Raman Media

Material	Δv (cm^{-1})		Gain $g \times 10^3$ (cm/MW)	
Liquids				
Benzene	992		3	
Water	3290		0.14	
N$_2$	2326		17	
O$_2$	1555		16	
Gases				
Methane	2916		0.66	(10 atm, 500 nm)
Hydrogen	4155	(vibrational)	1.5	(above 10 atm)
	450	(rotational)	0.5	(above 0.5 atm)
Deuterium	2991	(vibrational)	1.1	(above 10 atm)
N$_2$	2326	0.071 (10 atm, 500 nm)		
O$_2$	1555	0.016 (10 atm, 500 nm)		

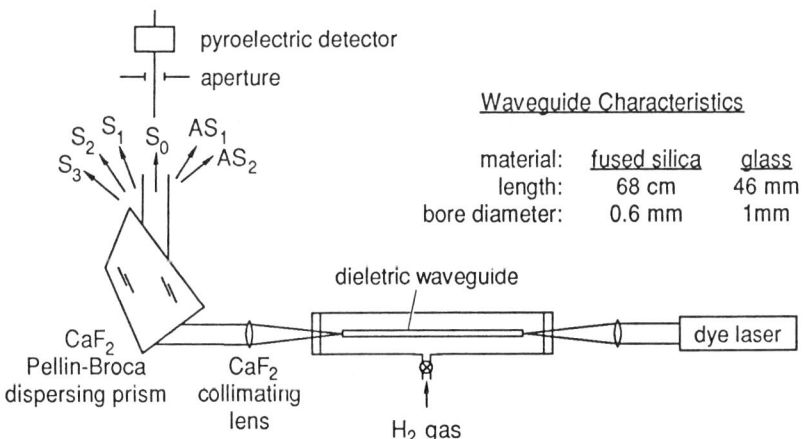

FIG. 17. Raman waveguide shifter using a high-pressure H$_2$ cell. The output of a frequency-doubled Nd:YAG laser pumps a dye oscillator and amplifier configuration whose output is focused into the waveguide containing 30-atm H$_2$ [173]. Reproduced with permission from Elsevier Science B.V.

infrared energy of 2.5 mJ generated in the third-order Stokes wave S_3 at 2 μm by guided-wave Raman shifting a 568-nm dye laser (62 mJ) with an energy conversion efficiency of ~4%. Using a Raman oscillator/waveguide-amplifier scheme to control the divergence of the Stokes beams, Hanna and Pacheco [176] demonstrated ~9% energy conversion efficiency to S_3 (1.58 μm) from 20 mJ input energy (532 nm) with H_2 as the Raman medium.

CO_2 lasers (9–11 μm), CO lasers (5 μm), or HF/DF lasers (2.5–4 μm) can also be used as intense pump sources for Raman-active media to cover large portions of the infrared wavelength region. Pumping liquid N_2 or O_2 as the Raman medium with the pressure-broadened tunable lines of HF and DF lasers allows the generation of tunable infrared coherent radiation from 5 to 10 μm for infrared spectroscopy [177, 178].

9.7 Up-Conversion Lasers

Up-conversion lasers [179] represent a class of optically pumped lasers that use nonlinear excitation processes to oscillate at frequencies higher than those used for pumping. Thus, up-conversion pumping is an alternative to using parametric nonlinear optical processes for converting infrared laser radiation to coherent output at shorter wavelengths. A number of up-conversion lasers has been demonstrated using trivalent rear-earth ions doped into crystals and glasses at sites lacking inversion symmetry as the laser medium. The interest in up-conversion lasers is motivated, to a substantial degree, by the availability of high-power red and near-infrared III–V semiconductor diode lasers that operate in spectral regions required for pumping compact solid-state up-conversion systems with wavelengths from the UV to the red for a wide range of applications.

By making use of a long-lived intermediate energy level positioned between the ground state and the upper laser level, nonlinear excitation can proceed in steps, with the intermediate state acting as a reservoir for the pump energy. In suitable laser materials, these metastable intermediate levels can be populated effectively so that they can hold as much population as, or more than, the ground state. Particularly attractive level structures for up-conversion lasing are found in Pr^{3+}, Nd^{3+}, Ho^{3+}, Er^{3+}, and Tm^{3+}. Three mechanisms can lead to efficient up-conversion excitation (Fig. 18):

a. A sequential two-step photon absorption process. In this case, up-conversion excitation is achieved with a first photon populating an intermediate metastable state and a second photon, which in the most general case would be of different energy, pumping the upper laser level via excited-state absorption [180, 181].

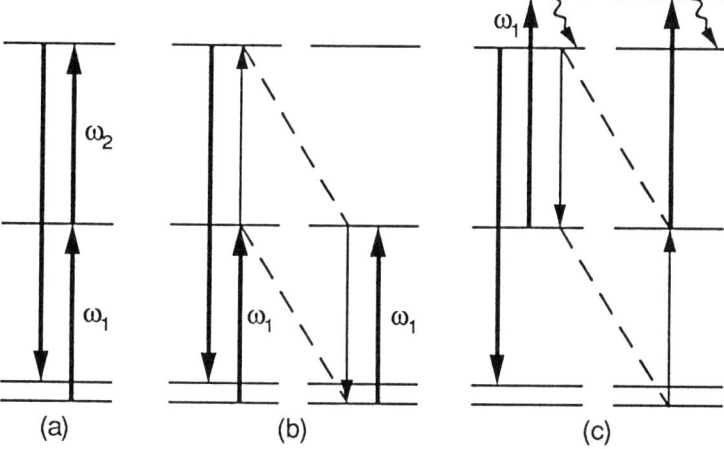

FIG. 18. Schematic of three up-conversion processes: (a) sequential two-step photon absorption, (b) the cross-relaxation process, and (c) the avalanche absorption process [179]. Reproduced with permission from the Optical Society of America.

b. *A cross-relaxation process.* The energy transfer between excited ions can result in efficient up-conversion when a sufficiently large number of ions has been excited to an intermediate state. Two ions in close physical proximity are coupled by a nonradiative process (electrical dipole–dipole interaction) in which one ion returns to the ground state while the other ion is promoted to the upper level. Phonons can participate in these energy transfer processes to make up for any mismatch between the energy of the donor and acceptor ions. Energy transfer processes can permit up-conversion excitation using a single pump laser that populates a long-lived intermediate state [182–184].

c. *An avalanche absorption process.* This process involves excited-state absorption of the pump light as well as interionic cross-relaxation and can provide efficient up-conversion pumping with a single laser. It is characteristic of this up-conversion process that the pump radiation is predominantly absorbed by excited-state absorption from an intermediate state that is populated via cross-relaxation by energy transfer [185].

The majority of the up-conversion lasers are operated at cryogenic temperatures, which is a practical disadvantage. At higher temperatures, spectral broadening leads to a reduction in the peak emission and absorption cross-sections, which in turn requires a higher inversion population to overcome laser losses and, at the same time, decreases pump absorption efficiency. The most successful approach to room-temperature up-conversion laser operation has been the use of fluoride glass optical fibers as laser media. Fibers offer important advantages for operating CW up-conversion lasers at ambient temperature on the basis of tight optical confinement and long interaction lengths (typically on the order of 1 m) that result in high excitation density and efficient pump absorption. In addition, depletion of the population of the ground-state manifold can lead to significant reduction of self-absorption losses. The use of CW pump sources is particularly effective for populating the metastable intermediate states due to the integrating effect associated with the long lifetimes of these states. In most cases, the rear-earth ion concentration is low (several percent), and sequential two-step photon absorption rather than energy transfer processes are used for up-conversion excitation. The first CW up-conversion laser operating at room temperature was demonstrated in 1990 by Allain et al. [186] by pumping an Ho_3^+-doped fluorozirconate fiber with the 647-nm line of a Kr^+ laser using doubly resonant sequential two-step photon absorption. Up to 10 mW of green 550-nm output was obtained from 300 mW of red pump power. By using an intracavity prism, the green output could be tuned over ~10 nm. A Pr^{3+}-doped 1.2-m fluorozirconate fiber laser pumped by two $Ti:Al_2O_3$ lasers at 1.01 and 0.835 μm was used to demonstrate a room-temperature CW up-conversion laser operation at blue, green, orange, and red wavelengths [181]. For 1 W at 1.01 μm, the slope efficiency of generated red light (605 nm) versus pump light (835 nm) was 7%, with a maximum red power output of 30 mW.

References

1. Shen, Y. R. (1984). *The Principles of Nonlinear Optics*, Wiley, New York.
2. Zernike, F. (1975). In *Methods of Experimental Physics*, Academic Press, Boston.
3. Franken, P. A., Hill, A. E., Peters, C. W., and Weinreich, G. (1961). *Phys. Rev. Lett.* **7**, 118.
4. Scifres, D. R., Burnham, R. D., and Streifer, W. (1982). *Appl. Phys. Lett.* **41**, 118.
5. Kane, T. J., and Byer, R. L. (1985). *Opt. Lett.* **10**, 65.
6. Walpole, J. N., Kintzer, E. S., Chin, S. R., Wang, C. A., and Missagia, L. J. (1992). *Appl. Phys. Lett.* **61**, 710.
7. Goldberg, L., Mehuys, D., Surette, M. R., and Hall, D. C. (1993). *IEEE J. Quantum Electronics* **Q-29**, 2028.
8. Pelouch, W. S., Powers, P. E., and Tang, C. L. (1992). *Opt. Lett.* **17**, 1070.
9. Koechner, W. (1992). *Solid-State Laser Engineering*, Springer-Verlag, Berlin.
10. Kato, K. (1986). *IEEE J. Quantum Electronics* **QE-22**, 1013. A new nonlinear material, CBO, that might be able to push the short-wavelength SHG limit to 170 nm is under investigation, as described in Wu, Y., Sasaki, T., Nakai, S., Tokotani, A., Tang, H., and Cheng, C. (1993). *Appl. Phys. Lett.* **62**, 2614.
11. Boyd, G. D., and Kleinman, D. A. (1968). *J. Appl. Phys.* **39**, 3597.

12. For short pulses, however, a large acceptance bandwidth ($\Delta v \approx 1/(\Delta n L)$; Δn = birefringence; L = crystal length) is required in order to sustain the short pulselength in the conversion process. In this case, the bandwidth can be increased, however, by using a dispersively compensated phase-matching scheme. See Cheville, R. A., Reiten, M. T., and Halas, N. J. (1992). *Opt. Lett.* **17**, 1343; Hofmann, Th., Mossavi, K., Tittel, F. K., and Szabo, G. (1992). *Opt. Lett.* **17**, 1691, and the references therein.
13. Yariv, A. (1990). *Quantum Electronics*, New York, Wiley.
14. Baer, T. (1986). *J. Opt. Soc. Am. B* **3**, 1175.
15. Dixon, G. J. (1989). Paper FP4 in *Dig. OSA Annual Meeting*, Washington, DC.
16. Band, Y., Ackerhalt, J. R., Chin, T., Krasinski, J. S., and Papanestor, P. (1989). In *OSA Proc. Tunable Solid-State Users*, p. 101, Optical Society of America, Washington, DC.
17. Anthony, D. W., Sipes, D. L., Pier, T. J., and Ressl, M. R. (1992). *IEEE J. Quantum Electronics* **QE-28**, 1148.
18. Nagai, H., Kume, M., Otha, I., Shimizu, H., and Kazumura, M. (1992). *IEEE J. Quantum Electronics* **QE-28**, 1164.
19. Grubb, S. G., Cannon, R. S., and Dixon, G. J. (1991). Paper CFJ6 in *Tech. Dig. CLEO '91*, Anaheim, CA, Optical Society of America, Washington, DC.
20. Hemmati, H. (1992). *IEEE J. Quantum Electronics* **QE-28**, 1169.
21. Fan, T. Y., and Byer, R. L. (1987). *IEEE J. Quantum Electronics* **QE-23**, 605.
22. Ashkin, A., Boyd, G. D., and Dziedzic, J. M. (1966). *IEEE J. Quantum Electronics* **QE-2**, 109.
23. Kozlovsky, W. J., Nabors, C. D., and Byer, R. L. (1988). *IEEE J. Quantum Electronics* **QE-24**, 913.
24. Dixon, G. J., Tanner, C. E., and Wieman, C. E. (1989). *Opt. Lett.* **14**, 731.
25. Kozlovsky, W. J., Lenth, W., Latta, E. E., Moser, A., and Bona, G. L. (1990). *Appl. Phys. Lett.* **56**, 2291.
26. Drever, R. W. P., Hall, J. L., Kowalski, F. V., Hough, J., Ford, G. M., Munley, A. J., and Ward, H. (1983). *Appl. Phys. B* **31**, 97.
27. Zimmermann, C., Hänsch, T. W., Byer, R., O'Brien, S., and Welch, D. (1992). *Appl. Phys. Lett.* **61**, 2741.
28. Yang, S. T., Pohalski, C. C., Gustafson, E. K., Byer, R. L., Feigelson, R. S., Raymakers, R. J., and Route, R. K. (1991). *Opt. Lett.* **16**, 1493.
29. In type II phase-matching in a negative (positive) birefringent crystal, two incoming waves at frequency ω, one polarized as an extraordinary ray and the other as an ordinary ray, generate a wave at frequency 2ω, polarized as an ordinary (extraordinary) ray.
30. Ou, Z. Y., Pereira, S. F., Polzik, E. S., and Kimble, H. J. (1992). *Opt. Lett.* **17**, 640.
31. Nabors, C. D., Kozolovsky, W. J., and Byer, R. L. (1988). In *Proc. SPIE*, Vol. 898, *Miniature Optics and Lasers*.
32. Schiller, S., and Byer, R. L. (1993). *J. Opt. Soc. Am. B* **10**, 1696.

33. Persaud, M. A., Tolchard, J. M., and Ferguson, A. I. (1990). *IEEE J. Quantum Electronics* **QE-26**, 1253.
34. Malcolm, G. P. A., Ebrahimzadeh, M., and Ferguson, A. I. (1992). *IEEE J. Quantum Electron* **QE-28**, 1172.
35. Armstrong, J. A., Bloembergen, N., Ducuing, J., and Pershan, P. S. (1962). *Phys. Rev.* **127**, 1918; Somekh, S., and Yariv, A. (1972). *Opt. Comm.* **6**, 301.
36. Fejer, M. M. (1992). In *Guided Wave Nonlinear Optics*, D. B. Ostrowsky and R. Reinsch (eds.), p. 133, Kluwer, Dordrecht, The Netherlands.
37. Fejer, M. M., Magel, G. A., Jundt, D. H., and Byer, R. L. (1992). *IEEE J. Quantum Electronics* **QE-28**, 2631.
38. Piltch, M. S., Cantrell, C. D., and Sze, R. C. (1976). *J. Appl. Phys.* **47**, 3514.
39. Thompson, D. E., McMullen, J. D., and Anderson, D. B. (1976). *Appl. Phys. Lett.* **29**, 113.
40. Okada, M., Takizawa, K., and Ieiri, S. (1976). *Opt. Comm.* **18**, 331.
41. Mao, H., Fu, F., Wu, B., and Chen, C. (1992). *Appl. Phys. Lett.* **61**, 1148.
42. A domain is a region of one orientation direction of the spontaneous electric polarization P_S and is linked to the sign of the nonlinear coefficient.
43. Fejer, M. M., Nightingale, J. L., Magel, G. A., and Byer, R. L. (1984). *Rev. Opt. Instrum.* **55**, 1791.
44. Xue, Y. H., Ming, N. B., Zhu, J. S., and Feng, D. (1983). *Chinese Phys.* **4**, 554; Feisst, A., and Koidl, P. (1985). *Appl. Phys. Lett.* **47**, 1125.
45. Magel, G. A., Fejer, M. M., and Byer, R. L. (1990). *Appl. Phys. Lett.* **56**, 108.
46. Jundt, D. H., Magel, G. A., Fejer, M. M., and Byer, R. L. (1991). *Appl. Phys. Lett.* **59**, 2657.
47. Khanarian, G., Norwood, R. A., Haas, D., Feuer, B., and Karim, D. (1990). *Appl. Phys. Lett.* **57**, 977.
48. Levine, B. F., Bethea, C. G., and Logan, R. A. (1975). *Appl. Phys. Lett.* **26**, 375.
49. Lim, E. J., Fejer, M. M., Byer, R. L., and Kozlovsky, W. J. (1989). *Electron. Lett.* **25**, 731.
50. Webjorn, J., Laurell, F., and Arvidsson, G. (1989). *IEEE Photon. Technol. Lett.* **1**, 316.
51. Matsumoto, S., Lim, E. J., Hertz, H. M., and Fejer, M. M. (1991). *Electron. Lett.* **27**, 2040.
52. Mizuuchi, K., and Yamamoto, K. (1992). *Appl. Phys. Lett.* **60**, 1283.
53. Makio, S., Nitanda, F., Ito, K., and Sato, M. (1992). *Appl. Phys. Lett.* **61**, 3077.
54. van der Poel, C. J., Bierlein, J. D., Brown, J. B., and Colak, S. (1990). *Appl. Phys. Lett.* **57**, 2074.
55. Mizuuchi, K., Yamamoto, K., and Taniuchi, T. (1991). *Appl. Phys. Lett.* **58**, 2732.
56. Stegeman, G. I., and Seaton, C. T. (1985). *J. Appl. Phys.* **58**, R57.
57. Yamamoto, K., Mizuuchi, K., and Taniuchi, T. (1991). *Opt. Lett.* **16**, 1156; Yamamoto, K., and Mizuuchi, K. (1992). *IEEE Photon. Technol. Lett.* **4**, 435;

Yamamoto, K., Mizuuchi, K., Kitaoka, Y., and Kato, M. (1993). *Appl. Phys. Lett.* **62**, 2599.
58. Yamada, M., Nada, N., Saitoh, M., and Watanabe, K. (1993). *Appl. Phys. Lett.* **62**, 435.
59. Lim, E. J., Hertz, H. M., Bortz, M. L., and Fejer, M. M. (1991). *Appl. Phys. Lett.* **59**, 2207.
60. Laurell, F., Brown, J. B., and Bierlein, J. D. (1992). *Appl. Phys. Lett.* **60**, 1064.
61. Kato, K. (1986). *IEEE J. Quantum Electronics* **QE-22**, 1013; Kato, K. (1986). *IEEE J. Quantum Electronics* **QE-17**, 1566.
62. Hemmati, H., Bergquist, J. C., and Itano, M. W. (1983). *Opt. Lett.* **8**, 73.
63. Watanabe, M., Hayasaka, K., Imajo, H., and Urabe, S. (1991). *Appl. Phys. Lett. B* **53**, 11; Watanabe, M., Hayasaka, K., Imajo, H., and Urabe, S. (1992). *Opt. Lett.* **17**, 46.
64. Bosenberg, W. R., Pelouch, W. S., and Tang, C. L. (1989). *Appl. Phys, Lett.* **55**, 1952.
65. Watanabe, M., Hayasaka, K., Imajo, H., and Urabe, S. (1993). *Opt. Comm.* **97**, 225.
66. Mückenheim, W., Lokai, P., Burghardt, B., and Basting, D. (1988). *Appl. Phys. B* **45**, 261.
67. Heitmann, U., Kötteritzsch, M., Heitz, S., and Hese, A. (1992). *Appl. Phys. B* **55**, 419.
68. Schomburg, H., Döbele, H. F., and Rückle, B. (1983). *Appl. Phys. B* **30**, 131.
69. Hilbig, R., and Wallenstein, R. (1982). *Appl. Opt.* **21**, 913.
70. Kean, P. N., Standley, R. W., and Dixon, G. J. (1993). *Appl. Phys. Lett.* **63**, 302.
71. Risk, W. P., and Kozlovsky, W. J. (1992). *Opt. Lett.* **17**, 707.
72. Nebel, A., and Beigang, R. (1992). *Opt. Comm.* **94**, 369.
73. Pine, A. S. (1974). *J. Opt. Soc. Am. B* **64**, 1683.
74. Bawendi, M. G., Rehfuss, B. D., and Oka, T. (1990). *J. Chem. Phys.* **93**, 6200; Xu, L. W., Gabrys, C., and Oka, T. (1990). *J. Chem. Phys.* **93**, 6210.
75. Canarelli, P., Benko, Z., Curl, R. F., and Tittel, F. K. (1992). *J. Opt. Soc. Am. B* **9**, 197; Hielscher, A. H., Miller, C. E., Bayard, D. C., Simon, U., Smolka, K. P., Curl, R. F., and Tittel, F. K. (1992). *J. Opt. Soc. Am. B* **9**, 1962.
76. Miller, C. E., Simon, U., Eckhoff, W. C., Tittel, F. K., and Curl, R. F. (1993). *AIP Conf. Proc., ELICOLS '93*, Hot Springs, VA.
77. Wieman, C. E., and Hollberg, L. (1991). *Rev. Sci. Instrum.* **62**, 1.
78. Simon, U., Miller, C. E., Bradley, C. C., Hulet, R. G., Curl, R. F., and Tittel, F. K. (1993). *Opt. Lett.* **18**, 1062.
79. Simon, U., Tittel, F. K., and Goldberg, L. (1993). *Opt. Lett.* **18**, 1931.
80. Simon, U., Waltman, S., Loa, I., Tittel, F. K., and Hollberg, L. (1995). *J. Opt. Soc. Am. B* **12**, 323.
81. Effenberger, F. J., and Dixon, G. J. (1992). *Advanced Solid-State Lasers*, Topical Meeting, New Orleans, LA.
82. Ludwig, C., Frey, W., Woerner, M., and Elsaesser, T. (1993). *Opt. Comm.* **102**, 447.
83. Hamm, P., Lauterwasser, C., and Zinth, W. (1993). *Opt. Lett.* **18**, 1943.

84. Shen, Y. R., ed. (1977). *Nonlinear Infrared Generation*, Topics in Applied Physics, Volume 16, Springer, Heidelberg.
85. Evenson, K. M., Inguscio, M., and Jennings, D. A. (1985). *J. Appl. Phys.* **57**, 956.
86. Evenson, K. M., Jennings, D. A., Leopold, K. R., and Zink, L. R. (1985). *Laser Spectroscopy*, Vol. VII, T. W. Hänsch and Y. R. Shen (eds.), Springer Series on Optical Science, Volume 49, p. 366, Springer-Verlag, Heidelberg; Evenson, K. M., Jennings, D. A., and Vanek, M. D. (1988). In *Frontiers of Laser Spectroscopy of Gases*, A. C. P. Alves (eds.), p. 43, Kluwer, Dordrecht, The Netherlands.
87. Inguscio, M. (1988). *Phys. Scripta* **37**, 699; Inguscio, M., Zink, P. R., Evenson, K. M., and Jennings, D. A. (1987). *Opt. Lett.* **12**, 867.
88. Nolt, I. G., DiLonardo, G., Evenson, K. M., Hinz, A., Jennings, D. A., Leopold, K. R., Vanek, M. D., Radistitz, J. V., and Zink, L. R. (1987). *J. Mol. Spectrosc.* **125**.
89. Brown, J. M., Zink, L. R., Jennings, D. A., Evenson, K. M., Hinz, A., and Nolt, I. G. (1986). *Appl. Phys.* **307**, 210; Leopold, K. R., Zink, L. R., Evenson, K. M., and Jennings, D. A. (1986). *J. Chem. Phys.* **84**, 1935; Jennings, D. A., Evenson, K. M., Zink, L. R., Demuynck, C., Destombes, J. L., Lemoine, B., and Johns, J. W. (1986). *J. Mol. Spectrosc.* **122**, 477.
90. Jennings, D. A., Evenson, K. M., Vanek, M. D., Nolt, I. G., Radistitz. J. V., and Chance, K. V. C. (1987). *Geophys. Res. Lett.* **14**, 722.
91. Craxton, R. S. (1980). *Opt. Comm.* **34**, 474.
92. Craxton, R. S. (1981). *IEEE J. Quantum Electronics* **QE-17**, 1771.
93. Nebel, A., and Beigang, R. (1991). *Opt. Lett.* **16**, 1729.
94. Reintjes, J. F. (1990). "Nonlinear Optical Processes," in *Encyclopedia of Modern Physics*, R. A. Meyers (ed.), Academic Press, New York.
95. Demtröder, W. (1991). *Laserspektroskopie*, Springer-Verlag, Berlin.
96. Hilbig, R., and Wallenstein, R. (1982). *Appl. Opt.* **21**, 913; Hilbig, R., and Wallenstein, R. (1982). *Appl. Opt. B* **28**, 202.
97. Harris, S. E. (1969). *Proc. IEEE* **57**, 2096.
98. Byer, R. L. (1975). "Optical Parametric Oscillators," in *Treatise on Quantum Electronics*, Vol. 1: *Nonlinear Optics*, Parts A and B, H. Rabin and C. L. Tang (eds.), Academic Press, New York.
99. Tang, C. L. (1975). "Spontaneous and Stimulated Parametric Processes," *Treatise on Quantum Electronics*, Vol. 1: *Nonlinear Optics*, Parts A and B, H. Rabin and C. L. Tang (eds.), Academic Press, New York.
100. Cheng, L. K., Rosker, M. J., and Tang, C. L. (1987). In *Tunable Lasers*, L. F. Mollennauer and J. C. White (eds.), Springer-Verlag, Berlin.
101. Yariv, A. (1990). Quantum Electronics, 3rd edition, Wiley, New York.
102. Ebrahimzadeh, M. E., Henderson, A. J., and Dunn, M. H. (1990). *IEEE J. Quantum Electronics* **QE-26**, 1241.
103. Bosenberg, W. R., Cheng, L. K., and Tang, C. L. (1989). *Appl. Phys. Lett.* **54**, 13.
104. Barnes, N., and Murray, K. (1990). Unpublished lecture.

105. Guyer, D. R., Hamilton, C., Braun, F., Lowenthal, D., and Ewing, J. (1991). In *Proceedings of the Conference on Lasers and Electro-Optics*, Baltimore, MD, Technical Digest Series, Optical Society of America, Washington, DC.
106. Wang, C. C., and Racette, G. W. (1965). *Appl. Phys. Lett.* **6**, 169.
107. Giordamaine, J. A., and Miller, R. C. (1965). *Phys. Rev. Lett* **14**, 973.
108. For a list of references, see Dmitriev, V. G., Gurzadyan, G. G., and Nikogosyan, D. N. (1911). In *Handbook of Nonlinear Optical Crystals*, Vol. 64, pp. 181–191, Springer, New York.
109. Byer, R. L. (1988). *Science* **239**, 742.
110. Miyabke, C. I., Braun, F., and Guyer, D. R. (1992). In *Proc. Advanced Solid-State Laser Conference*, Optical Society of America, Washington, DC.
111. Pinard, J., and Young, J. F. (1972). *Opt. Comm.* **4**, 425.
112. Bordui, P. F., and Fejer, M. M. (1993). *Annu. Rev. Mater. Sci.* **23**, 321.
113. Fan, Y. X., Eckardt, R. C., Byer, R. L., Nolting, J., and Wallenstein, R. (1988). *Appl. Phys. Lett.* **53**, 2014; Fan, Y. X., Eckardt, R. C., Byer, R. L., Chen, C., and Jiang, A. D. (1989). *IEEE J. Quantum Electronics* **QE-25**, 1196.
114. Cheng, L. K., Bosenberg, W. R., and Tang, C. L. (1989). *Appl. Phys. Lett.* **54**, 13.
115. Wang, Y., Xu, Z., Deng, D., Zheng, W., Liu, X., Wu, B., and Chen, C. (1991). *Appl. Phys. Lett.* **58**, 1461.
116. Fix, A., Schröder, T., Wallenstein, R., Haub, J. G., Johnson, M. J., and Orr, B. J. (1993). *J. Opt. Soc. Am. B* **10**, 1744.
117. Withers, D. E., Robertson, G., Henderson, A. J., Tang, Y., Cui, Y., Sibett, W., Sinclair, B. D., and Dunn, M. H. (1993). *J. Opt. Soc. Am. B* **10**, 1737.
118. Komine, H. (1993). *J. Opt. Soc. Am. B* **10**, 1751.
119. Zhang, J. Y., Huang, J. Y., Shen, Y. R., and Chen, C. (1993). *J. Opt. Soc. Am. B* **10**, 1758.
120. Robertson, G., Henderson, A., and Dunn, M. H. (1992). *Appl. Phys. Lett.* **60**, 271.
121. Ebrahimzadeh, M. E., Hall, G. J., and Ferguson, A. I. (1992). *Opt. Lett.* **17**, 652; Ebrahimzadeh, M. E., Hall, G. J., and Ferguson, A. I. (1992). *Appl. Phys. Lett.* **60**, 1421.
122. Vanherzeele, H. (1990). *Appl. Opt.* **29**, 2246.
123. Powers, P. E., Ellington, R. J., Pelouch, W. S., and Tang, C. L. (1993). *J. Opt. Soc. Am. B* **10**, 2162.
124. Hall, G. J., Ebrahimzadeh, M., Robertson, A., Malcolm, G. P. A., and Ferguson, A. I. (1993). *J. Opt. Soc. Am. B* **10**, 2168.
125. McCarthy, M. J., and Hanna, D. C. (1993). *J. Opt. Soc. Am. B* **10**, 2180.
126. Lotshaw, W. T., Unternahrer, J. R., Kukla, M. J., Miyake, C. I., and Brown, F. D. (1993). *J. Opt. Soc. Am. B* **10**, 2191.
127. Nebel, A., Fallnich, C., Beigang, R., and Wallenstein, R. (1993). *J. Opt. Soc. Am. B* **10**, 2195.
128. Chung, J., and Siegman, A. E. (1993). *J. Opt. Soc. Am. B* **10**, 2201.

129. Apesi, A., Reali, R. C., Kubecek, V., Kumazaki, S., Takagi, Y., and Yoshihara, K. (1993). *J. Opt. Soc. Am. B* **10**, 2211.
130. Gräser, Ch., Wang, D., Beigang, R., and Wallenstein, R. (1993). *J. Opt. Soc. Am. B* **10**, 2218.
131. Vodopyanov, K. L. (1993). *J. Opt. Soc. Am. B* **10**, 1723.
132. Budni, P. A., Schunemann, P. G., Knights, M. G., Pollak, T. M., and Chicklis, E. P. (1992). In *Digest of Advanced Solid-State Lasers*, Optical Society of America, Washington, DC.
133. Fan, Y. X., Eckardt, R. C., Byer, R. L., Route, R. K., and Feigelson, R. S. (1984). *Appl. Phys. Lett.* **45**, 313.
134. Bakker, H. J., Kennis, J. T. M., Kop, H. J., and Lagendijk, A. (1991). *Opt. Comm.* **86**, 58.
135. Krause, H.-J., and Daum, W. (1993). *Appl. Phys. B* **56**, 8.
136. Eckardt, R. C., Fan, Y. X., Byer, R. L., Marquardt, C. L., Storm, M. E., and Esterowitz, L. (1986). *Appl. Phys. Lett.* **49**, 608.
137. Catella, G. C., Shiozawa, L. R., Hietanen, J. R., Eckardt, R. C., Route, R. K., Feigelson, R. S., Cooper, D. G., and Marquardt, C. L. (1993). *Appl. Opt.* **32**, 3948.
138. Budni, P. A., Knights, M. G., Chicklis, E. P., and Schepler, K. L. (1993). *Opt. Lett.* **18**, 1068.
139. Gordon, L. A., Woods, G. L., Eckardt, R. C., Route, R. K., Feigelson, R. S., Fejer, M. M., and Byer, R. L. (1993). *Electron. Lett.* **29**, 1942.
140. Gordon, L. A., Eckardt, R. C., and Byer, R. L. (1994). In *Proc. OE/LASE'94*, SPIE, Los Angeles.
141. Zyss, J. (1994). *Molecular Nonlinear Optics: Materials, Devices, and Physics*, Academic Press, Boston.
142. Eckardt, R. C., Nabors, C. D., Kozlovsky, W. J., and Byer, R. L. (1991). *J. Opt. Soc. Am. B* **8**, 646.
143. Parent, A., and Lavigne, P. (1989). *Opt. Lett.* **14**, 399.
144. Kane, T. J., and Byer, R. L. (1985). *Opt. Lett.* **10**, 65.
145. Kane, T. J., Nilsson, A. C., and Byer, R. L. (1987). *Opt. Lett.* **12**, 175.
146. Day, T., Gustafson, E. K., and Byer, R. L. (1991). *Opt. Lett.* **15**, 221.
147. Kane, T. J., Kozlovsky, W. J., and Byer, R. L. (1986). *Opt. Lett.* **11**, 216.
148. Nabors, C. D., Farinas, A. D., Day, T., Yang, S. T., Gustafson, E. K., and Byer, R. L. (1989). *Opt. Lett.* **14**, 1189.
149. Komine, H. (1988). *Opt. Lett.* **13**, 643.
150. Ebrahimzadeh, M. E., Robertson, G., and Dunn, M. H. (1991). *Opt. Lett.* **16**, 767.
151. Brosnan, S. J., and Byer, R. L. (1979). *IEEE J. Quantum Electronics* **QE-15**, 415.
152. Bosenberg, W. R., Pelouch, W. S., and Tang, C. L. (1989). *Appl. Phys. Lett.* **55**, 1952.
153. Bosenberg, W. R., and Guyer, D. R. (1992). *Appl. Phys. Lett.* **61**, 387.
154. Haub, J. G., Johnson, M. J., Orr, B. J., and Wallenstein, R. (1991). *Appl. Phys. Lett.* **58**, 1718.

155. Bosenberg, W. R., and Tang, C. L. (1990). *Appl. Phys. Lett.* **56**, 1819.
156. Johnson, M. J., Haub, J. G., Barth, H.-D., and Orr, B. J. (1993). *Opt. Lett.* **18**, 441.
157. Marshall, L. R., Kansinski, J., Hays, A. D., and Burnham, R. (1991). *Opt. Lett.* **16**, 681.
158. Tang, C. L., Bosenberg, W. R., Ukachi, T., Lane, R. J., and Cheng, L. K. (1992). *Proc. IEEE* **80**, 365.
159. Thresholds for nanosecond OPOs are higher than for CW OPO operation, which can result in crystal surface damage before the threshold intensity is reached (Section 9.5.6).
160. Guyer, D. R., and Lowenthal, D. D. (1990). *Proc. SPIE 1220, Nonlinear Optics* **41**.
161. Simultaneously, for short pulses a large acceptance bandwidth ($\Delta\nu \approx 1/(\Delta nL)$; Δn = birefringence; L = crystal length) is required in order to sustain the short pulselength in the conversion process.
162. Edelstein, D. C., Wachman, E. S., and Tang, C. L. (1989). *Appl. Phys. Lett.* **54**, 1728; Wachman, E. S., Edelstein, D. C., and Tang, C. L. (1990). *Opt. Lett.* **15**, 136.
163. Pelouch, W. S., Powers, P. E., and Tang, C. L. (1992). *Opt. Lett.* **17**, 1070.
164. Nabors, C. D., Eckardt, R. C., Kozlovsky, W. J., and Byer, R. L. (1989). *Opt. Lett.* **14**, 1134.
165. Lee, D., and Wong, N. C. (1992). *Opt. Lett.* **17**, 13.
166. Smith, R. G. (1973). *IEEE J. Quantum Electronics* **QE-9**, 530.
167. Nabors, C. D., Yang, S. T., Day, T., and Byer, R. L. (1990). *J. Opt. Soc. Am. B* **7**, 815.
168. Yang, S. T., Eckardt, R. C., and Byer, R. L. (1993). *Opt. Lett.* **18**, 971.
169. Guha, S., Wu, F. J., and Falk, J. (1982). *IEEE J. Quantum Electronics* **QE-18**, 907.
170. Colville, F. G., Henderson, A. J., Padgett, M. G., Zhang, J., and Dunn, M. H. (1993). *Opt. Lett.* **18**, 205; Colville, F. G., Padgett, M. G., Henderson, A. J., Zhang, J., and Dunn, M. H. (1993). *Opt. Lett.* **18**, 1065.
171. Bosenberg, W. R., Guyer, D. R., Lowenthal, D. D., and Moody, S. E. (1992). *Laser Focus World*, **165**, May.
172. White, J. C. (1987). In *Tunable Lasers*, L. F. Mollennauer and J. C. White (eds.), Topics in Applied Physics, Springer-Verlag, Berlin.
173. Mannik, L., and Brown, S. K. (1986). *Opt. Comm.* **57**, 360.
174. Wilke, V., and Schmidt, W. (1979). *Appl. Phys.* **18**, 177.
175. Hartig, W., and Schmidt, W. (1979). *Appl. Phys.* **18**, 235.
176. Hanna, D. C., and Pacheco, M. T. T. (1986). *Opt. Comm.* **60**, 107.
177. Grasiuk, A. Z., and Zuharev, I. G. (1978). *Appl. Phys.* **17**, 211.
178. Wellegehausen, B., Ludewigt, K., and Welling, H. (1985). *Proc. SPIE* **492**, 10.
179. Lenth, W., and Macfarlane, R. M. (1992). *Optics and Photonics News*, **8**.
180. Macfarlane, R. M., Tong, F., Silversmith, A. J., and Lenth, W. (1988). *Appl. Phys. Lett.* **52**, 1300.
181. Smart, R. G., Hanna, D. C., Tropper, A. C., Davey, S. T., Carter, S. F., and Szebesta, D. (1991). *Electron. Lett.* **27**, 1307.

182. Lenth, W., Silversmith, A. J., and Macfarlane, R. M. (1988). In *Advances in Laser Science*, Vol. 3, A. C. Tarn, J. L. Gole, and W. C. Stwalley (eds.), *AIP Confer. Proc.* **172**, 8.
183. Hebert, T., Risk, W. P., Macfarlane, R. M., and Lenth, W. (1990). In *Proc. Adv. Solid State Lasers*, Vol. 6, p. 379, H. J. Jenssen and G. Dube (eds.), Optical Society of America, Washington, DC; Hebert, T., Wannemacher, R., Lenth, W., and Macfarlane, R. M. (1990). *Appl. Phys. Lett.* **57**, 1727.
184. Macfarlane, R. M., Robinson, M., and Pollack, S. A. (1990). *Proc. SPIE* **1223**, 294.
185. Macfarlane, R. M., Wannemacher, R., Hebert, T., and Lenth, W. (1990). Paper CWF-1 in *Tech. Dig. CLEO '91*, Anaheim, CA, Optical Society of America, Washington, DC.
186. Allain, J. Y., Monerie, M., and Poignant, H. (1990). *Electron. Lett.* **26**, 166; Allain, J. Y., Monerie, M., and Poignant, H. (1990). *Electron. Lett.* **26**, 261; Allain, J. Y., Monerie, M., and Poignant, H. (1991). *Electron. Lett.* **27**, 189.

10. OPTICAL WAVELENGTH STANDARDS

Jürgen Helmcke

Physikalisch-Technische Bundesanstalt
Braunschweig, Germany

10.1 Introduction

Frequency-stabilized lasers operating in the visible range are widely used as wavelength standards in length metrology and precision laser spectroscopy. In modern length metrology, we rely on reference wavelengths generated by lasers for which the frequency is stabilized to a suitable absorption line. With the definition of the SI base unit of length, the meter, "the length of the path traveled by light in vacuum during a time interval of 1/299,792,458 of a second" [1], the wavelength λ of a stabilized laser is determined by a measurement of its frequency and results from the relation $\lambda = c/\nu$. Consequently, the value of the speed of light, $c = 299,792,458$ m/sec is fixed by the definition of the length unit. Frequency-stabilized lasers can therefore be regarded as wavelength standards or optical frequency standards, depending on their application.

The development of novel methods of nonlinear laser spectroscopy like cooling and trapping of atoms or ions and optical frequency measurements has made a strong impact on length/frequency metrology and atomic physics. The application of these methods may eventually lead to the development of optical frequency standards and optical clocks with unprecedented low uncertainty.

The major advantage of a wavelength standard operating in the *optical* range is provided by its high frequency, which is more than four orders of magnitude higher compared to the present primary standard of time and frequency, the cesium atomic clock. Consequently, for a given interaction time of the absorbers with the exciting field, the spectral resolution can be increased by several orders of magnitude. Such ultrahigh resolution has already been used to investigate line distortions caused by broadening due to the second-order Doppler effect [2] in a thermal ensemble of atoms. It was shown that second-order Doppler broadening ultimately limits the uncertainty of frequency standards based on thermal absorbers.

This chapter is organized as follows. The next section describes the basic scheme of a laser wavelength standard. In Section 10.3 we discuss as an important practical example an He–Ne laser whose frequency can be stabilized to the center of an absorption line in molecular iodine. Such iodine-stabilized lasers represent

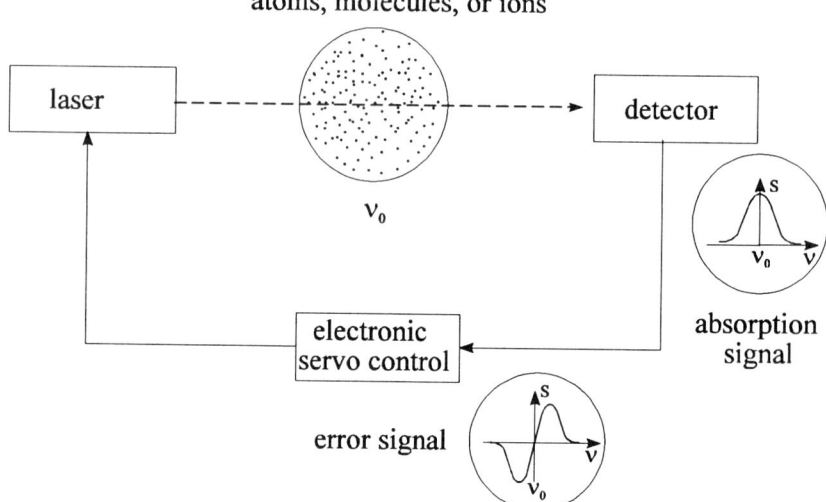

FIG. 1. Basic concept of an optical frequency standard.

the "working horses" for precision interferometric measurements in many laboratories. The relative frequency reproducibility of these I_2-stabilized lasers is limited to approximately 2.5×10^{-11}.

Better reproducibilities and lower values of the frequency uncertainty can be obtained with reference frequencies provided by extremely narrow absorption lines of laser-cooled and trapped ions or atoms. As a typical example of such a precision wavelength standard, we describe in Section 10.4 a laser that utilizes the narrow intercombination line ($\lambda \approx 657$ nm) of laser-cooled trapped Ca atoms. Finally, in Section 10.5, we discuss the measurement of optical frequencies and present a phase-coherent frequency measurement of the Ca-stabilized laser.

10.2 Basic Scheme of an Optical Wavelength Standard

The basic scheme of a laser wavelength standard is shown schematically in Fig. 1. The radiation of a single-frequency laser passes through an ensemble of atoms, molecules, or ions, providing an absorption line suitable as a reference frequency for stabilization. The ensemble of absorbers can be contained in a cell, in an atomic beam, or in a trap. If the laser frequency is tuned across the absorption line, the power of the transmitted laser beam changes due to the absorption, and an absorption feature is detected versus the laser frequency. The stabilization circuit converts the detected absorption signal to an error signal that is proportional to the frequency offset $\Delta\nu$ from the reference frequency. This error signal in turn drives a servo system that controls the laser frequency such that the error signal vanishes at its zero-crossing. The realization of a precision wavelength standard requires

- a narrow linewidth and precise tunability of the probing laser field,
- a narrow symmetric reference line that should be independent of external fields and operation parameters and that should provide a high signal-to-noise ratio for stabilization,
- a long interrogation time of the absorbers with the light field to achieve a high spectral resolution,
- a low velocity of the absorbers to reduce residual shifts and broadenings by the Doppler effect and to increase the interrogation time, and
- a precise knowledge of the stabilized frequency.

The generation of laser radiation with narrow linewidth and precise tunability is discussed in this volume by M. Zhu and J. L. Hall [3]. The observed width of the reference line is usually broadened by the Doppler effect, by collisions between the particles, and by the limited interaction time of the moving particles with the laser radiation. The Doppler broadening can be suppressed in the first order by nonlinear Doppler-free methods that are discussed in Chapter 13 of Volume 29B (Atoms and Molecules). The influence of collisions can be reduced by utilizing absorption cells with low vapor pressure or atomic beams. The residual linewidth is then determined by the natural linewidth, that is, the lifetimes of the corresponding energy levels, by the limited interaction time of the absorbers with the light field, and by the laser power. These line-broadening effects have to be addressed for the development of a precision optical wavelength standard (see Section 10.4).

In this section we present the basic components of a laser wavelength standard. The following paragraph (10.2.1) shows examples of how the error signal can be extracted from a Doppler-free absorption feature. Section 10.2.2 analyzes the electronic servo loop that converts the error signal to a control signal, stabilizing the laser frequency to the center of an absorption line. In Section 10.2.3 we discuss experimental methods used to determine the stability and reproducibility of the stabilized laser frequency.

10.2.1 Generation of the Error Signal

There are several methods to observe narrow Doppler-free absorption lines as references for the frequency stabilization. These methods have been described in previous chapters in this volume. Most wavelength standards utilize the methods of saturated absorption. The signal of the saturated absorption has to be converted

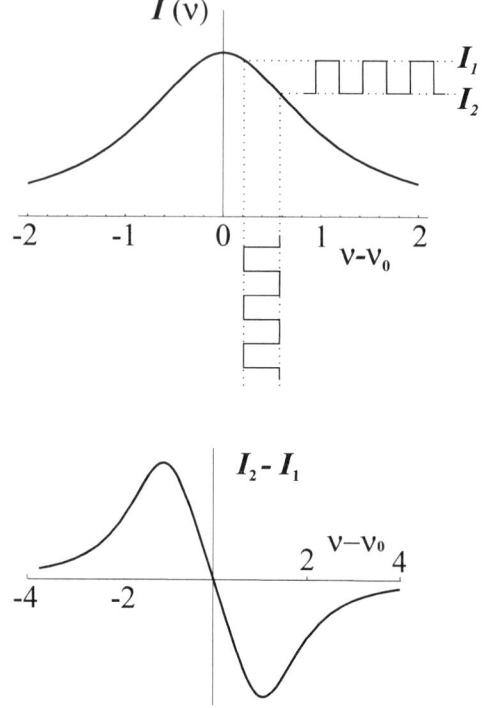

FIG. 2. Generation of the error signal by square-wave modulation and phase-sensitive detection.

to an error signal that is antisymmetric with the detuning $\Delta\nu = \nu - \nu_0$ of the laser frequency ν from the line center ν_0 and that changes linearly with $\Delta\nu$ near line center.

A simple concept to generate the error signal is to modulate the laser frequency between two discrete values (Fig. 2a) and detect the two corresponding absorption signals. The difference of the two signals leads to an error signal that is zero at the line center and changes linearly with detuning close to the line center (Fig. 2b). For small modulation amplitudes $\vartheta\nu$ the error signal is proportional to the first derivative of the absorption signal.

Instead of square-wave modulation, most lasers use a harmonic frequency modulation, and the modulated absorption signal consequently contains harmonics of the modulation frequency. The amplitudes of these harmonics change with detuning and therefore contain information on the offset $\Delta\nu = \langle\nu\rangle - \nu_0$ of the mean laser frequency $\langle\nu\rangle$. The amplitude of a particular harmonic signal in the power of the transmitted laser beam can be detected phase-sensitively by a corresponding bandpass filter followed by a phase-sensitive detector (psd) that is gated by the corresponding harmonic of the modulation frequency. For a symmetric absorption line, the amplitudes of the odd harmonics generated at the absorption profile are antisymmetric with the detuning from the line center and have a zero-crossing at $\Delta\nu = 0$. These odd harmonics are suitable as error signals for the frequency stabilization. If the modulation width is small compared to the linewidth, the derived signal represents the corresponding derivative of the saturation peak, that is, the first derivative for the first harmonic and the third derivative for the third harmonic. In the following we calculate the amplitudes of these harmonics [4]. The saturation peak can be described in good approximation by a Lorentzian line profile, $I(\nu)$:

$$I(\nu) = \frac{A}{1 + [(\nu - \nu_0)/\nu_{1/2}]^2}, \quad (1)$$

which is superimposed to the tuning curve of the laser. In Eq. (1), A and $\nu_{1/2}$ are the height and the width of the absorption feature, respectively. The tuning curve of the laser, that is, the laser power versus the frequency, is usually much wider than the saturation dip and can be approximated near the saturation line by a linear frequency dependence of constant slope. The modulated laser frequency $\nu(t)$ can be written as

$$\nu(t) = \nu_0 + \Delta\nu + \vartheta\nu \cdot \cos(\omega \cdot t), \quad (2)$$

where $\Delta\nu$ is the detuning of the mean laser frequency $\langle \nu \rangle = \nu_0 + \Delta\nu$ from the line center ν_0 and $\vartheta\nu$ is the modulation amplitude. If the modulation frequency $\omega/2\pi$ is small compared to the linewidth, the detected signal follows the laser frequency $\nu(t)$ adiabatically and we can replace ν in Eq. (1) by $\nu(t)$ from Eq. (2), leading to a time-dependent absorption signal:

$$I(t) = \frac{A}{1 + \{[\Delta\nu + \vartheta\nu \cdot \cos(\omega \cdot t)]/\nu_{1/2}\}^2}.$$

If we relate $\Delta\nu$ and $\vartheta\nu$ to the halfwidth $\nu_{1/2}$ with $d_D = \Delta\nu/\nu_{1/2}$ and $d_A = \vartheta\nu/\nu_{1/2}$, this equation transforms to

$$I(t) = \frac{A}{1 + [d_D + d_A \cdot \cos(\omega \cdot t)]^2}. \quad (3)$$

The signal $I(t)$ is periodic in time and can be expressed by a Fourier series:

$$S(t) = \frac{A_0}{2} + \sum_{m=1}^{\infty} A_m \cdot \cos(m\omega \cdot t), \quad (4)$$

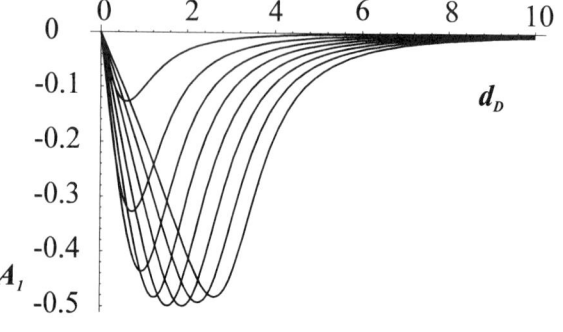

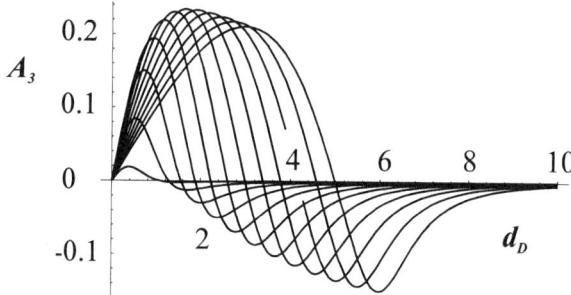

FIG. 3. Amplitudes of the first harmonic A_1 and the third harmonic A_3 versus the detuning of the laser frequency d_D. The modulation amplitude was chosen as parameter $d_A = 0.2$, 0.6, 1.0, ..., 3.0 for A_1, and $d_A = 0.5, 1.0, 1.5, ..., 6.0$ for A_3.

where the Fourier coefficient A_m represents the signal amplitude in the mth harmonic of the modulation frequency $\omega/(2\pi)$. It can be calculated by the relation

$$A_m = \frac{2}{\pi} \int_0^\pi S(\tau) \cdot \cos(m\tau)\, d\tau .$$

In the case of a symmetric profile $S(\nu)$, the odd harmonics A_{2m+1} are antisymmetric with the detuning d_D. Consequently, they have a zero-crossing at the line center ($d_D = 0$):

$$A_{2m+1}(d_D) = -A_{2m+1}(-d_D) \quad \text{and} \quad A_{2m+1}(d_D = 0) = 0.$$

The odd harmonics can therefore be utilized as error signals for stabilization. The even harmonics, on the other hand, are symmetric with a horizontal slope at the center. For the coefficients A_1 and A_3 we get [4]

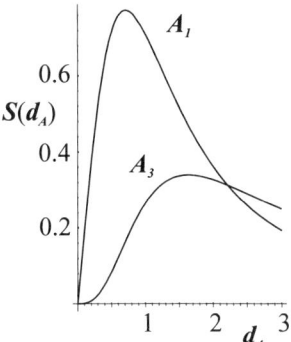

FIG. 4. Slope $S(d_A)$ of the error signal at the line center versus the modulation width for the first- and the third-harmonic signals, A_1 and A_3, respectively.

$$A_1 = \frac{1}{d_A}\{(\text{sign } d_D)P_- - d_D P_+\}, \tag{5a}$$

$$A_3 = \frac{1}{d_A^3}\{(\text{sign } d_D)[4(1-3d_D^2) + 3d_A^2]P_- + [4(3-d_D^2) + 3d_A^2]d_D P_+ - 16d_D\}, \tag{5b}$$

with

$$P_\pm = \frac{1}{\rho}\sqrt{2(\rho \pm \alpha)} \quad \text{and} \quad \rho = \sqrt{\alpha^2 + 4d_D^2}, \quad \alpha = 1 + d_A^2 - d_D^2.$$

A_1 and A_3 are shown versus d_D with the modulation amplitude d_A as a parameter in Figs. 3a and 3b, respectively. Both curves have zero-crossings at $d_D = 0$. In this central range their amplitudes depend linearly on the detuning d_D. Correspondingly, they can be used as error signals for the servo control. The slopes S at the line center of the curves in Figs. 3a and 3b are characteristic quantities influencing the gain of the servo control system. They are shown in Fig. 4 versus the modulation amplitude d_A. For a given saturation signal $I(\nu)$, the slope depends on the modulation width d_A. For small amplitudes $d_A \ll 1$, the amplitude of the first harmonic signal increases with the first power and that of the third harmonic with the third power of d_A. The largest slope, and therefore the highest sensitivity, can be obtained with the first harmonic $A_1(d_D)$. In many cases, however, the saturation signal $I(d_D)$ is superimposed to a frequency-dependent background, which can be represented near the line center by a baseline of constant slope. Thus, a constant signal would be added to the first harmonic signal $A_1(d_D)$ and the zero-crossing would be shifted, leading to a frequency offset of the stabilized laser frequency. This shift can be strongly reduced if we use the third harmonic or another odd harmonic of higher order for stabilization. If the amplitude of the background changes linearly or quadratically with laser frequency, the background will only generate harmonics up to the second order. Therefore, the third harmonic A_3 is not affected and the influence of the background is strongly reduced [5]. Since the maximum slope of the error signal decreases with increasing harmonic order, the third harmonic represents a good compromise between the maximum possible servo gain and suppression of the background. The third harmonic is therefore frequently used for frequency stabilization of lasers. It has three zero-crossings, of which the central one coinciding with the line center is used for stabilization.

So far, the signals A_1 and A_3 were calculated under the assumption of a perfect harmonic frequency modulation. However, if the modulation contains harmonic distortions, spurious signals will be generated in the detection band by harmonic mixing. For example, a second harmonic distortion will mix with the fundamental frequency to give a signal at the fundamental frequency and at the third harmonic [6]. This signal will be superimposed to the error signal A_1 or A_3 and consequently cause a frequency offset. If the modulation contains third-harmonic components, these harmonics will cause offsets that are, in the case of third harmonic detection, proportional to the slope of the background. It is therefore important to generate a frequency modulation of low distortion. In most servo systems, higher harmonic distortions are suppressed to better than -70 dB.

10.2.2 Electronic Frequency Servo Control

This section describes how the error signal is used to control the laser frequency close to the center of the absorption line. If the servo control is operating, the initial frequency v_i of the free-running laser is corrected to v_s close to v_0 by the servo loop (Fig. 5). This frequency v_s is compared with the frequency v_0 at the line center. The difference $\delta v = v_s - v_0$ represents the residual frequency deviation of the stabilized laser from the line center v_0. This deviation is converted to an amplitude $U = C\delta v$, where the constant C is proportional to the slope of the error signal versus the detuning (see Fig. 2b). This difference signal $C\delta v$ is amplified in the following control circuit by a factor $g(f)$ and is then transferred to a transducer, which corrects the laser frequency v_i by $-CDg(f)\delta v$. Here, D is the sensitivity of the transducer that transforms the control voltage to a corresponding frequency shift of the laser and $g(f)$ characterizes the frequency dependence of the electronic servo loop. The servo loop provides a negative feedback to the laser frequency, and the corrected value v_s can be written as

$$v_s = v_i - C \cdot D \cdot g(f) \cdot \delta v. \qquad (6)$$

If we subtract v_0 from both sides, Eq. (6) can be rewritten as

$$\frac{\delta v}{\Delta v} = \frac{1}{1 + C \cdot D \cdot g(f)}. \qquad (7)$$

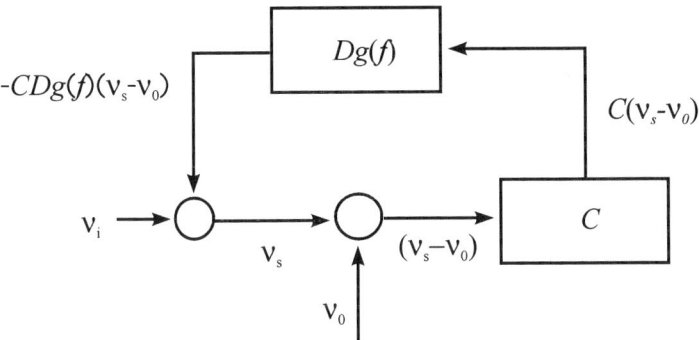

FIG. 5. Equivalent circuit of the frequency stabilization.

We see that the frequency deviation $\Delta \nu$ of the free-running laser from the line center ν_0 is reduced by a factor of $1/[1 + C \cdot D \cdot g(f)]$, where $C \cdot D \cdot g(f)$ represents the gain of the open servo loop. To achieve a negligible small value of the residual frequency deviation $\delta \nu$, the servo gain should be as high as possible. In particular, the deviation of the mean frequency averaged over long times should be close to zero. This requires that $g(f = 0) \to \infty$. Phase shifts and time delays of the servo loop put limitations on the maximum gain and the bandwidth of the servo loop. A simple technique to provide high servo gain at low frequencies and to avoid the influence of phase shifts and time delays at high frequencies is to have a decreasing servo gain with increasing frequency [3]. Such characteristics can be obtained by an integrating behavior of the servo gain, that is, if $g(f)$ is proportional to $1/f$. For an integrating servo control, the total transfer function of the control system can be characterized by the unity gain frequency f_C at which the gain of the open servo loop $|C\,D\,g(f_C)| = 1$. At fluctuation frequencies $f \ll f_C$, the frequency deviations will be reduced by a factor of f/f_C. For an integrating transfer function of the servo loop, an increase in the servo gain at a certain servo frequency f corresponds to an increase in the unity gain frequency by the same factor. Consequently, in an integrating servo loop a high servo gain requires a large servo bandwidth f. In cases where large frequency fluctuations at low frequencies have to be suppressed at a limited servo bandwidth f, an additional integration can be introduced into the servo loop, leading to a double integrating behavior and consequently to higher servo gain at frequencies $f \ll f_C$. The stability of the servo loop requires that this double integration is effective for frequencies well below the unity gain frequency f_C. A suitable maximum crossover frequency f_{di} between double and single integrating behavior is at approximately $f_{di} \approx f_C/4$ [3].

10.2.3 Stability, Reproducibility, and Uncertainty

The stability, reproducibility, and uncertainty of the stabilized laser frequency are important characteristics of any frequency or wavelength standard. The stability and reproducibility can be investigated by measuring the beat frequency $v_B = |v_1 - v_2|$ between two identical stabilized laser systems 1 and 2 (see Fig. 6). In this setup, the two laser beams are combined by a beam splitter to be coaxial and focused on a fast photodetector. The photocurrent contains a signal oscillating with the beat frequency $|v_1 - v_2|$. This signal is amplified and fed to a frequency counter. The instability can be described by the Allan standard deviation [7]:

$$\sigma_y(\tau) = (1/v_0) \sqrt{(1/2n) \sum_n (v_{B,n+1} - v_{B,n})^2},$$

where the frequency difference $v_{B,n+1}$ is determined by two successive measurements of the beat frequency v_B using an integrating time τ of the frequency counter. Supposing that both lasers contribute equal amounts to the frequency fluctuations and that their fluctuations are not correlated, $\sigma_y(\tau)$ of one laser is given by

$$\sigma_y(\tau) = (1/2v_0) \sqrt{(1/n) \sum_n (v_{B,n+1} - v_{B,n})^2}. \tag{8}$$

The dependence of $\sigma_y(\tau)$ versus the integrating time τ of the counter depends on the frequency noise of the laser. For example, if white frequency noise is dominant, $\sigma_y(\tau)$ decreases with the square root of the averaging time τ [7].

The reproducibility of a wavelength standard can be investigated by measuring the dependence of its frequency on the various operation parameters and by frequency intercomparisons between lasers of different designs.

The uncertainty of a wavelength standard in principle contains two contributions. The first one is the realization uncertainty, that is, the uncertainty to which the line center of an unperturbed atom at rest can be realized. The second contribution stems from the uncertainty of the determination of its wavelength. For many wavelength standards, the uncertainty of their wavelength values is larger than the realization uncertainty, and the latter gives an estimation of the potential of the standard. As a typical example, we will describe in Section 10.4.4 the various effects contributing to the uncertainty of a Ca optical frequency standard.

10.3 Iodine-Stabilized Lasers

Most wavelength standards of visible radiation utilize reference frequencies that are provided by absorption lines of molecular iodine [8]. The *B–X* electronic

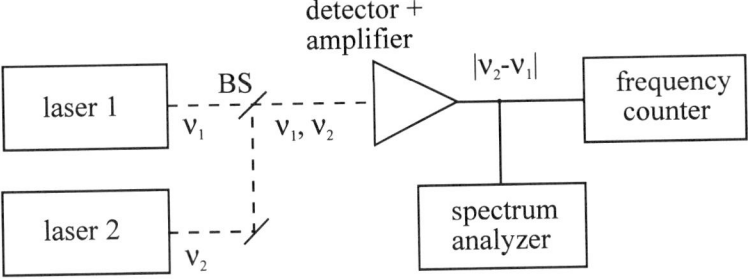

FIG. 6. Typical setup for the measurement of beat frequencies.

transition of molecular iodine vapor has a very rich absorption spectrum almost over the entire visible spectrum up to the dissociation limit at approximately 500 nm [9]. The different absorption lines belong to the various vibrational and rotational levels. Due to the nuclear spin of iodine, each of the absorption lines is split into several hyperfine components. Natural iodine consists of only one isotope: $^{127}I_2$. However, the slightly radioactive isotope ($^{129}I_2$) is also used as a reference spectrum for frequency stabilization. Iodine vapor has been used for frequency stabilization in atomic beams [10] and in absorption cells [11]. Many applications use iodine absorption cells, and the linewidth as well as the position of the line center slightly depends on the iodine vapor pressure. The vapor pressure can be adjusted by controlling the temperature of the cooling finger that is usually attached to the cell.

10.3.1 Iodine-Stabilized He–Ne Laser ($\lambda \approx 633$ nm)

A wavelength standard most commonly used in length metrology and precision spectroscopy is the iodine-stabilized He–Ne laser operating at a wavelength of approximately 633 nm. This laser belongs to the wavelengths standards recommended by the International Committee of Weights and Measures (Comité International des Poids et Mesures, CIPM) for the realization of the length unit [1, 8].

Figure 7 shows schematically a typical setup of an iodine-stabilized He–Ne laser. The laser head consists of an He–Ne plasma tube and an iodine absorption cell mounted inside the laser resonator. The lengths of the plasma tube and the absorption cell are typically 20 and 10 cm, respectively, corresponding to a minimum length of the resonator of approximately 35 cm and a free spectral range of ≈430 MHz. The laser resonator should be made as rugged as possible, and the spacers should consist of a low-thermal-expansion material. The radii of curvature of the laser mirrors are usually in the range between $r = 60$ cm and $r = 4$ m, with the radius $r = 1$ m most commonly used. The mirrors have reflectivities of approximately 98%, allowing single-frequency operation over most of the free

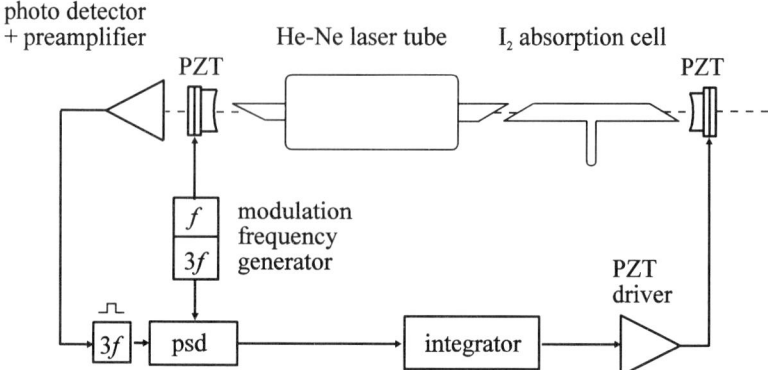

FIG. 7. Schematic setup of an I_2-stabilized He–Ne laser (psd = phase-sensitive detector).

spectral range. Output powers up to 300 µW have been achieved, being sufficient for many applications in spectroscopy and interferometry. The two mirrors are mounted on piezoelectric transducers (PZTs), which makes it possible to change the resonator length and thereby the laser frequency by applying tension to the PZTs. Figure 8a shows a typical tuning curve of the output power of the laser versus the PZT voltage if the laser frequency is not stabilized. We observe peaks in the output power whenever the laser frequency coincides with an absorption line of iodine vapor. The four peaks shown in Fig. 8a correspond to saturated absorption features of the hyperfine structure components d, e, f, and g of the transition $R(127)$, $v' = 5$, $v'' = 11$ of $^{127}I_2$. The separation between the peaks is approximately 13 MHz (see [1, 8]), and the full width at half maximum of each line is ≈5 MHz, corresponding to a line $Q = v/\delta v \approx 10^8$. The height of each peak is approximately 0.1% of the laser power.

To generate the error signal for stabilization, the length and thereby the frequency of the laser is modulated harmonically (Fig. 7). Third-harmonic detection is used for stabilization (see Section 10.2.1). If the laser frequency is tuned close to the absorption frequency v_0, the harmonics of the modulating frequency are generated at the absorption feature in the output power of the laser and the third-harmonic signal is selected for stabilization. This signal is phase-sensitively detected and, after passing through a lowpass filter, serves as an error signal for stabilization. A typical "third-harmonic" spectrum of molecular iodine is shown in Fig. 8b for an I_2-stabilized He–Ne laser ($\lambda \approx 633$ nm). Each of the seven absorption features is suitable for stabilization. With the servo loop operating, the length of the laser is stabilized such that the third-harmonic signal of the corresponding hfs component vanishes at its central zero-crossing.

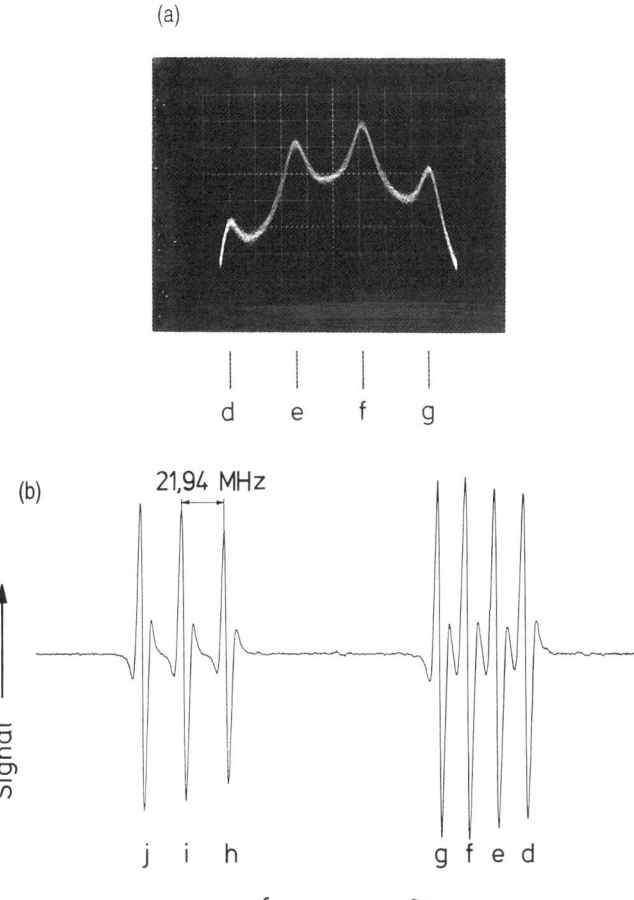

FIG. 8. Output power (a) and third harmonic spectrum (b) of an He–Ne laser with intracavity I_2 absorption cell versus laser frequency.

10.3.2 Stability and Reproducibility of the I_2-Stabilized He–Ne Laser

The stability and reproducibility of the I_2-stabilized He–Ne laser were investigated by beat frequency measurements of two independent lasers, as described in Section 10.2.3. Figure 9 shows the measured Allan standard deviation of an iodine-stabilized laser versus the integrating time τ. In the range between $\tau = 10$ msec and $\tau = 100$ sec, $\sigma_y(\tau)$ decreases approximately, with the square root of τ corresponding to white frequency noise. The minimum instability of about $2 \cdot 10^{-13}$ is observed at an integrating time $\tau \approx 1000$ sec.

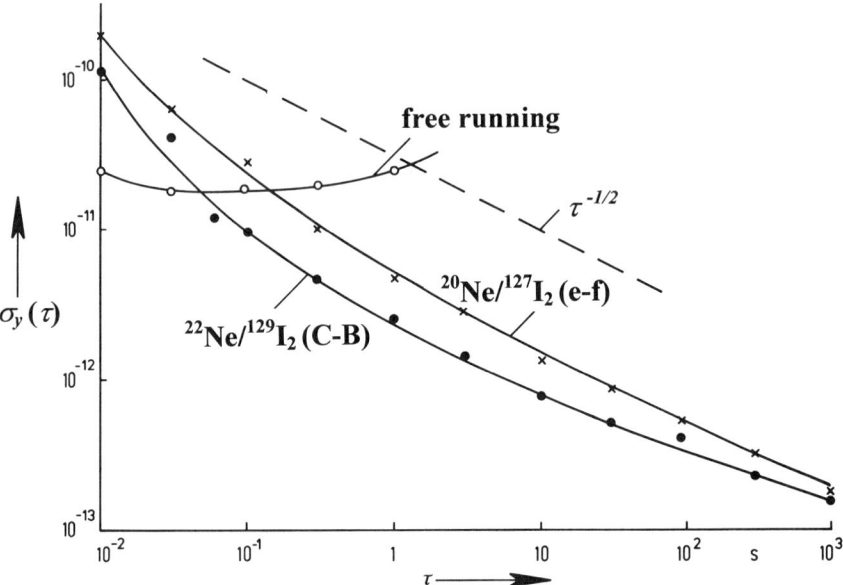

FIG. 9. Typical Allan standard deviation of He–Ne lasers ($\lambda \approx 633$ nm) stabilized to hyperfine structure components of $^{127}I_2$ and $^{129}I_2$.

The reproducibility of the iodine-stabilized He–Ne laser was investigated by measuring the dependence of the laser frequency on the various operation parameters and by frequency intercomparisons between lasers of different institutes. Figure 10 shows the measured dependence of the stabilized laser frequency on the iodine pressure and the width of the frequency modulation. The laser frequency depends on the modulation width, the iodine vapor pressure, and slightly on the laser power. It is therefore difficult to realize the true line center of the absorption line with very low uncertainty. The coefficients are approximately –6 kHz/Pa and –10 kHz/MHz for the pressure and modulation dependence, respectively. International frequency comparisons between iodine-stabilized lasers have shown that the stabilized frequencies of the majority of iodine-stabilized He–Ne lasers coincide within a range of approximately 12 kHz, corresponding to $2.5 \times 10^{-11} \nu$. The frequency of the I_2-stabilized He–Ne laser has recently been measured, leading to precise values of the frequency and the wavelength of the laser [12]. The recommended frequency and wavelength values for the hfs component i of the transition B–X, 11–5, $R(127)$ of $^{127}I_2$ are

$$\nu = 473,612,214,705 \text{ kHz} \quad \text{and} \quad \lambda_{vac} = 632.991398220 \text{ nm},$$

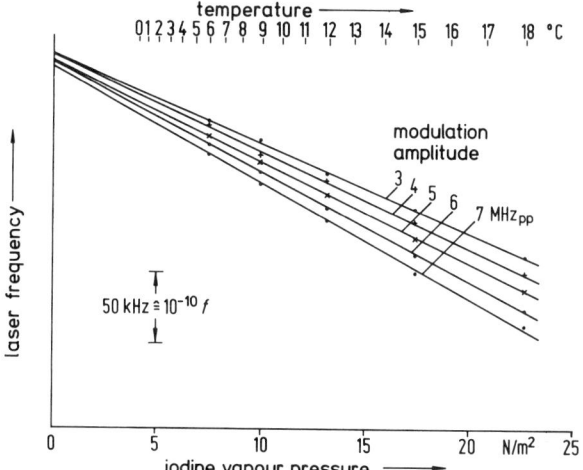

FIG. 10. Pressure dependence of the stabilized frequency of an I_2-stabilized He–Ne laser for different modulation widths.

with an estimated relative standard uncertainty of 2.5×10^{-11}. This low value for relative uncertainty can be obtained only if the operation parameters are held at the following values:

cell wall temperature = $25 \pm 5°C$
cold point temperature = $15 \pm 0.20°C$
frequency modulation width between the peaks = 6 ± 0.3 MHz
one-way intracavity beam power = 10 ± 5 mW.

The iodine-stabilized He–Ne laser ($\lambda \approx 633$ mn) represents a simple precision wavelength standard that can conveniently be used in length metrology and that provides a precise optical reference frequency for laser spectroscopy. Using an intracavity dispersion prism and mirrors with suitable coatings, the He–Ne laser can also operate at a variety of other wavelengths (e.g., 543, 612, and 640 nm). For all emission ranges, the laser frequency can be stabilized to iodine hfs lines, leading to further wavelengths standards recommended for realization of the meter [8]. The frequency-doubled YAG laser pumped by a diode laser ($\lambda \approx 532$ nm) provides another very promising candidate for an I_2-stabilized wavelength standard. This laser has very low intrinsic frequency noise and operates at an output power of several tens of milliwatts, which is orders of magnitude higher than that of He–Ne laser wavelength standards.

10.4 Wavelength Standards Utilizing Narrow Resonances of Laser-Cooled Absorbers

The iodine-stabilized He–Ne laser described in the previous section utilizes the coincidence of a narrow iodine absorption line with the emission wavelength of an He–Ne laser. This laser is small, transportable, and easy to operate and provides an uncertainty that is sufficient in most practical applications. Still lower uncertainties can be achieved if absorption lines are used that are even better suited as reference frequencies. Such lines are provided by narrow (forbidden) absorption lines in atoms or ions that can be trapped and laser cooled.

Unfortunately, the frequencies of these lines do not coincide with the frequencies of simple gas lasers and need tunable lasers like dye or diode lasers to excite these transitions. New generations of such wavelength standards are developed in several laboratories. For example, the intercombination lines $^3P_1$–$^1S_0$ of the alkaline earths are excellent reference lines for the development of a precision optical frequency standard [13, 14]. Further reference lines are also provided, for example, in silver atoms [14] and in Yb$^+$ and Hg$^+$ ions [15, 16]. There are several groups investigating these lines for development of optical frequency standards of the highest possible accuracy. As a typical example of such an optical frequency standard, we will describe in the following section a laser whose frequency is stabilized to the intercombination line of ^{40}Ca.

10.4.1 Ca-Stabilized Laser

The frequency of the intercombination transition of calcium at $\lambda \approx 657$ nm has a natural linewidth of $\delta\nu \leq 400$ Hz, corresponding to a line Q of $\nu/\delta\nu > 10^{12}$. ^{40}Ca has no hyperfine structure, and the ground state is not degenerate. Correspondingly, each atom in the ground state is available for excitation. The (clock) transition ($m_j = 0 \rightarrow m_j = 0$) has a very small quadratic dependence on electric and magnetic fields of only 30 mHz/(V/cm)^2 and about 10^8 Hz/T^2 [17], respectively. It can be excited with diode lasers, allowing for development of a precision transportable optical Ca frequency standard. Furthermore, the velocity of the Ca atoms can be reduced by laser cooling and trapping [18, 19] using the transition $^1P_1$–$^1S_0$ at $\lambda \approx 423$ nm. A laser stabilized to the Ca intercombination transition was recommended for realization of the SI unit of length, the meter [8].

The transit time broadening of the "clock transition" observed with *thermal* Ca atoms can be strongly decreased by separated-field excitation, leading to narrow (Ramsey) structures [20, 21] that can be interpreted as atom interferences [22]. They are useful as ultranarrow discriminant signals for frequency stabilization. Section 10.4.2 will discuss the generation of atom interferences by separated-field excitation for an effusive Ca beam and for Ca atoms stored in a magnetooptical

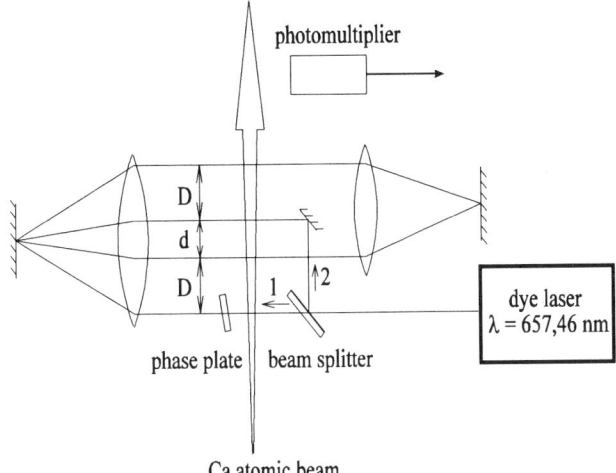

FIG. 11. Separated-field excitation of a Ca atomic beam. The atoms interact with four coherent traveling waves obtained from a single laser beam by the use of two cat's eye retroreflectors. The phase plate allows introducing a phase difference between the first and subsequent laser beams. A photomultiplier detects the excited atoms by their fluorescence approximately one decay length downstream.

trap (MOT). Section 10.4.3 describes the frequency stabilization scheme. The influences affecting the uncertainty of the standard are discussed in Section 10.4.4.

10.4.2 Atom Interference Generated by Separated-Field Excitation

10.4.2.1 Separated-Field Excitation of an Effusive Beam (Bordé Atom Interferometer)
In this section we describe the generation of atom interferences in an effusive atomic beam. The atoms sequentially pass two pairs of equally spaced traveling laser beams (Fig. 11). Two "cat's eye" retroreflectors, each consisting of a lens and a mirror in the focal plane, are employed to get four parallel laser beams, with the first two beams running in the same direction and the second pair being counterpropagating. The frequency of the laser is tuned close to that of the $^3P_1$–$^1S_0$ transition of ^{40}Ca. Fluorescent decay of the excited atoms is detected by a photomultiplier about one decay length (20 cm) downstream. Figure 12 shows the observed interference signal versus the laser frequency.

The method of separated-field excitation was first introduced by Ramsey for the microwave range [23]. According to Bordé [22], separated-field excitation using four interaction zones can be interpreted as a (Bordé) atom interferometer that provides state labeling of the output ports. If an atom is excited from the ground state |a> to the excited state |b> in one of the four excitation zones (Fig.

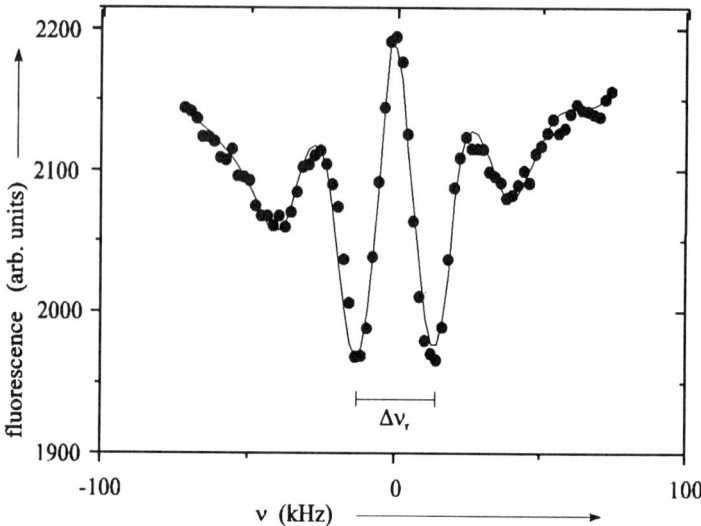

FIG. 12. Interference signal of the Bordé atom interferometer versus the laser frequency, observed in an effusive Ca atomic beam. The interference signal is generated in the center of the Doppler-free saturation dip. At a separation of $2D = 3.5$ cm, the recoil splitting Δv_r is observed. The line represents a fit to the measured data.

13), it is deflected due to momentum transfer from the photon to the atom. The probability to excite the atom to the $^3P_1$ state can be adjusted by the field strength and the interaction time, leaving the atom in a coherent superposition of the $^1S_0$ ground state and the $^3P_1$ state. Since the atoms in both states have different momenta, it is more adequate to use the picture of wave packets.

The calcium wave packets enter the first of the four interaction zones in the ground state $^1S_0$. In the first (and subsequently in each other) zone, the matter wave is coherently split into two partial waves with internal states $^3P_1$ and $^1S_0$. If an absorption or emission of a photon is induced, the transverse momentum of the Ca atom is changed by the momentum of the photon $\hbar k$ ($\hbar$ is Planck's constant/2π, k is the laser wavenumber $2\pi/\lambda$). There are two different possibilities (marked by two trapezoids of Fig. 13) for which two partial waves are combined in the fourth interaction zone. Both cases differ in the excited state by their transverse momenta caused by the photon recoil, $+\hbar k$ and $-\hbar k$, respectively. The different signs of these photon recoils lead to a frequency splitting [13], and each trapezoid corresponding to one recoil component can be regarded as a separate Mach–Zehnder interferometer. There are two exit ports where the calcium matter waves leave the interferometer either in the excited state (port I) or in the ground state (port II). The probability to find the matter wave in the excited state contains an interference term W that harmonically oscillates with the phase difference $\Delta\Phi$ between the partial waves

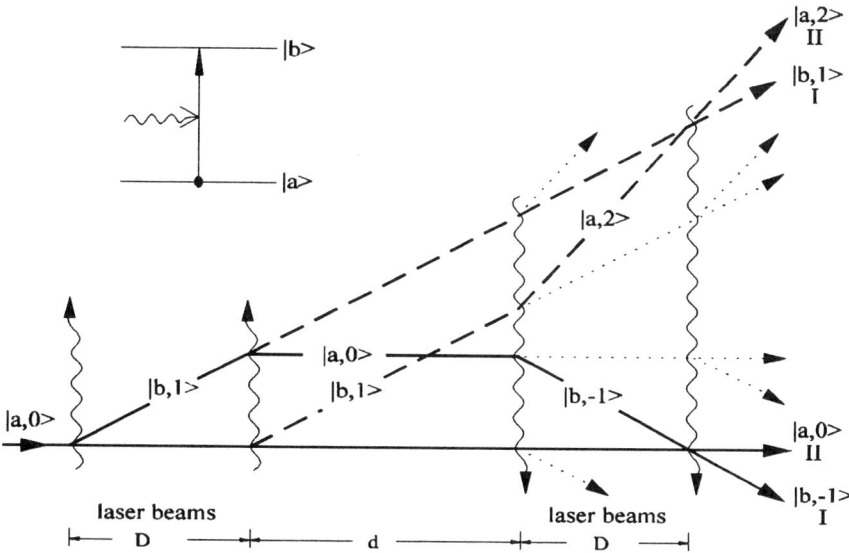

FIG. 13. Separated-field excitation of an atomic beam by four traveling laser fields (Fig. 11) interpreted as atom interferometer. Full lines = high-frequency recoil component; dashed lines = low-frequency component.

according to $W = W_0 \cos \Delta\Phi$. The phase difference can be expressed as follows [20]:

$$\Delta\Phi = 2\pi[\Delta\nu \pm h \cdot \nu^2/(2Mc^2) + v \cdot v^2/(2c^2)] \cdot 2D/v + (\varphi_2 - \varphi_1 + \varphi_4 - \varphi_3). \quad (9)$$

where D denotes the distance between the parallel laser beams and v is the velocity of the atoms in the beam (Fig. 13). The first term $\Delta\nu = \nu - \nu_0$ describes detuning of the laser frequency ν from the frequency of the clock transition ν_0 and leads to the oscillation of W with the laser frequency (Fig. 12). This oscillation is frequently called optical Ramsey interference in analogy to Ramsey resonances observed in the microwave range [23]. The width of its period $\delta\nu = v/(2D)$ is inversely proportional to the transit time D/v of the atoms between two parallel laser beams. With thermal atoms, the interference fringes wash out rapidly with increasing detuning, leaving only the central fringes. This situation is similar to white light interferences in an optical Mach–Zehnder interferometer, and mainly the achromatic "white light" interference fringe survives. The second term in Eq. (9) results from the photon momentum changing the velocity of the wave packet (recoil) and leading to a recoil splitting of the interferences $\Delta\nu_r = h \cdot \nu^2/(Mc^2) \approx 23.08$ kHz. The sign of the recoil term is positive for the low-frequency recoil component (Fig. 13, trapezoid with dashed lines) and negative for the high-frequency component (solid lines). The third term results from the second-order Doppler effect. For thermal atoms with a wide velocity distribution, it results in a shift and an asymmetry of the interference structure [2]. An additional phase shift $\varphi_2 - \varphi_1 + \varphi_4 - \varphi_3$ is introduced between the two arms of the interferometers by the phases φ_i of the four laser beams ($i = 1, 2, 3, 4$). The use of well-aligned cat's eyes compensates this phase shift close to zero. All phase shifts cause frequency shifts of the interference fringes (Eq. (9)), which may increase the uncertainty of the standard. For thermal atoms, optical phase errors (corresponding to a residual first-order Doppler effect) and the second-order Doppler effect limit the uncertainty to a few parts in 10^{12} [24].

10.4.2.2 Separated-Field Excitation of Laser-Cooled and Trapped Atoms
To further reduce this uncertainty level, we have applied a scheme for laser cooling and magnetooptical trapping some 10^6 Ca atoms [25]. With this scheme, the mean velocity of the effusive beam (≈ 700 m/sec) is reduced to an rms velocity of the trapped atoms of $|v| < 1$ m/sec, and the influence of the Doppler effect is negligible (see Section 10.4.4).

Since the atoms are confined in a small volume, separated-field excitation has to be performed in the time domain where the atomic waves are excited by short pulses of coherent laser radiation. In contrast to conventional continuous saturated absorption, where a linewidth of 1 kHz is generated by atoms within a velocity range $|v_z| < 0.7$ mm/sec, this range can be increased by excitation with short pulses since the velocity interval δv increases with inverse pulsewidth. The spectral resolution is then determined by the time-separation T between the coherent laser pulses. Pulsed (time) separated-field excitation is therefore advantageous to increase the number of atoms contributing to the narrow Doppler-free signal.

To avoid systematic shifts and broadenings due to the Zeeman effect and the ac Stark effect, all trapping fields have to be turned off before the reference (clock) transition can be probed. The corresponding time sequence consists of atom trapping ($t_1 \approx 13$ msec), turning off the trapping fields ($t_2 \approx 100$ μsec), separated-field excitation of the reference transition ($t_3 \approx 0.5$ msec), and detection ($t_4 \approx 0.5$ msec) is shown in Fig. 14. Separated-field excitation was performed by three pulses of standing laser fields [26]. The phase difference $\Delta\Phi$ can be calculated from Eq. (9) if the transit time D/v is replaced by the pulse separation T. Due to the fixed time separation T, the coherence length of the atomic wave has no influence on the interference contrast and the envelope of the interference structure is determined by the pulsewidth, leading to an increased number of fringes for short pulses. Figure 15 shows the measured interference signal versus the laser frequency for three different pulse separations T. In these experiments, T was chosen such that the recoil splitting Δv_r was equal to or a multiple of the period width of the interference $1/(2T)$. The narrowest fringe width $1/(4T)$ obtained with trapped atoms was limited by residual phase noise of the laser to approximately 2 kHz.

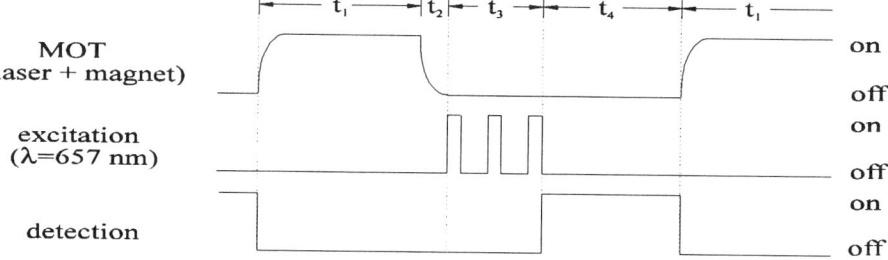

FIG. 14. Time sequence for trapping, separated-field excitation, and detection of Ca atoms. Separated-field excitation and detection of interferences are performed when the trapping fields are turned off ($t_1 \approx 13$ msec = trapping time; $t_2 \approx 0.1$ msec = allowed decay time for trapping fields; $t_3 \approx 0.5$ msec = excitation time of the clock transition; $t_4 \approx 0.5$ msec = detection time).

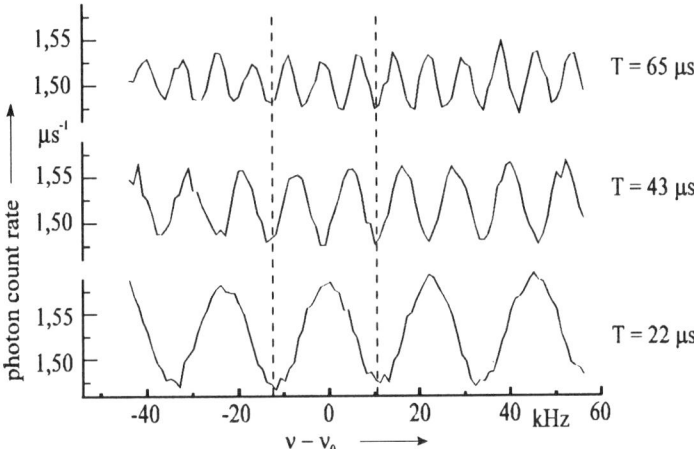

FIG. 15. Atom interferences versus the laser frequency observed with trapped Ca atoms for three different pulse separations T. The dashed lines mark the frequencies of the recoil components.

10.4.3 Frequency Stabilization

The interferometer signals discussed in the preceding section are used to stabilize the frequency of a dye laser to the center of the Ca intercombination line. The stabilization scheme consists of three stages (Fig. 16): (a) prestabilization to a resonance of an optical cavity, (b) stabilization to the center of the intercombination line in an effusive beam, and (c) with trapped atoms.

For prestabilization to an eigenfrequency of an optical resonator we use the method of phase modulation [27]. The fast frequency fluctuations are controlled

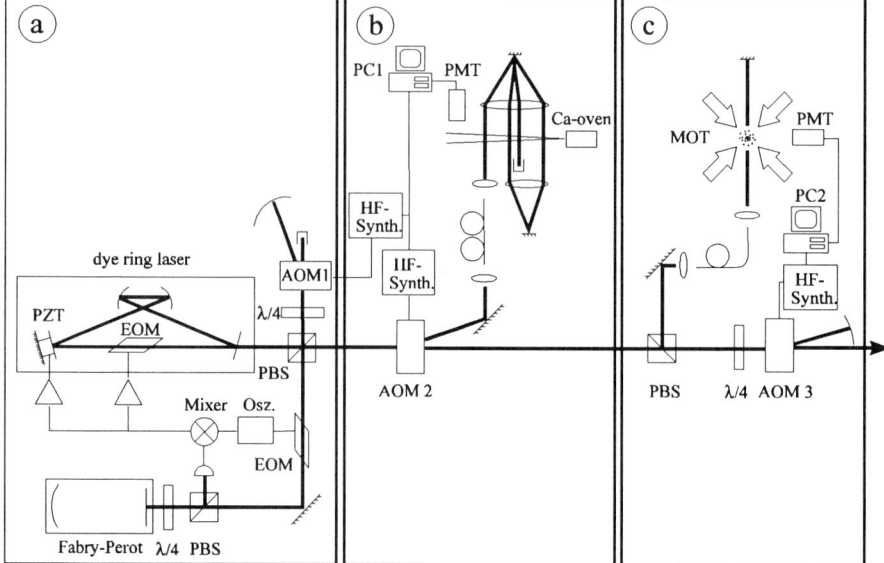

FIG. 16. Schematics of the Ca optical frequency standard. (a) Prestabilization to a resonance of an optical cavity. Stabilization to the Ca intercombination line (b) in an effusive beam and (c) in an MOT.

by an intracavity electrooptic modulator (EOM), whereas the slow ones and the drift are compensated by a piezoelectric transducer (PZT). The resonator consists of a Zerodur spacer, with Zerodur mirrors optically contacted to the spacer. Frequency tuning is performed by an acoustooptic modulator (AOM1) with the driving frequency controlled by a computer [28]. The observed linewidth is a few kilohertz, and the drift of the stabilized laser system is less than 100 Hz/sec. Stabilization to the center of the Ca intercombination line is performed by the following two steps (Fig. 16b,c). For excitation of the atomic beam we use laser radiation diffracted in the first order of an acoustooptic modulator (AOM2). The error signal for stabilization is generated by a square wave modulating the laser frequency ν (diffracted at AOM2) using a total width of 10 kHz (see Section 10.2.1). The difference between the signal amplitudes at ν + 5 kHz and ν − 5 kHz is calculated, and a signal proportional to this error signal is generated and added to the signal controlling the laser frequency at AOM2, leading to a digital integrating servo control. With this stabilization system, the laser frequency stayed in lock for several hours.

To minimize frequency shifts caused by the second-order Doppler effect and by optical phase errors in the excitation, the laser frequency is further stabilized to interference fringes generated in the MOT [25]. For this purpose we use the

TABLE I. Contributions to the Frequency Uncertainty of an Optical Frequency Standard Based on Laser-Cooled and Trapped Ca Atoms [40]

Effect	Achieved uncertainty	Attainable uncertainty
First-order Doppler effect	<5 Hz	<1 Hz
Second-order Doppler effect	<3 MHz	<3 mHz
Magnetic fields	<0.5 Hz	<0.5 Hz
Stark effect, 24 mHz/(V/cm)^2	45 MHz	45 mHz
ac Stark effect	<5 Hz	<0.1 Hz
Black-body radiation (300 K)	<30 mHz	<30 mHz
Collision of cold atoms	<100 Hz	<1 Hz
Influence of second recoil component	<7 Hz	7 mHz
Stabilization scheme	<300 Hz	<0.4 Hz
Counting errors	<100 Hz	<0.1 Hz
H-maser	15 Hz	<0.5 Hz
Cs-clock	7 Hz	<0.5 Hz
Total uncertainty $\Delta\nu$	<0.35 kHz	<1.7 Hz
Total relative uncertainty $\delta\nu/\nu$	$<8 \times 10^{-13}$	$<4 \times 10^{-14}$

radiation passing AOM2 at the zeroeth order that is already prestabilized to the Ca beam. The use of this radiation facilitates finding the correct interference fringe. The servo control is similar to that used with the Ca beam, except that modulation is applied to AOM3, resulting in a frequency-modulated laser frequency.

10.4.4 Estimation of Realization Uncertainty

The frequency uncertainty of the standard is determined by the uncertainty to realize the line center of the undisturbed atom at rest and by that of the frequency measurement. In this section we estimate the various contributions to realization uncertainty. The uncertainty of frequency measurement is discussed at the end of Section 10.5. The contributions to uncertainty are summarized in Table I.

10.4.4.1 Phase Shifts Generated by the Phase of the Laser Fields In the four-beam geometry of Fig. 11, phase differences in optical excitation show up directly as phase differences between the partial atomic waves and result in frequency offsets of the interference signal (Eq. (9)). The optical phase differences can be adjusted by a glass plate behind the first interaction. Since the sign of the phase shift changes if the direction of the laser beams is reversed, the method of laser beam reversal is a convenient means to adjust the phase difference in the matter-wave interferometer [24]. The residual contribution to the uncertainty is a few parts in 10^{12} for the effusive beam. In the case of trapped atoms, the influence of the Doppler effect is strongly reduced. After switching off the trapping fields, the atomic cloud is expanding and the atoms are accelerated by gravity. If the laser beam probing the clock transition is reflected under a small angle α, a residual first-order Doppler shift $\Delta v_D \approx v_0 \cdot (v/c) \cdot \alpha$ will occur. Gravitational acceleration leads to a velocity <1 cm/sec during the elapsed time of 1 msec. With α controlled to ≤0.3 mrad, this effect contributes to less than $10^{-14}v$. Since the velocity of the atoms is reduced to less than 1 m/sec, the corresponding second-order Doppler shift is reduced by six orders of magnitude to <3 mHz.

10.4.4.2 Phase Shifts Generated by the ac Stark Effect In the regions between the first and second and between the third and fourth interaction zones, the Ca partial waves are in different states in each arm of the interferometer (Fig. 16). A change in the potential energy of one of the partial waves accomplished by additional laser radiation tuned close to the $^1P_1$–$^1S_0$ transition leads to an additional phase difference caused by the ac Stark effect [29]. For the development of an optical frequency standard based on cold atoms, it is therefore important that the cooling radiation be turned off when the clock transition is probed.

10.4.4.3 Phase Shifts Generated by a Static Electric Field A static electric field E generated for example by surface charges in a vacuum chamber causes additional phase shifts $\delta\Phi_E$ in the corresponding partial waves due to the quadratic Stark effect and the Aharonov–Casher effect [30]:

$$\delta\Phi_E = -\frac{2D}{\hbar N}\left[\frac{1}{2}\Delta\alpha \cdot E^2 + \frac{v}{c^2}(\mu \times E)\right]. \tag{10}$$

Since the partial waves are in different states with different polarizabilities α and different magnetic moments μ between the first and second and between the third and fourth interactions (Fig. 13), a phase difference is induced by the electric field. The first term in Eq. (10) describing the quadratic Stark effect leads to a frequency shift of approximately 0.03 Hz/(V/cm)^2 [17]. Assuming a maximum field strength of 1 V/cm, the corresponding frequency shift of 30 mHz can be neglected at the present level of uncertainty.

The second term in (10) corresponds to a phase shift caused by the Aharonov–Casher effect. It is generated by an interaction of the magnetic dipole moment μ (in the excited state) with the field E. The phase is shifted, even though no classical force is acting on the atom. The dependence of the atomic phase on field strength was measured in our interferometer, verifying the Aharonov–Casher effect with a relative uncertainty of 2.2% [31]. The corresponding frequency dependence is 16 mHz/(V/cm), which can be neglected at the present state of optical frequency standards.

10.4.4.4 Influence of Collisions of Cold Atom Collisions of the cold atoms represent another potential source of systematic frequency shifts. With at most 10^7 atoms stored in a volume of approximately 1 mm^3, the distance between the atoms is about 5 μm, which is large compared to the wavelength of the laser and to the size of the wave packets. For measurements with different numbers of stored atoms, we have found no shift within an uncertainty of 100 Hz. At ultimate precision, however, collision-induced shifts may eventually limit the accuracy of optical frequency standards.

10.4.4.5 Frequency Shifts by a Superposition of Both Recoil Components If the recoil splitting δv_r does not coincide with the interference period δv_i or with an integer multiple $m \cdot v_i$ adjusted by the pulse separation T of the excitation (Section 10.2.2), the signals of both recoil components do not superimpose constructively. Even though the center frequency of each recoil component does not change, the sum of both signals shifts the observed superposition of fringes by an amount corresponding to the difference $\delta v_r - m \cdot v_i$. In the present experiment, the limited time resolution of the pulse generator used to gate the laser pulses causes a shift at each recoil component of +70 Hz for the high-frequency recoil component and –70 Hz for the low-frequency one. For the mean frequency of both recoil components, we estimate the contribution to the uncertainty to be less than 10% of this shift (<7 Hz).

10.4.4.6 Influence of Magnetic Fields Magnetic fields shift the frequency of the clock transition. Using the splitting between the σ components, the magnetic field can be determined with a relative uncertainty of <1%. For a residual field strength of 0.5 mT, the corresponding contribution to the uncertainty is 0.5 Hz.

10.5 Optical Frequency Measurement

For any wavelength standard, the value of its wavelength has to be determined. With the value of the speed of light fixed by the definition of the meter, the wavelength is derived from the frequency of the laser by the relation $\lambda = c/v$. In turn, the frequency of the standard has to be related to the primary standard of time and frequency, the Cs atomic clock. In order to avoid uncertainties additional to those of the Cs clock and the optical frequency standard, the large gap of frequencies has to be bridged by a *phase-coherent* measurement chain. Several concepts have been proposed for the synthesis and measurement of optical frequencies [32,

33, 34]. In this section we describe an optical frequency measurement which utilizes the principle of harmonic mixing, which is performed in several steps leading to a "frequency chain." In each step of the chain, the frequencies of two oscillators are compared. The lower frequency of the one oscillator is multiplied in a nonlinear device and the corresponding harmonic is compared to the higher frequency of the second oscillator that is operating close to the harmonic. The corresponding beat frequency is detected and can be phase-locked, resulting in a phase-coherent determination of the frequency ratio of the two oscillators. The nonlinear devices used in such a chain are Schottky diodes, metal–insulator–metal (MIM) diodes, and nonlinear crystals. The principle of harmonic mixing has been applied for frequency measurements of He–Ne lasers stabilized to absorption lines in methane at $\lambda \approx 3.39$ μm [35] and iodine at $\lambda \approx 633$ nm [12]. In the following we present a frequency measurement of the Ca optical frequency standard that represents the first phase-coherent frequency measurement of visible radiation [36].

Figure 17 shows the scheme of the chain used to measure the frequency of the optical Ca frequency standard. The high-frequency part of this chain is down-locked from the Ca optical frequency standard to the color-center laser (CCL). The lower part is locked from the 100 MHz standard frequency output of a hydrogen maser up to the methanol laser at ≈ 4.2 THz. The intermediate part of the chain, consisting of all CO_2-lasers, is locked to a methane-stabilized He–Ne laser to improve the frequency stability. To obtain the value of the Ca transition frequency, we simultaneously counted the beat signals of the methanol laser with the backward wave oscillator (carcinotron) and with the CO_2-laser, and the beat signal of the two CO_2-lasers with the CCL using totalizing counters. Combining the beat signals yields a frequency ratio independent of fluctuations of the intermediate transfer oscillators. This method allows one to track the phase of all intermediate oscillators and therefore leads to a truly phase-coherent measurement. We have operated the frequency measurement chain in the phase-coherent mode for an integrated time of about 3 hours. The dye laser was subsequently stabilized to the high- and low-frequency recoil component of the Ca atoms in the effusive beam. Frequency shifts due to optical phase errors were largely compensated by the method of laser beam reversal [24]. For each recoil component, the measured frequency values have a Gaussian distribution with an FWHM of approximately 900 Hz. The standard deviation of the mean is of the order of 20 Hz.

To determine the frequency of the intercombination transition of an unperturbed Ca atom at rest, we have simultaneously measured the frequency offset between the effusive beam and the trapped atoms. The remaining frequency difference is attributed to a residual first-order Doppler shift (optical phase errors) and the second-order Doppler shift in the thermal beam.

Figure 18 shows the Allan standard deviation of the measured laser frequency stabilized to the effusive beam (dots) and the MOT (squares) together with a fit of

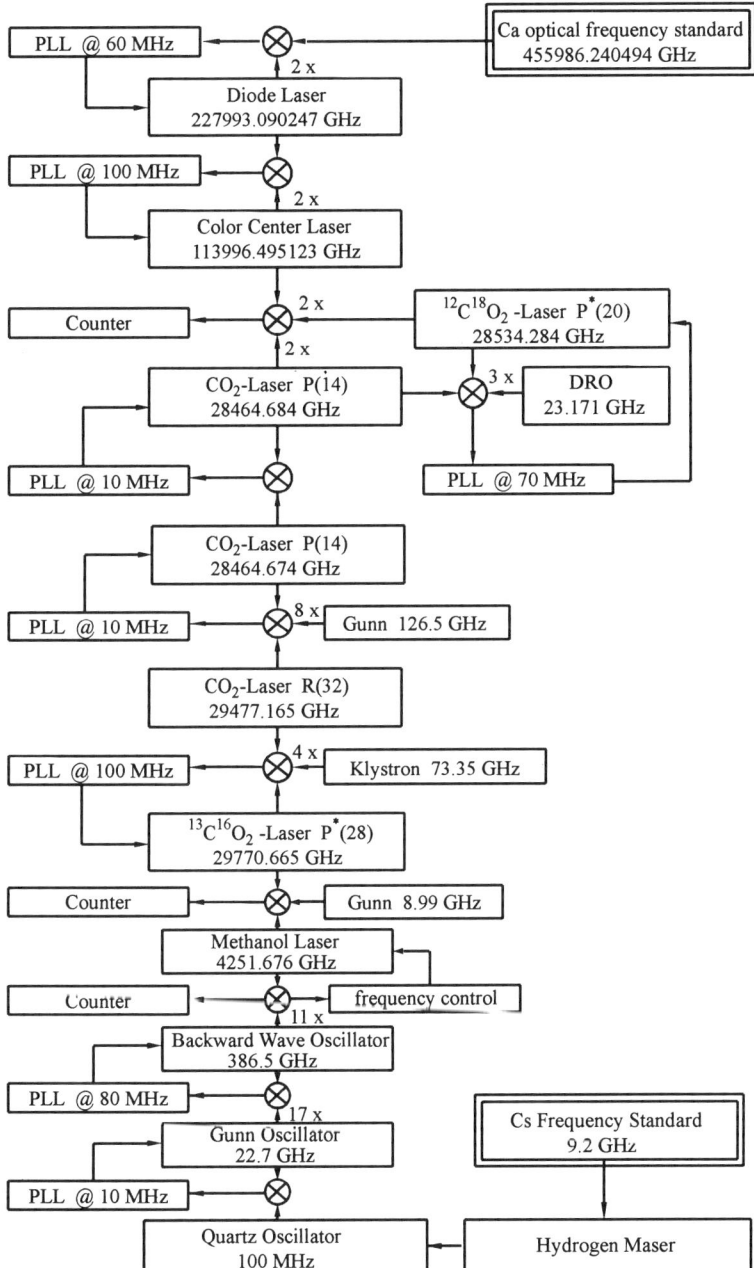

FIG. 17. Scheme of the PTB frequency chain. Each auxiliary oscillator is phase-locked to the Cs clock.

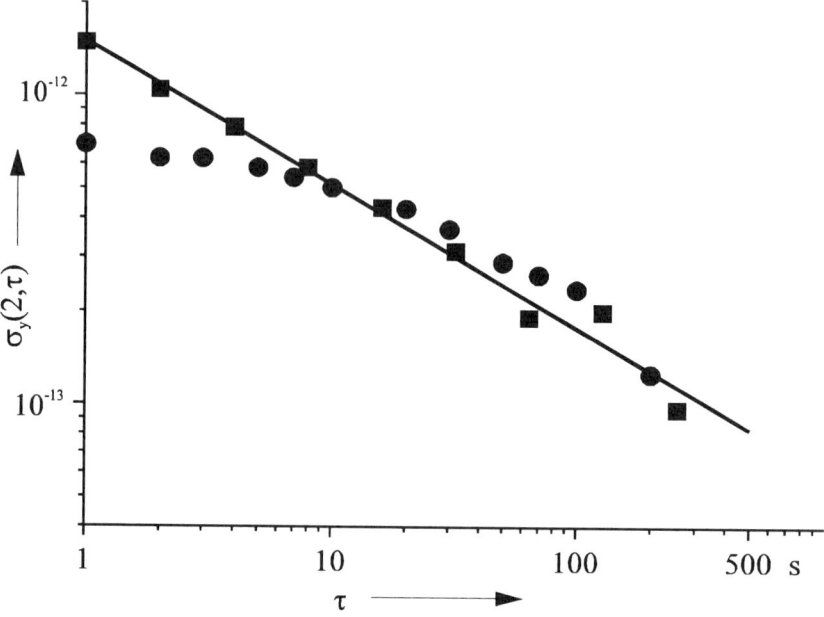

FIG. 18. Allan standard deviation $\sigma_y(\tau)$ of the measured frequency versus the integration time.

the data of the MOT (straight line) versus the integration time τ. The frequency instability of the MOT is dominated by white frequency noise, as can be derived from the fact that it is proportional to $\tau^{-1/2}$. It is limited by the signal-to-noise ratio of the trapped atoms (S/N ≈ 10) and the selected spectral resolution ($\delta\nu$ ≈ 10 kHz) given by the pulse separation. An estimation of the fractional instability $\Delta\nu/(\nu \cdot$ S/N) = 2.2×10^{-12} for τ = 1 sec agrees well with the fitted value of 1.6×10^{-12}. The mean of the high- and low-frequency recoil components is

$$455\ 986\ 240\ 494.07\ (0.35)\ \text{kHz.}$$

The uncertainty of the frequency value consists of the realization uncertainty estimated in Section 10.4.4 and the uncertainty of the frequency measurement. The corresponding contributions to the uncertainty are summarized in Table I. In principle, the phase-coherent frequency measurement should not contribute to the uncertainty. However, cycle slips may occasionally occur that lead to counting errors. The possible occurrence of such cycle slips depends critically on the signal-to-noise ratio of the relevant beat signals [37]. We estimated the rate of cycle slips in the most crucial stages of the chain and conclude that their contributions (<100 Hz) can be neglected at the present uncertainty level.

10.6 Conclusions

It was the aim of this chapter to discuss the basic design principles and the behavior of optical wavelength standards. As typical examples, we have described an iodine-stabilized He–Ne laser ($\lambda \approx 633$ m) as a traditional wavelength standard and a Ca-stabilized laser ($\lambda \approx 657$ nm) as a representative of a new generation of standards based on laser-cooled absorbers. Regarding the large variety of different laser wavelength standards, this chapter cannot be comprehensive. For example, we did not discuss lasers stabilized to single trapped ions [15, 16] or to other trapped atoms. For example, very narrow atomic interferences in trapped Mg atoms have also been observed and used for a frequency stabilization by the group of W. Ertmer [38]. We have chosen the Ca wavelength standard as a representative of an optical frequency standard based on cold absorbers since it is recommended for realization of the length unit and since it provides the lowest frequency uncertainty in the visible range due to precision measurement of its frequency.

Frequently, realization of optical frequency standards utilizing laser-cooled and trapped absorbers requires dye lasers for cooling and for probing the reference (clock) transition. With the development and refinement of diode lasers and diode laser pumped solid-state lasers, it is envisaged that dye lasers can soon be replaced by these new lasers. The introduction of laser diodes would greatly reduce the size, the complexity, and the power consumption of precision wavelength standards. With the use of a small atomic beam or a small trap for atoms or ions, the development of a transportable optical frequency standard is envisaged for the near future. For example, using a diode laser spectrometer, we have already observed Ca interference structures as narrow as 2 kHz [39]. Furthermore, the availability of powerful diode lasers at $\lambda \approx 846$ nm and of efficient nonlinear crystals for this wavelength opens the possibility to generate radiation with sufficient power at $\lambda = 423$ nm to cool and trap Ca atoms using diodes. Cooling and trapping of atoms or ions using radiation synthesized from diode lasers or diode laser-pumped lasers would be an important step on the way to a transportable precision optical frequency standard of unprecedented low uncertainty.

Acknowledgments

It is my pleasure to thank Dr. Fritz Riehle and Dr. Harald Schnatz for many helpful discussions during the preparation of the manuscript. I am also grateful to Tillman Trebst for his help in preparing the figures and for careful reading of the manuscript. The work reported in Sections 10.4 and 10.5 was supported by the Deutsche Forschungsgemeinschaft.

References

1. Editor's note: "Documents Concerning the New Definition of the Meter" (1984). *Metrologia* **19**, 163–177.

2. Barger, R. L. (1981). *Opt. Lett.* **6**, 145.
3. Zhu, M., and Hall, J. L. (1997). Chapter 5 in *Experimental Methods in the Physical Sciences: Atomic, Molecular, and Optical Physics*, Volume 29C (this volume), Academic Press, New York.
4. Bayer-Helms, F., and Helmcke, J. (1977). In *PTB-Report ME-77*, F. Bayer-Helms (ed.), pp. 85–109.
5. Wallard, A. J. (1972). *J. Phys. Electron.* **5**, 927.
6. Helmcke, J., and Bayer-Helms, F. (1977). In *PTB-Report ME-77*, F. Bayer-Helms (ed.), pp. 111–131.
7. Barnes, J. A., Chi, A. R., Cutler, L. S. Healey, D. J., Leeson, D. B., McGunigal, T. E., Mullen Jr., J. A., Smith, W. L., Snydor, R. L., Vessot, R. F. C., and Winkler, G. M. R. (1971). *IEEE Trans. Instrum. Meas.* **IM-20**, 105–120.
8. Quinn, T. J. (1993/94). *Metrologia* **30**, 523–541.
9. Gerstenkorn, S., and Luc, P. (1978). *Atlas du spectre d'absorption de la molécule d'iode*, Édition du CNRS, Paris.
10. See, for example, Hackel, L. A., Hackel, R. P., and Ezekiel, S. (1977). *Metrologia* **13**, 141–143.
11. Hanes, G. R., and Baird, K. M. (1969). *Metrologia* **5**, 32.
12. Acef, O., Zondy, J. J., Abed, M., Rovera, D. G., Gérard, A. H., Clairon, A., Laurent, Ph., Millerioux, Y., and Juncar, P. (1993). *Opt. Comm.* **97**, 29.
13. Bergquist, J. C., Barger, R. L., and Glaze, D. J. (1979). In *Laser Spectroscopy*, Vol. IV, Springer Series on Optical Science No. 21, H. Walther and K. W. Rothe (eds.), p. 120, Springer, Berlin.
14. See, for example, Ertmer, W., Blatt, R., and Hall, J. L. (1983). In *Laser Cooled and Trapped Atoms*, W. D. Phillips (ed.), U.S. National Bureau of Standards Special Publication No. 653, pp. 154–161.
15. See, for example, Tamm, Ch. (1993). *Appl. Phys. B* **56**, 240; Bell, A. S., Gill, P., Klein, H. A., Levick, A. P., Tamm, Ch., and Schnier, D. (1991). *Phys. Rev. A* **44**, R20.
16. Bergquist, J. C., Diedrich, F., Itano, W. M., and Wineland, D. J. (1989). In *Laser Spectroscopy*, Vol. 9, p. 274, M. S. Feld, J. E. Thomas, and A. Moradian (eds.), Academic Press, San Diego.
17. Zeiske, K. (1995). Ph.D. thesis, University of Hannover.
18. Beverini, N., Giammanco, F., Maccioni, E., Strumia, F., and Vassani, G. (1989). *J. Opt. Soc. Am. B* **6**, 2188–2193; Witte, A., Kisters, Th., Riehle, F., and Helmcke, J. (1992). *J. Opt. Soc. Am. B* **9**, 1030–1037.
19. Kurosu, T., and Shimizu, F. (1992). *Jpn. J. Appl. Phys.* **31**, 908.
20. Bordé, Ch. J., Avrillier, S., van Lerberghe, A., Salomon, Ch., Bassi, D., and Scoles, G. (1981). *J. Phys., Colloque C-8*, supplement au n° 12, pp. C8-15–C8-19; Bordé, Ch. J., Salomon, Ch., Avrillier, S., van Lerberghe, A., Bréant, Ch., Bassi, D., and Scoles, G. (1984). *Phys. Rev. A* **30**, 1836.
21. Helmcke, J., Zevgolis, D., and Yen, B. Ü. (1982). *Appl. Phys. Lett. B* **28**, 83.
22. Bordé, Ch. J. (1989). *Phys. Lett. A* **140**, 10.

23. Ramsey, N. F. (1950). *Phys. Rev. Lett.* **78**, 695.
24. Morinaga, A., Riehle, F., Ishikawa, J., and Helmcke, J. (1989). *Appl. Phys. B* **48**, 165–171.
25. Kisters, Th., Zeiske, K., Riehle, F., and Helmcke, J. (1994). *Appl. Phys. B* **59**, 89–98.
26. Baklanov, Ye. V., Dubetsky, B. Ya., and Chebotayev, V. P. (1976). *Appl. Phys.* **9**, 171.
27. Drever, R. W. P., Hall, J. L., Kowalski, F. V., Hough, J., Ford, G. M., Munley, A. J., and Ward, H. (1983). *Appl. Phys. B* **31**, 97–105.
28. Helmcke, J., Snyder, J. J., Morinaga, A., Mensing, F., and Gläser, M. (1987). *Appl. Phys. B* **43**, 85–91.
29. Riehle, F., Kisters, Th., Witte, A., and Helmcke, J. (1992). In *Laser Spectroscopy*, pp. 246–251, M. Ducloy, E. Giacobino, and G. Camy (eds.), World Scientific, Singapore.
30. Aharonov, Y., and Casher, A. (1984). *Phys. Rev. Lett.* **53**, 319.
31. Zeiske, K., Zinner, G., Riehle, F., and Helmcke, J. (1995). *Appl. Phys. B* **60**, 205–209.
32. Telle, H. R., Meschede, D., and Hänsch, T. W. (1990). *Opt. Lett.* **15**, 532.
33. Wong, N. C., and Lai, B. (1996). In *Frequency Standards and Metrology*, J. C. Bergquist (ed.), pp. 319–326, World Scientific, Singapore.
34. Kourogi, M., Ohtsu, M., and Saito, T. (1996). In *Frequency Standards and Metrology*, J. C. Bergquist (ed.), pp. 327–332, World Scientific, Singapore.
35. Kramer, G., Weiss, C. O., and Lipphardt, B. (1989). In *Frequency Standards and Metrology*, A. de Marchi (ed.), pp. 181–186, Springer-Verlag, Berlin.
36. Schnatz, H., Lipphardt, B., Helmcke, J., Riehle, F., and Zinner, G. (1996). *Phys. Rev. Lett.* **76**, 18.
37. Telle, H. (1996). In *Frequency Control of Semiconductor Lasers*, M. Ohtsu (ed.), pp. 137–172, Wiley, New York.
38. See, for example, Sengstock, K., Sterr, U., Hennig, G., Bettermann, D., Müller, J. H., and Ertmer, W. (1993). *Opt. Comm.* **103**, 73.
39. Kersten, P., Celikov, A., D'Evelin, L., Zinner, G., and Riehle, F. (1997). To be published.
40. Riehle, F., Schnatz, H., Lipphardt, B., Kersten, P., Trebst, T., Zinner, G., and Helmcke, J. (1996). In *Proceedings of the Workshop on "Frequency Standards Based on Laser-Manipulated Atoms and Ions,"* J. Helmcke and S. Penselin (eds.), pp. 11–20, PTB-Report, ISBN 3-89429-719-0.

11. PRECISE WAVELENGTH MEASUREMENT OF TUNABLE LASERS

Miao Zhu

Hewlett-Packard Laboratories
Palo Alto, California

John L. Hall

JILA, National Institute of Standards and Technology
and University of Colorado
Boulder, Colorado

11.1 Introduction

For many applications involving tunable lasers, an adequate knowledge of the optical frequency is of great importance. However, measuring the optical frequency directly is not a trivial task [1–3]. Alternatively, one measures the (plane wave) laser wavelength $\lambda = c/f$, where c is the speed of light, and f is the laser frequency. For low-resolution applications, a monochromator often gives enough resolution. For high-resolution applications, interferometric measurements of the laser wavelength are commonly used. While frequency-stabilized tunable lasers can give sub-Hertz linewidths, their frequencies measured via interferometric methods are less accurate by many orders. Therefore, in these ultrahigh-resolution measurements, say higher than 10^{-10}, direct optical frequency measurement, either directly linked to the microwave primary frequency standard or to a well-documented atomic transition, is probably a better choice.

A variety of interferometers have been developed for measuring optical wavelength. The λ-meter [4, 5], the Fizeau interferometer [6], the Fabry–Perot cavity [7], and the σ-meter [8] are probably the ones most often used. The properties of these wavemeters are summarized in Table I.

In the following sections we discuss the λ-meter and the Fizeau wavemeter in some detail. Rather less detail is given for the Fabry–Perot cavity, as the techniques are closely related to the Fizeau system. The reader is referred to the literature for the σ-meter approach [8, 9].

TABLE I. Summary of Interferometer Properties

Laser type	λ-meter	Fizeau interferometer	Fabry–Perot cavity	σ-meter
Laser type	CW only	CW or pulse	CW or pulse	CW or pulse
No. of interferometers	one	one	multi	multi
Additional low-resolution device	no	no	no	yes
Calibration	no	yes	yes	yes
Reference laser	yes	no	no	yes
Resolution	10^{-9}	10^{-7}	10^{-11}	10^{-9}
Power spectrum measurement	yes	no	no	no

11.2 The λ-Meter (Scanning Michelson Interferometer)

11.2.1 Basic Principles of the λ-Meter

The λ-meter is basically a scanning Michelson interferometer [10, 11]. The principle of the λ-meter has been described elsewhere in detail [4]. At its heart it consists of measuring the same length-by-displacement using two wavelengths. The two wavelengths give rise to two different multipliers (fringe numbers) to express the same physical distance. The ratio of two fringe numbers is then inversely related to the wavelength ratio. We will look at how the data are obtained during the motion.

For the basic arrangement shown in Fig. 1a, a monochromatic plane wave input laser field generates an interference term in the output signal, which is proportional to

$$\left[E(t)E^*\left(t - \frac{L}{c}\right) + \text{c.c.} \right] \propto \cos\left(\frac{2\pi L}{\lambda_{un}}\right) = \cos\left(\frac{4\pi v t}{\lambda_{un}}\right), \qquad (1)$$

where L is the optical path difference of the two arms, c is the speed of light, λ_{un} is the laser wavelength to be measured, and v is the speed of the moving mirror. Another interference term generated by a reference laser determines the displacement of the moving mirror in real time. The wavelength of this reference laser, λ_{ref}, is well documented.

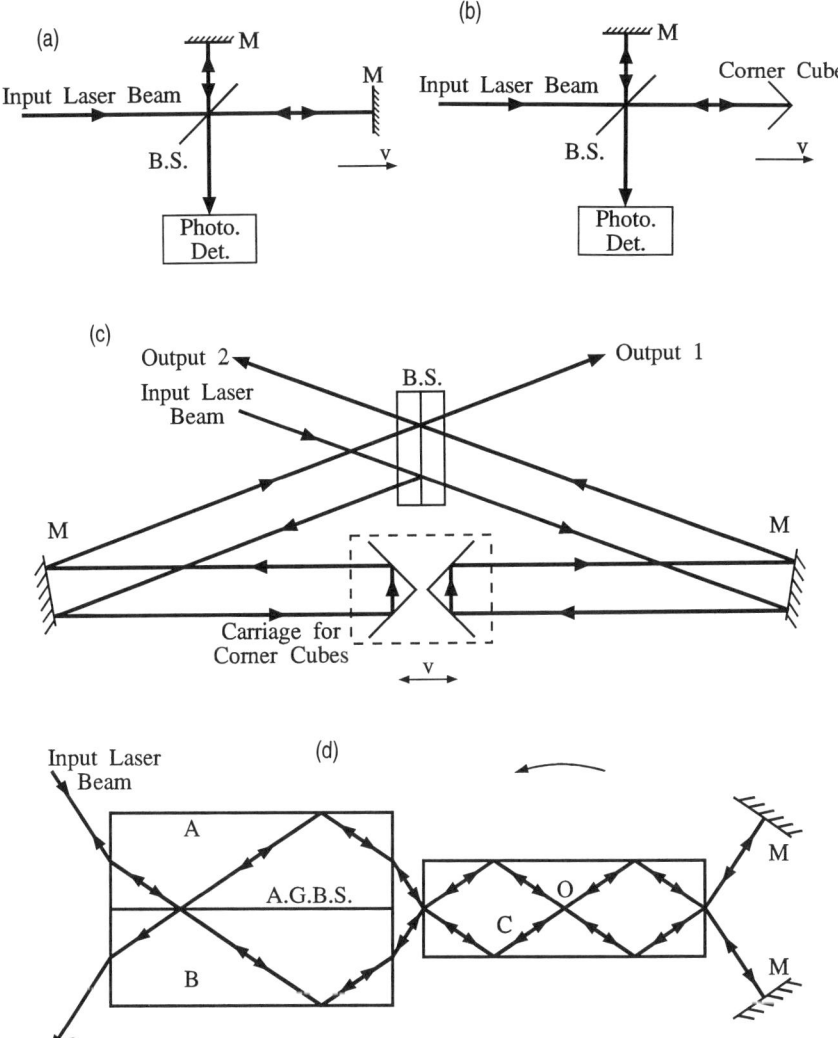

FIG. 1. λ-meter configurations: B.S. = beam splitter, M = mirror, A.G.B.S. = air gap beam splitter. (a) Basic scanning Michelson interferometer. (b) The moving mirror in (a) is replaced by a corner cube. Alternatively, a cat's eye can be used to replace the moving mirror. (c) Both mirrors in (a) are replaced by corner cubes. The symmetric arrangement cancels the misalignment due to the dispersion of the beam splitter. (d) Rotating parallelepiped configuration. A, B, and C are three rectangular parallelepipeds. The axis, marked as O, of rotation of parallelepiped C passes the center of the parallelepiped C, and is normal to the plane of the figure. See the text for detailed discussion.

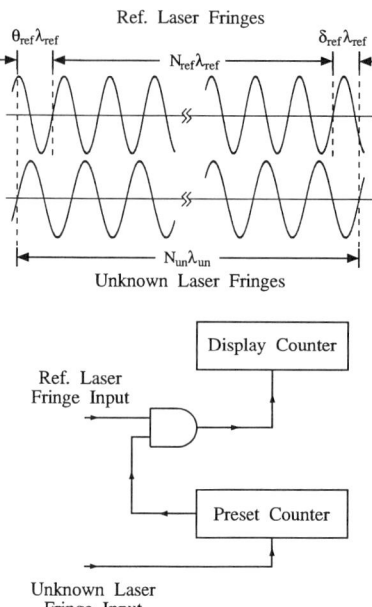

FIG. 2. Fringe counting for λ-meter data acquisition. Preset counter opens the gate for N_{un} unknown laser fringes. Only direct counting is shown here. See the text for more discussion.

Figure 2 illustrates one of the data acquisition methods. During the data acquisition time, the scanned interferometer arm difference is equal to N_{un} (integer), unknown laser wavelengths. There are $(N_{ref} + \theta_{ref} + \delta_{ref})$ reference laser wavelengths for the same interferometer arm difference scan. The unknown laser wavelength, λ_{un}, and the reference laser wavelength, λ_{ref}, are related by

$$\lambda_{un} = \frac{\lambda_{ref}}{N_{ref}} (N_{ref} + \theta_{ref} + \delta_{ref}) . \qquad (2)$$

To read the unknown laser wavelength directly, one typically will count the reference laser fringes, generating triggers for example at the positive-going zero-crossings. The duration of the gate opening time is controlled by another counter counting N_{un} unknown laser fringes. This direct counting scheme has a counting uncertainty of ±1 count, which sets the scale of the instrument resolution to $1/N_{ref}$. The interpolation task is then to learn the fringe fractions $\theta_{ref}\lambda_{ref}$ and $\delta_{ref}\lambda_{ref}$ at the instants when the gate is open and closed.

The principle of measuring the length-by-displacement with λ_{un} and λ_{ref} simultaneously eliminates many influences, for example, the static optical phase shift of all the optical elements, the mechanical imperfections, the varying mirror speed. Consequently, the instrument itself does not require calibration. Since only a single-element photodetector is needed for each set of fringes, there is no tedious detector uniformity calibration involved. A commercially available frequency counter can be used for fringe counting. As an alternative, a fringe counter can be readily built for this application. It is not necessary to have a computer dedicated for the measurement.

This time-varying interference term shown in Eq. (1) can also be seen as the autocorrelation function of the input laser field with a time delay of L/c. If this signal is properly sampled and scaled, a Fourier transform of the sampled data is basically the input laser's power spectrum.

11.2.2 Physical Limits on Accuracy and Resolution: Setting the Scale

The λ-meter can be used for highly accurate measurements of laser wavelengths. For instance, a measurement with 10^{-9} accuracy at a 10-μm wavelength has been reported [12]. For common laboratory lasers in the visible and near-infrared range, $1-2 \times 10^{-7}$ (Δf on the order of 50–100 MHz) accuracy could be a rational goal for resolution. While this resolution may not be satisfactory in a metrology laboratory, it can usually guide the laser frequency to the place of interest, for example, within the central 10% of the Doppler profile of an atomic or molecular transition, where the more detailed spectroscopic features can be obtained by scanning the laser frequency. The higher resolution could be then obtained using these spectroscopic features with, for example, one of the methods of Doppler-free spectroscopy [13]. Here we discuss some of the limits of the achievable accuracy and resolution of the λ-meter.

One ultimate accuracy limit of laser wavelength measurement by using a scanning Michelson interferometer arises from the diffraction of laser beams [12]. Using the full beam of a TEM_{00} mode in near-field, the error due to diffraction of the Gaussian beam is

$$\frac{\lambda_m - \lambda}{\lambda} = \frac{\lambda^2}{4\pi^2 w_0^2}, \tag{3}$$

where λ_m is the measured wavelength, λ is the laser wavelength, and w_0 is the waist radius of the beam. In the visible and near-infrared wavelength region this limit is less serious. Using a large beam waist size to circumvent this limit requires expensive large-size/high-quality optical elements. Still, for a red beam ($\lambda = 633$ nm) of $w_0 = 2$ mm, the diffraction correction is 2.5×10^{-9}. Note that, if a matching value of λ/w_0 is chosen for the reference beam, there will be a very strong cancellation of this already small systematic offset. So diffraction does not need to be our first-order limitation.

At least a few papers using λ-meter techniques have contained carefully measured precise results that are wrong by a scale factor. The displacement measured can easily be in error if the laser propagation direction is not parallel to the mirror's motion. If the angle between the laser propagation direction and the mirror's motion direction, θ, is 1 milliradian, the fractional error will be $1 - \cos\theta \approx \theta^2/2 = 5 \times 10^{-7}$. In practice, the mirror displacement is represented by the reference laser fringes; then θ is the angle between the two laser propagation directions. Fortunately, there are powerful and simple alignment methods to be discussed momentarily.

Dispersion of the air's refractive index ($n_{air} - 1 \approx 2.7 \times 10^{-4}$ at 10^5 Pa) will also affect the measured wavelength, assuming the measurement is done in the air. This error can be basically eliminated by putting the interferometer into a vacuum enclosure maintained below 0.1 Pa. The error can be reduced about tenfold if the chamber is filled with an atmosphere of helium rather than of air. Ordinarily the refractive dispersion is compensated by calculating the air's refractive index according to the measured laboratory conditions: air composition, total pressure, temperature, etc. For nearby wavelengths the dispersion of the basic 270-ppm air index may be only some 0.5%, leading to ppm level corrections of the apparent measured wavelength ratios to convert them to a vacuum basis. However, unless the wavelengths are very similar, it rapidly becomes troublesome to make the air dispersion corrections adequately to reach an inaccuracy below 10^{-8}.

The accuracy of the reference laser frequency, which determines the scanning distance of the interferometer, obviously limits the accuracy obtainable by the measurements. As will be discussed subsequently, with a frequency-stabilized reference laser, for example, an He–Ne laser stabilized to an iodine hyperfine transition, this limit is usually not a problem.

In data acquisition the direct fringe counting method is seldom used alone in practice because of its ±1 counting uncertainty. The maximum scanning distance is usually limited by the practical interferometer size and the data acquisition time. Since the obtained fringes can show a very high signal-to-noise ratio, they offer us a clear invitation into the art of subdividing a single fringe. Thus, a more practical resolution limit is set by how well the fraction of a fringe can be determined. Several different methods have been developed, as we will discuss in the later sections.

So the bottom-line summary is that laser beam collimation, alignment, refractive index correction, and fractional fringe determination stand as our problem areas. All these problems need to be addressed properly before accurate measurement results can be obtained.

11.2.3 Structure of Scanning Michelson Interferometers

11.2.3.1 Optical Configurations
There are two major limitations for the basic arrangement of a scanning Michelson interferometer using plane mirrors, as

shown in Fig. 1a. First, it is necessary that the moving mirror be maintained in perfect alignment. This is a stringent requirement if the mirror is moved over a distance of more than a couple of centimeters. The second difficulty is the strong optical feedback, whose amplitude and phase depend on the changing arm difference returned back to the laser. To relax the stringent requirement for perfect alignment of the moving mirror, one can replace the moving mirror with a retroreflector, as shown in Fig. 1b. Either a corner cube or a cat's eye can be used as the retroreflector. The cat's eye maintains the polarization if a small numerical aperture is used. The corner cube can be used respecting its intrinsic polarization eigendirections [14], or more often it is metal-coated on the roof facets to minimize depolarization effects. The gap, a_0, between the edges of the corner cube introduces an additional correction [12], which can be estimated as

$$\frac{\text{gap correction}}{\text{diffraction correction}} = -\frac{9}{2\sqrt{\pi}} \frac{a_0}{w_0}. \tag{4}$$

Hollow (or open) corner cube retroreflectors with first surface mirrors and adequately small gaps are ideal and are widely used for this λ-meter application since they are commercially available at reasonable cost. Consequently, other types of polarization-preserving retroreflectors [15] are seldom used for laser wavelength measurements. Alternatively, one can use a PZT-driven flat mirror with a quadrant detector and a servo system to maintain the perfect alignment [16]. The advantage of this method is that it is easier to obtain a high-quality large-size mirror than a corner cube with similar quality and size.

One advantage of the Michelson arrangement with plane mirrors is that the interferometer can be used as a shearing interferometer to test and establish laser beam collimation and alignment. One plane mirror versus a corner cube functions similarly. Further, when the reference and unknown laser beams are aligned with this interferometer to give a flat interference pattern, both laser beams are aligned to be perpendicular to a common mirror and hence are parallel to each other [17].

Figure 1c shows another commonly used arrangement with both mirrors replaced by retroreflectors. Although this configuration loses the advantage of autoalignment, this is ordinarily provided by temporarily inserting a wedge-free partially reflecting mirror perpendicular to one of the beams. The other laser input is then aligned to be perpendicular to this plate (both faces of this plate), and so one has generated the parallel beams needed for good accuracy. (This procedure is described more fully below in Section 11.2.4.) A major advantage of the dual corner cube configuration is that there is an input/output spatial offset, and hence there is no direct optical feedback to the laser. Also, the configuration can be made symmetric so that the dispersion of the beam splitter does not introduce a wavelength-dependent misalignment. It can be arranged so that the light beams stay basically within the individual sectors of the corner cube, and so the gaps of the corner cube do not strongly affect the laser beams. If the laser beam to be measured has a serious amount of amplitude noise, the offset configuration with two corner retroreflectors has another very important advantage: it generates a complementary output signal. This complementary signal nominally has a phase shift of π when a low-loss dielectric coating is used for the beam splitter. In this case, the signal-to-noise ratio can be increased by dc-subtracting the two signals. One will find an interferometric phase in which the output is zero, basically independent of the amplitude of the laser. The discriminator is then carefully set to trigger at this level, thus gaining a tremendous suppression of the laser's amplitude noise. If a slightly absorbing metallic coating is used for the beam splitter, this complementary output, whose phase shift now can be significantly away from π, can be used for autocorrection of the occasional backward motion of the moving retroreflectors [18]. But for any reasonably well-designed mechanical system, it is not necessary to have this autocorrection.

Returning to our consideration of various interferometric approaches, Fig. 1d shows a quite different configuration. It consists of three identical fused silica blocks and two mirrors. An air gap between the first two blocks, A and B, serves as a beam splitter. The scan of the optical path difference is accomplished by rotating the third fused silica block, C. The axis of rotation passes through the center of the block and is normal to the plane of the figure. This optical configuration provides a higher data acquisition rate and is less sensitive to translation motion. Detailed discussion on this arrangement can be found in references [19–21]. In the following sections we limit our discussion to a scanning interferometer with linear mechanical translation, such as the one shown in Fig. 1c, but the basic principles apply to all the other arrangements.

11.2.3.2 Mechanical Design Guidelines
In order to change the optical path difference of two arms of the interferometer while maintaining the alignment, the mechanical design should be made so that the motion of the carriage, upon which the retroreflectors are mounted, is as straight and smooth as possible. Smoothness of the motion is important because irregularities can directly limit measurement of the fringe fraction. The general guidelines for construction of the carriage are: (1) the carriage's center of mass should be at or just below the points where the vertical supporting forces are applied; (2) the translating force exerted on the carriage should pass the carriage's center of the mass to reduce the rotational torque on the carriage; and (3) the apices of the retroreflectors should coincide with each other and coincide with the center of the mass of the carriage, because the rotation of a hollow retroreflector around its apex does not result in displacement of the reflected beam. Practically, these requirements are difficult to meet simultaneously, especially for hollow corner cube retroreflectors where the structural material is actually external. But efforts should be made to approach these guidelines in the design.

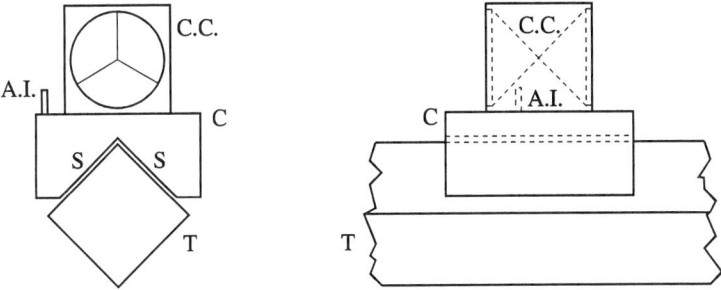

FIG. 3. Corner cube carriage for an air track system. A.I. = compressed air inlet, C = carriage, C.C. = corner cubes, S = surfaces the compressed air comes through to levitate the carriage, T = track.

There are two ways to support the moving carriage. One approach uses an air track to support the carriage [4]. The compressed air can travel inside the track and come out on the small holes on the track surface, although the later design refinements by Drullinger lead to improved performance by suppressing the unnecessary air flows [22]. Figure 3 shows one contemporary design for an air-floating system. A precisely ground steel bar with a 40×40 mm^2 square cross is used as a track to support the carriage. The necessary air pressure to support this slider is supplied via a thin flexible air supply hose, from a pressure $\sim 3 \times 10^5$ Pa. The air comes out of the bottom of the carriage through a number of small holes. Because the carriage experiences only very small air friction, the main and still very small dragging force is due to flexing of the ~3-mm diameter gum–rubber compressed air supply tubing. The carriage, when supplied with adequate compressed air, is completely levitated from the track and requires only a slight push at each end of the track to maintain the motion. The smooth motion makes fringe interpolation as painless as possible. The main drawback of the air-floating support system is that it cannot be used in the vacuum to circumvent the influence of the air refractive index. Since there is a horizontal driving force exerted on the carriage only when it is at the ends of the track, the system does not work very well when the traveling length of the carriage is excessively long: it works well at 1.3 m, but some slowdown is evident. This slowdown requires a larger dynamic stability range for the phase-locked loop (PLL) if this approach is used for the fringe fraction interpolations. This also indicates that the speed of the carriage cannot be servoed.

Higher resolution measurements, say better than 10^{-7}, require a longer arm difference interferometer, perhaps operation in a vacuum, and maybe even a servo control for the carriage motion. This leads to an alternative support system for the carriage. Figure 4 shows the basic design used by Hall and Lee [23]. Other

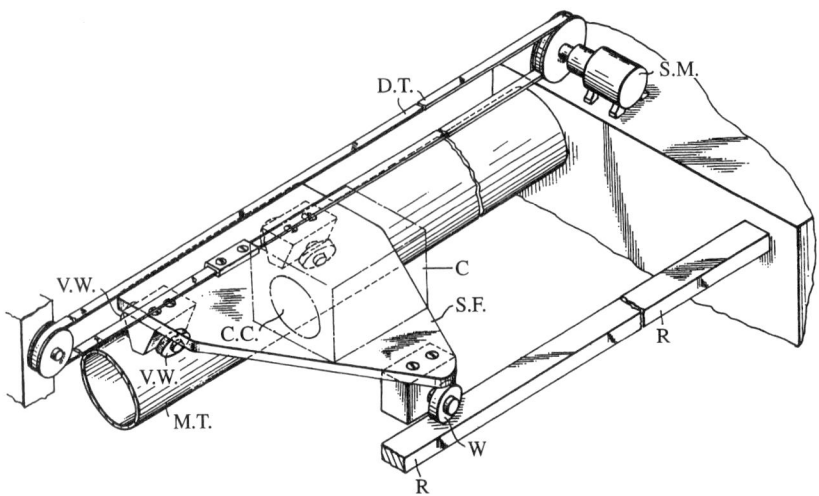

FIG. 4. Structure of JILA's first λ-meter. C = carriage, C.C. = corner cube, D.T. = driving tape, M.T. = metallic tube, R = railway, S.F. = support frame, S.M. = synchronous motor, V.W. = V-shaped wheel mount, W = wheel mount. See reference [23].

variations of this design we have tried include replacing the rectangular track with a round track, and using narrow sets of wheels attacking the support rods at ±45° from above, mimicking a kinematic mount. A V-shaped contacting surface with a teflon coating was successfully used [12]. One finds it useful to smooth the motion with a flywheel [12, 24]. Either a mylar tape from a cassette or a reel-to-reel tape recorder, or a special thin steel tape, or a braided steel wire can be used to transmit the driving force to the carriage. One or more strong springs are usually used to tighten the driving tape or wire. Either a dc motor or an ac motor (preferably a synchronous motor) with proper gear reduction can be used as the driving source. Some vibration isolation may be required to reduce the motor's perturbation to the laser systems that are to be used for measurement.

Further control of the speed of the carriage in this kind of support system could be obtained using servo control to phase lock the reference laser fringe to a suitable signal generated by a quartz oscillator. The rotation speed of the motor could be readily controlled either by changing the motor power supply voltage (for a dc motor) or frequency (for an ac synchronous motor driven by a stereo amplifier). To increase the unity gain frequency of this servo loop, the retroreflectors can be mounted on a PZT tube or PZT stacks instead of mounted on the carriage directly. But attention should then be paid to the carriage recoil. In fact, with a simple mechanical system using a flywheel and a wire for carriage pulling, when things

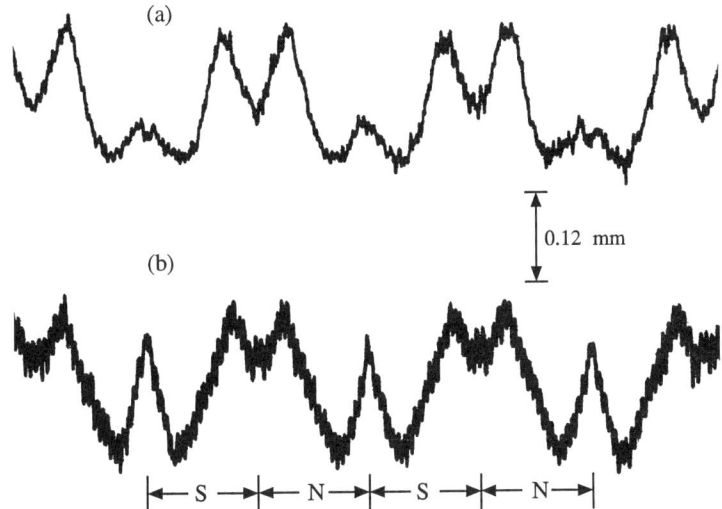

FIG. 5. Straightness of the tracks in JILA's 2.5-m λ-meter. Corner cube carriage travel direction: N = north, S = south. The data shown here are the lateral displacement of the laser beam reflected from the corner cube. The track distortion is half of this displacement. (a) Vertical. (b) Horizontal.

are working normally one finds a speed variation below 5% with Fourier frequencies below some tens of hertz. These speed variations are basically adiabatic with respect to the dual channel phase-locked fringe multipliers and affect both channels equally, so that there is little incentive to work harder on the constant-velocity issue unless some very demanding data acquisition methods are used.

In the 2.5-m JILA evacuable λ-meter, the transverse displacement of the laser beam reflected from the corner cube is controlled within 0.24 mm peak-to-peak, as shown in Fig. 5, limited by the straightness of the tracks. If this is due purely to the lateral displacement of the carriage, the two output beams will experience the same transverse displacement and result in no effect on the measurement. However, if this is partly due to the rotation of the carriage, the displacements of two output beams are not quite the same, giving an additional term which shows in the diffraction correction

$$\frac{\lambda_m - \lambda}{\lambda} = \frac{\lambda^2}{4\pi^2 w_0^2}\left(1 + \frac{x_0^2}{2w_0^2}\right), \qquad (5)$$

where x_0 is the small distance between two parallel beams. Basically, this effect arises because the beams of finite aperture cannot have flat wavefronts everywhere in the optical path. There is some evidence that the rotation effect dominates in our system.

11.2.4 Alignment

11.2.4.1 Collimation of Laser Beams
The first step in aligning laser beams to a λ-meter is to collimate both laser beams to the desired waist sizes. One can determine the existing beam parameters by measuring the beam sizes at two places, then making use of the Gaussian beam propagation formula. Or one of the commercially available automated beam measurement systems may be used. A beam expander (a reversed telescope) is usually needed for collimation. The ratio of the focal lengths of the telescope lenses can be readily calculated from the existing beam waist and the desired beam waist. Either Keplerian (two positive lenses) or Galilean (one negative lens and one positive lens) telescope configuration can be used. The Galilean configuration is more compact than the Keplerian version, but the Keplerian configuration offers a waist between two lenses for an optional spatial filter. Lenses designed for infinite conjugate ratio are preferred for this application. The degree of collimation can be tested either with a parallel plate or a wedged plate working as a shearing interferometer [25]. This shearing interferometer not only tests the beam collimation but also the aberrations of the whole optical system. In fact, one can just use this shearing interferometer to test the lenses used for the beam expander if it is necessary. As noted earlier, a stronger discrimination can be obtained by setting up a temporary Michelson interferometer that uses a plane mirror against a retroreflector (these are intrinsically image-inverting, thus maximizing the wavefront shear). The interference fringes from the shearing interferometer can be observed by eye or by using a CCD camera. For simplicity one usually uses a plane-parallel plate tipped at an appropriate small angle to induce wavefront shear while still providing some overlap between the two reflected spots. In order to avoid inducing birefringence and/or distortion, a minimum force should be used to mount the solid interferometer plate. In the case of a parallel plate, the interference fringe pattern along the shear direction, x, is given by

$$I(x) \approx I_0(x) \cos\left(\frac{2\pi}{\lambda} \frac{x_0 x}{R} + \varphi_0\right), \tag{6}$$

where $I_0(x)$ is the intensity envelope, x_0 is the shear distance between two reflected beams, R is the radius of curvature of the wavefront, and φ_0 is an initial phase. A change in the incidence angle changes the shear distance x_0, which is proportional to the sensitivity of this interferometer. It also changes the initial phase φ_0, so that a dark fringe pattern can be reached for easy adjustment. For scale, if the change in the fringe across a 20-mm diameter beam with 2 mm shear is less than $\lambda/10$, at

$\lambda = 600$ nm the radius of curvature is larger than 600 m. In the case of a slightly wedged shear plate, the wedge should be aligned perpendicular to the shear direction. There are always interference fringes across the beam. The number of fringes depend on the wedge angle and the wavelength. The rotation direction of the fringes relative to the shear direction indicates the convergence or divergence of the beam. It is somewhat difficult to determine the direction of the fringes for a perfectly collimated beam unless the wedged plate is premounted and a reference line is made with a perfectly collimated beam.

11.2.4.2 Making Two Beams Parallel (and Overlapped) It is important to align the reference laser beam and the unknown laser beam such that they are parallel and overlapped spatially. The relative measurement error is $(1 - \cos \theta) \propto (\theta^2/2)$, where θ is the angle between the two laser beams. The best solution for this alignment part is to choose the interferometer configuration, for example, as shown in Fig. 1b or Fig. 1c, with a temporary flat plate as described earlier, such that this alignment is automatically accomplished when the beams are aligned to the interferometer to generate a flat interference pattern. In the case this interferometer arrangement is not chosen, the following guidelines can be used. If both unknown and reference laser beams can propagate in an optical fiber with single modes, one can combine both beams into a suitable optical fiber that serves as a spatial filter. A polarization-preserving fiber should be used with two laser beams orthogonally polarized, as this allows polarization to be used to separate these two beams for detection. One disadvantage is the low coupling efficiency of the optical fiber (typically <50%). To reduce the interference due to fiber end reflections, the fiber output end could be cut somewhat away from 90° relative to the fiber axis. However, because of Snell's law and the dispersion of the optical fiber material, one prefers the fiber output end to be cut perfectly 90° relative to the fiber axis. One choice to reduce interference then is to cut the input end of the fiber a few degrees away from 90° relative to the fiber axis while the output end is cut as close to 90° relative to the fiber axis as possible. An experimental difficulty is that the angle of the fiber end relative to the fiber axis is not a very well-controlled parameter. On some occasions, the "test and fail" procedure has to be used several times before a suitable cut is obtained. We are also getting reasonable results by attaching an antireflection-coated window to the well-cleaved fiber end with index-matching grease or cement. Lateral expansion of the diverging beam greatly reduces the coherent feedback from the finite reflection of the window's exit face. However, the fiber axis still needs to be normal to the exiting window surface, and some wavefront aberrations may be introduced if the fiber's numerical aperture (NA) is quite large.

The collimation of both laser beams exiting the optical fiber is ideally accomplished with a single set of optics at the output of the fiber. Achromatic lenses or an all-reflection system should be used. Because both beams have the same waist sizes when they leave the optical fiber [26], collimation of both beams usually cannot be perfect simultaneously even with perfect achromatic optics. Sometimes one has to compromise this fact and include it into the measurement error budget. On the other hand, it is not impossible to design a custom collimation system compensating this fact. We are planning to try an all-reflective long-focal-length telescope system someday soon.

When an optical fiber cannot be used as the spatial combiner/filter, the two laser beams have to be aligned and overlapped by other means. If a single detector can be used for both the reference and unknown lasers, then a quadrant detector can be positioned at two places to test the overlap of two laser beams when the co-propagating configuration is used. Otherwise, the laser beams have to be retroreflected before alignment. A weaker alternative method uses two separated mechanical apertures, but a long distance between these two apertures is needed for the required accuracy. Sometimes the length of the λ-meter can be used as part of this baseline distance for alignment. This requires that one of the laser beams (usually the reference beam) be aligned with the λ-meter before the other beam is aligned, but this usually does not cause extra difficulties.

Another accuracy limitation is dispersion of a wedged beam splitter, which would cause two laser beams to angularly deviate from each other, introducing an uncorrectable cosine error. This could be readily corrected using a compensating plate with identical material and wedge angle as the beam splitter, but a judicious choice of a small wedge is more typically employed. Of course the beam splitter's rear face should be antireflection coated.

11.2.4.3 Alignment of the Interferometer
The following instructions can be understood with reference to Fig. 1c:

1. Align the input beam such that the transmitted beam is approximately at the correct position on the right-hand mirror.

2. Use the beam splitter to position the reflected beam on the left-hand mirror.

3. Use the right-hand mirror to align the beam such that the wave vector **k** is parallel to the carriage velocity **v**. If **k** is not parallel to **v**, the output beam from the retroreflector will move laterally when the retroreflector moves longitudinally. This part of alignment can be readily accomplished by aligning the laser beam iteratively when the retroreflector is placed at two extreme positions. For precision work, this return spot can be detected with a four-quadrant position detector and displayed in an x–y mode oscilloscope. Because of the available electronic magnification, the track straightness defects and the error of the retroreflector certainly can show up in this part of the alignment.

4. Use the beam splitter and the left-hand mirror to align the left part of the interferometer such that **k** is parallel to **v** AND so that the output beam is overlapped with the output beam of the right part of the interferometer.

5. Since the movement of the beam splitter in step 4 will change the transmitted beam spatially, minor realignment of the right part of the interferometer is necessary.

6. Several iterations of the above procedure may be required before the whole system is aligned. As noted, a quadrant detector is a very useful tool in this alignment procedure.

11.2.4.4 Stray Light and Feedback Isolations Stray light from all undesired surfaces, even after multiple reflection/transmission paths, can introduce systematic error in the measurement. Light baffles can be used to keep stray light from reaching the detectors in the whole carriage traveling distance. Another source of measurement error can arise from optical feedback to the laser, especially in the interferometer arrangement from which there is a direct output beam propagating back to the laser. A combination of a linear polarizer and a Faraday rotator can be used as an isolator between the laser and the λ-meter. A carefully aligned Faraday rotator type isolator can provide more than 30 dB isolation at one particular wavelength. Acoustooptic modulators (AOMs) can also be used for isolation [27]. The reflected beam would have twice the frequency shift, say 160 MHz if an 80-MHz AOM is used. This frequency-shifted backscattered beam is drastically weakened inside the laser cavity (at least when the laser cavity has a narrow enough linewidth). This frequency shift also prevents the feedback beam from interacting with the same atoms that provide gain for the lasing frequency. So the laser is happy. Unfortunately, with no further isolation precautions, this frequency-shifted beam will return into the interferometer only somewhat weakened and will set up a measurable fringe pattern for an optical frequency shifted by 160 MHz. Its harmonics will also appear in a geometric progression. In our experience, isolation is never better than just marginally sufficient if one is interested in 10^{-9} wavelength measurement accuracy.

11.2.5 Reference Laser

The accuracy of the reference laser frequency affects the measurement result directly. The He–Ne laser at 633 nm is usually chosen because of low cost and convenience. The gain profile width limits the wavelength accuracy of a free-running 633-nm He–Ne laser to about 10^{-6}. To improve the accuracy of this reference laser one can use a well-designed polarization-stabilized He–Ne laser or a Zeeman-stabilized He–Ne laser to approach about 10^{-9} accuracy by locking to the Ne gain curve [28, 29]. Changes in the gas pressure over the operating life can cause a frequency shift of a few tens of megahertz, so that calibration is necessary for accurate work [29]. For ultimate accuracy one can use well-documented hyperfine transitions in iodine molecules as references to which an He–Ne laser frequency can be locked. He–Ne lasers ($\lambda = 612$ nm) stabilized to iodine transitions are also used in some laboratories. As diode laser systems become better developed, other choices appear to be reasonable, such as the Rb two-photon line at 778 nm ($5s \leftrightarrow 5d$ transitions) [30] or the Rb D_2 line at 780 nm [31]. Very soon there will be a plethora of stabilized diode lasers that may serve as the reference laser.

11.2.6 Data Acquisition

11.2.6.1 Detect and Count Fringes As discussed earlier, the scanning Michelson interferometer generates two sets of interference fringes whose ratio determines the unknown laser wavelength according to Eq. (2). Measurement of the fringe ratio is analogous to period (or time interval) measurement in the electronic counter. The difference here is that the frequencies of both fringe sets are proportional to the carriage speed, as shown in Eq. (1), which usually varies with time. Some fringe interpolation methods may not work well due to this frequency variation.

Evidently, the leakage of one set of fringes into the other will add a second sinewave signal and cause systematic errors. In the case the two laser beams are counterpropagating, it is somewhat less difficult to separate the beams for detecting fringes. Polarizers or special coatings can be used to increase efficiency for the case when maximum signal isolation is required. In the case that the two laser beams are co-propagating, more attention is needed in efficiently separating two beams. Although polarization or a specially coated beam splitter can be used, it is difficult to completely avoid leakage of one laser fringe into the other. Interference filters (5–10 nm bandwidth, even a 0.3 nm bandwidth is possible [32]) can efficiently prevent the unwanted laser beam from reaching the detector. Colored glass or interference "edge" filters can be used in the case the unknown laser will scan over a broad range. If the unknown laser wavelength is too close to the reference laser wavelength to use an interference filter, a different reference laser wavelength may be used, for example, an He–Ne laser with $\lambda = 612$ nm instead of an He–Ne laser with $\lambda = 633$ nm, or one of the NIR Rb-stabilized diode lasers mentioned above.

If none of these optical methods can be used or they are insufficient for the most demanding measurements, some electronic methods may be helpful. One simple method is to add fuzz to the trigger level at the cost of some resolution. A very effective technique is to use a notch filter in the servo loop to filter out the unwanted fringe frequency. This method keeps the voltage-controlled oscillator (VCO) frequency in the PLL from being pulled. For high-enough beat frequencies (wavelength ratios) we have successfully used a sampling-type notch filter driven by the main fringe signal from the other channel.

It is straightforward to construct a photodetector for fringe counting using either an operational amplifier or a transimpedance amplifier and a photodiode fabricated from material suitable for the wavelength under study. An ac-coupled amplifier after the photodetector removes the dc component and amplifies the fringe signal. This signal is ready for use as the input to a commercial frequency counter.

Using a commercial dual-channel frequency counter to count the ratio of the reference laser fringe and the unknown laser fringe is the most convenient method for obtaining the unknown laser wavelength. Some commercial frequency

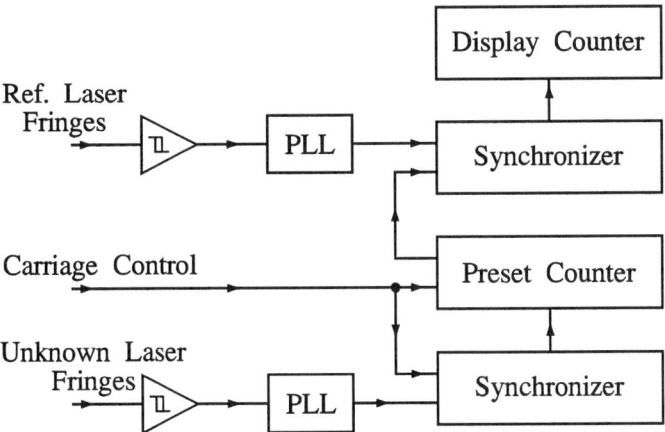

FIG. 6. Block diagram of λ-meter data acquisition system: PLL = phase-locked-loop. Synchronizers are used instead of the simple gates. See the text for discussion.

counters can display the product of the frequency ratio and a constant that is set by the thumbwheel switches on the front panel. Alternatively, the scale factor can be downloaded from the computer. So either the wavelength or the wavenumber can be displayed directly. The measuring result can be easily read by a computer, too. It is quite typical to find an obsolete but serviceable computer programmed to display the digital wavelength data in huge colored digits so that the mildly myopic laser user can read the λ value from across the lab!

If one decides not to use a commercial frequency counter, a λ-meter fringe counter such as the one shown in Fig. 6 can be constructed readily. The basic function of this preset counter is that the unknown laser fringe is used to control the gate of the counter for reference laser fringe counting, or vice versa. With correct setting of the gate (preset number of unknown laser fringes) from the thumbwheel or computer interface, the displayed result is the wavelength of the unknown laser directly. Similarly, the wavenumber of the unknown laser can be displayed if the reference laser fringe is used to control the gate for counting the unknown laser fringes. The fringe frequency is about 300 kHz for a typical λ-meter. Even when we use a phase-locked loop with a multiplication factor of 100 for fringe interpolation, the TTL circuit family (especially the faster newer versions) work adequately here, after respect is given to cable terminations.

The first stage shown in Fig. 6 is a Schmitter trigger that turns a sinusoidal fringe signal into a TTL compatible signal for the counters. The threshold voltage and the hysteresis of this Schmitter trigger should be chosen to accommodate the amplitude and signal-to-noise ratio of the input fringe signal [23]. This trigger stage is still needed before the phase/frequency detector when a phase-locked loop

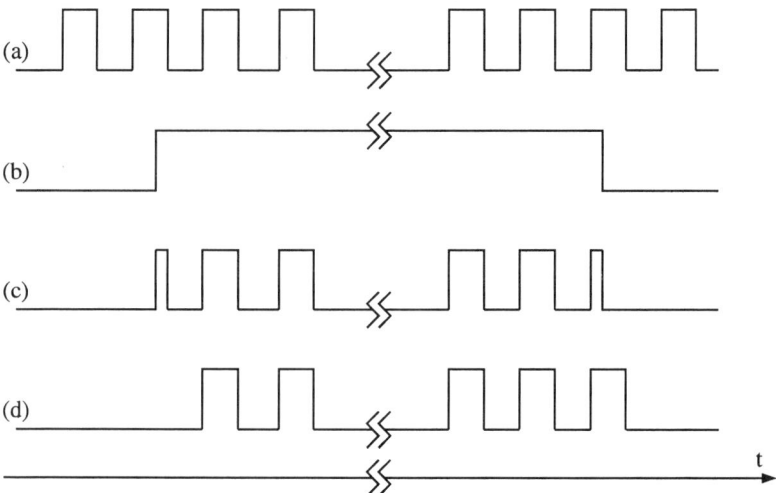

FIG. 7. Principles of synchronizer. (a) Input fringes. (b) Input control signal. (c) Output of a simple AND gate with (a) and (b) as the input signals. (d) Output of a synchronizer with (a) and (b) as the input signals.

is used for fringe interpolation. This requirement comes because we make use of such phase-lock chips as the MC4044, MC12040, or AD9901, which perform the phase/frequency lock function.

Because the counting circuit's response to partial clock pulses is not reliable, the partial clock pulses generated by the gate can cause a bias error in the measurement [33]. This bias error can be eliminated using a synchronizer. The synchronizer only allows the whole clock pulse to pass through as shown in Fig. 7. We find some older general-purpose synchronizer chips (MC74120) may not work as reliably as one expects.

11.2.6.2 Measurement of Fractional Fringe In order to increase the resolution of the λ-meter, the fraction of a fringe at each end of the measurement has to be determined. Using PLLs to multiply the fringe frequencies is a convenient method [23]. The principles of the PLL can be found in reference [34]. The advantage of using a PLL is threefold. First, multiplication of the fringe frequency effectively increases the counted fringe number such that the $\pm 1/N$ measurement error is reduced by the factor of the multiplication. Second, the PLL works as a flywheel to interpolate through the possible fringe dropouts, for example, in the case of an air bubble in the dye jet. Third, the PLL acts like a tracking filter to reduce the bandwidth and hence noise, and even the error due to the small leakage of the other fringe. The choice of the servo bandwidth for the PLL depends on how smooth the carriage motion is and how quiet the VCO is. One possible design is to use a broad servo bandwidth just after the carriage changes the direction of motion for fast acquisition, and to use a narrower servo bandwidth after the PLL is locked, to reduce noise, including the noise from leaking fringes. The irregularity of the carriage motion limits how well a fringe can be interpolated. Practically, a 50 to 100× multiplication factor is easily achieved for a well-designed towed-carriage type of mechanical system.

As we pointed out earlier, some powerful methods developed for time-interval measurement with an electronic counter [33, 35] may not work as well in the λ-meter application. The charge-pump type of analog vernier requires that the clock frequency (fringe frequency) be the same in the charge and discharge intervals. The locking range of the phase-startable/phase-lockable oscillator is usually less than 1%. Improvement of carriage motion smoothness certainly will open doors for these powerful methods in λ-meter measurements.

Another method to measure the fringe fraction was developed by Monchalin *et al.* [12]. This method uses reference laser ($\lambda = 633$ nm) fringe zero-crossing to sample the unknown laser ($\lambda = 10$ μm) fringe. First, the relative phase of the fringes at a number of locations along the scan is determined by either using least-squares fitting or Fourier summation scan based on the approximate value of the unknown wavelength (obtained from direct fringe counting). Then these relative phases are used to derive the more precise value of the unknown wavelength. Compared to the global fitting of all data points, this method is less sensitive to amplitude change along the scan and uses less computing time.

Kowalski *et al.* used a different method to measure the fringe fractions at both ends of the scan [36]. They measured the time interval between the zero-crossings of the two sets of fringes normalized by the two following periods of the reference laser fringes. Even though normalization is used, this method still relies on very smooth mechanical motion of the carriage. In principle, this method is as precise as the previous method. However, since the data are less redundant, this method is more sensitive to fluctuations and intensity noise: basically the PLL approach averaged over the previous 50 or 100 fringes.

Snyder developed yet another method to improve the resolution of the λ-meter [37]. Instead of measuring N unknown laser fringes obtained from a single interferometer arm difference scan, this method divides these N fringes into N_s subsets, as shown in Fig. 8. Each subset contains $N - N_s + 1$ fringes and shifts from the previous subset by 1 fringe. Each of these overlapping subsets is measured. Then the final results are obtained by averaging all these subsets measurements. Although the quantization error for each subset is $1/(N - N_s + 1)$, the error for the final result is improved by a factor of nearly $1/\sqrt{N_s}$ from averaging, if all the subsets can be treated as *independent* measurements. Remarkably, implementation of this scheme is straightforward, using hardware counters, and can also be easily per-

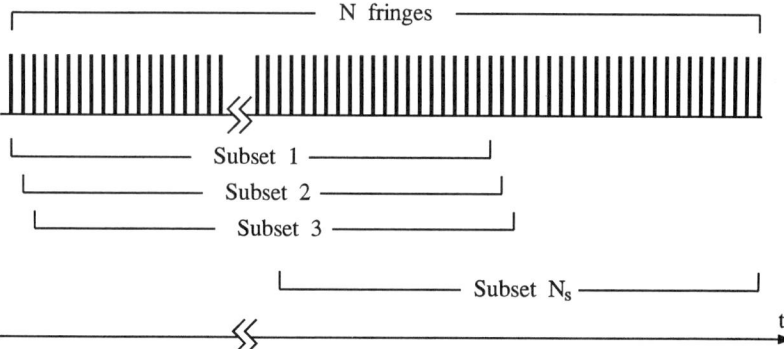

FIG. 8. Snyder's fringe counting scheme. See text for detailed discussion.

formed in software. Snyder showed this algorithm to be equivalent to a global least-squares fit of the entire wavelength pair fringe data set.

To guarantee that the averaged result approaches the true value, a synchronizer has to be used to eliminate the bias from partial pulses, as discussed before. Any coherence or rational frequency ratio between the two lasers can also degrade the measured result [33]. This coherence leads to total loss of averaging when the frequency of one laser is the (sub)harmonic of the other laser's frequency. Other frequency ratios may lead to a partially averaged result and a record length–dependent bias.

Random phase modulation of one set of fringes can be exploited to destroy this coherence [33]. Two important parameters for this method are the rms value of phase modulation, σ_{ph}, and the modulation bandwidth, f_c. A large σ_{ph} destroys the coherence more effectively. However, the rms value of the measurement error of each subset increases proportional to σ_{ph}. The choice of the modulation bandwidth depends on the frequency of the unmodulated fringe set. A large modulation bandwidth permits high fringe frequency. On the other hand, too large a bandwidth will also degrade the single subset measurement. The short summary is that 10^{-7} and 10^{-8} accuracy can come rather easily and that further progress requires care and patience.

11.2.6.3 Fourier Spectrometer Junttila *et al.* used the fast Fourier transform (FFT) of the interference term in the output signal of a scanning Michelson interferometer to measure the power spectrum [38]. This topic itself is worth a whole book. Readers interested in this topic can find a number of references (e.g., [39]). With the high-performance specialized integrated circuits and ever-increasing computing power, the measured result can be obtained in a fraction of a second.

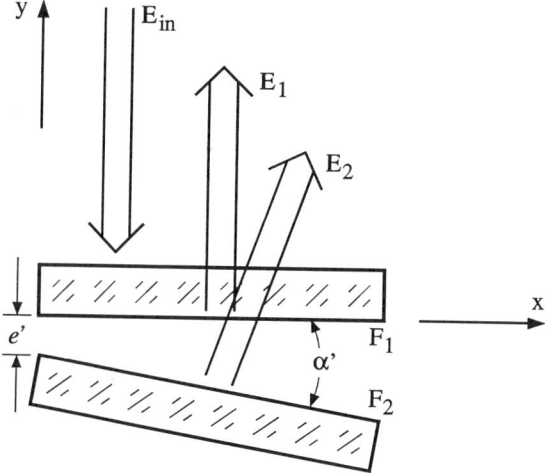

FIG. 9. Fizeau interferometer. F_1 and F_2 are two uncoated surfaces that form the interferometer. E_{in} is the input laser field. E_1 and E_2 are the fields reflected from the uncoated surfaces F_1 and F_2, respectively.

Some aspects of this approach were employed by Monchalain et al. [12] and by the famous interferometer group under P. Connes [40].

11.3 The Fizeau Wavemeter

11.3.1 Basic Principles of the Fizeau Interferometer

Another commonly used interferometer for wavelength measurement is the Fizeau wedge interferometer [41]. As shown in Fig. 9, a Fizeau interferometer consists of two optical flats with a wedge angle. The multibeam Fizeau interferometer, using reflective coatings on the optical flat surfaces, exhibits asymmetrical fringes because of the higher-order beam walkoff [42]. This fact and the wavelength-dependent optical phase shift of the coating [43, 44] make it difficult to interpret the fringes for wavelength measurement. Although multibeam Fizeau interferometers are used in some occasions, we will limit our discussion in the following sections to the two-beam Fizeau interferometer.

In the plane $y = d$, for a normal-incidence plane wave input laser field E_{in}, the optical interference of the fields E_1 and E_2, respectively reflected from two uncoated surfaces F_1 and F_2, is given by

$$I(x) = I_0 \left\{ 1 + A \cos \left[2\pi x \frac{\sin(2\alpha')}{\lambda} + 2\pi d \frac{\cos(2\alpha') - 1}{\lambda} + 2\pi \frac{2e' \cos^2 \alpha'}{\lambda} \right] \right\}, \quad (7)$$

where the contrast A (<1) results from the small imbalance in the two interfering field amplitude, λ is the input laser wavelength, α' is the wedge angle, and e' is the thickness of the Fizeau interferometer at $x = 0$. The effect of the thickness of the input optical flat is included in d. For a nonplane wave input laser beam, the envelope of the fringes is modulated by the laser beam intensity profile.

For fixed $y = d$, Eq. (7) can be further simplified as

$$I(x) = I_0 \left\{ 1 + A \cos\left[2\pi x \frac{2\alpha}{\lambda} + \frac{2e}{\lambda}\right]\right\}, \qquad (8)$$

where the effective wedge angle, $\alpha \equiv \sin(2\alpha')/2$ (for small angle $\alpha \approx \alpha'$), and the effective thickness, $e \equiv e'\cos^2\alpha' + (d/2)[\cos(2\alpha') - 1]$, are two quantities that characterize the Fizeau interferometer. These two quantities are measured during the calibration process (see the subsequent discussion in Section 11.3.4) and are later used for calculation of the wavelength under study.

Since there are no moving parts in a Fizeau interferometer, it can measure a pulsed laser's wavelength as well as that of a CW laser, provided the laser linewidth is less than the free spectral range of the interferometer $c/2e$, which is about 150 GHz for a typical design.

The interference fringe pattern shown in Eq. (8) is measured by a detector array. The analog output signal from the detector, after being digitized, is read by a computer for wavelength calculation.

11.3.2 Structure of the Fizeau Wavemeter

Figure 10 shows the setup of a Fizeau wavemeter [6]. The input laser beam is collimated using a beam expander with a spatial filter. The collimating result can be tested using a shearing interferometer, as discussed in Section 11.2.4.1. The size of the laser beam after collimation is larger than the aperture of the interferometer in order to obtain uniform illumination.

The interference fringes are generated by two uncoated inner surfaces of the optical flats separated by a wedged spacer (~20 arc-seconds). To avoid the stray fringes from the outer surfaces of the optical flats, these outer surfaces can be either AR-coated or made of a wedge (~30 arc-minutes) perpendicular to the spacer wedge. The spacer and the optical flats should be made of the same material, for example, commercially available high-quality fused silica, to avoid excess strain due to differential thermal expansion. The well-polished wedge spacer, with a channel for pumping the dispersive air out of the interferometer, can be optically contacted to the flats to form the interferometer. After the optical contact bond, a test of flatness of the inner surfaces can reveal the deformation of the flats due to surface nonuniformity [45]. Since the interferometer is usually kept in a vacuum

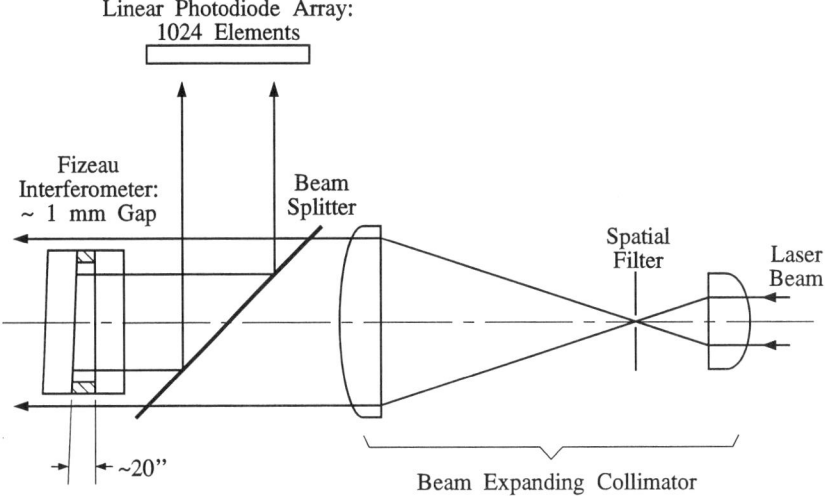

FIG. 10. Fizeau wavemeter. After reference [6].

enclosure, care should be taken to minimize the error due to deformation of the vacuum enclosure window.

The beam splitter in Fig. 10 steers the interference fringes to a linear photodiode array with 1024 elements for detection. The fringe period for visible light (600 nm) in this arrangement is about 3 mm, which covers about 120 pixels for a 25 μm/pixel photodiode array. The analog output signal from the photodiode array is digitized using a sample/hold amplifier and an analog-to-digital converter (ADC). A 14-bit ADC at 10 megasamples per second is now readily available for fast sampling requirements. The digital signal is then read into a computer for wavelength calculation.

11.3.3 Calculation of Wavelength

To calculate the unknown wavelength, λ_{un}, we can write the effective thickness, e, as

$$2e = \left(m + \frac{\Theta}{2\pi} \right) \lambda_{un}, \qquad (9)$$

where m (an integer) is the interference order, and Θ ($|\Theta| < \pi$) is the phase at $x = 0$. Then Eq. (8) becomes

$$I(x) = I_0 \left\{ 1 + A \cos \left[2\pi x \frac{2\alpha}{\lambda_{un}} + 2m\pi + \Theta \right] \right\} = I_0 \left\{ 1 + A \cos \left[2\pi \frac{x}{X_{un}} + \Theta \right] \right\}, \qquad (10)$$

where

$$X_{un} \equiv \frac{\lambda_{un}}{2\alpha}. \tag{11}$$

is the fringe period. The wavelength calculation process is described as follows. First, the fringe period X_{un} and phase Θ are calculated from fitting the data, and the approximate wavelength, $\lambda_{un}^{(a)}$, is extracted using Eq. (11). This approximate wavelength is used to calculate the exact interference order, m. For visible light, with a 1-mm gap, the interference order is about 3000. Thus, one wants to have

$$\frac{|\lambda_{un}^{(a)} - \lambda_{un}|}{\lambda_{un}} \approx 10^{-4}, \tag{12}$$

which requires that both $|\Delta X_{un}|/X_{un}$ and $|\Delta\alpha|/\alpha$ be smaller than 10^{-4}. The small uncertainty of the fringe period ($\sim 1/100$ of a pixel) is achieved by averaging all the fringe periods measured. After the interference order is unambiguously determined, the final wavelength, $\lambda_{un}^{(f)}$, is given by

$$\lambda_{un}^{(f)} = \frac{2e}{m + (\Theta/2\pi)}. \tag{13}$$

The uncertainty of $\lambda_{un}^{(f)}$ can be written as

$$\frac{|\Delta\lambda_{un}^{(f)}|}{\lambda_{un}^{(f)}} \approx \frac{|\Delta e|}{e} + \frac{|\Delta\Theta|}{2m\pi}. \tag{14}$$

The main concern for the wavelength calculation algorithm is that the resultant wavelength should be insensitive to various fluctuations such as the intensity profile of the input laser beam. The time needed to complete the calculation is relatively less important nowadays because of the availability of fast computers. Snyder developed a fast algorithm for deriving the wavelength from the Fizeau wavemeter fringes [46]. First, a symmetric digital filter with transfer function

$$f_{\text{filter}}(x) = \begin{cases} 0, & x = 0, \ |x| > b \\ 1, & -b \leq x < 0 \\ -1, & 0 < x \leq b \end{cases}, \tag{15}$$

is used to filter the raw data. The filter width, b, is chosen to be

$$b = 0.371 \frac{\lambda}{2\alpha} \tag{16}$$

to minimize the phase error in the fitting. The estimated fringe period is used for choosing b. After filtering, all the extrema of the fringes show as zeroes, which can be found using interpolation. Then least-squares fitting is used to fit a straight line to all these extrema. The slope is half the fringe period and the intercept is related to the initial phase of the fringes. This result is used to find out the interference order, and thus the wavelength. This more accurate λ is used in the next iteration to determine the filter width, b. Usually the iteration converges rapidly. This algorithm is fast because there are no trigonometric functions used and the least-squares fitting is linear.

For this particular algorithm the estimated fringe period uncertainty and the initial phase uncertainty are

$$|\Delta X_{un}| \approx \frac{2\sqrt{3}}{N^{3/2}} |\Delta x|, \qquad (17)$$

and

$$\frac{|\Delta \Theta|}{2\pi} \approx \frac{N}{2\sqrt{3}} \frac{|\Delta X_{un}|}{X_{un}} \approx \frac{1}{\sqrt{N}} \frac{|\Delta x|}{X_{un}}, \qquad (18)$$

respectively, where N is the number of the extrema used in the least-squares fitting, and $|\Delta x|$ is the uncertainty of each extremum. These two estimated uncertainties, together with Eq. (14), not only determine the achievable resolution, but also restrict the choice of α and e in the Fizeau wavemeter.

11.3.4 Calibrations

It is obvious that the wavelength derivation is based on a knowledge of the effective wedge angle, α, and the effective thickness, e, in Eq. (8). These two physical quantities need to be determined in the calibration process after the interferometer is assembled to obtain the required uncertainties Δα/α and Δe/e. Calibration needs to be done at several well-documented wavelengths. In the visible and near-infrared regions, transitions in molecular iodine and in some atoms are well documented [47]. Calibration could be performed using a tunable laser (or several separate lasers) locked to atomic/molecular transitions. The reference wavelengths chosen for calibration should cover both ends of the interested spectrum if possible.

The variation of the individual element responsivity of the detector array could easily exceed one least significant bit (LSB) of the digitizer and thus deform fringe shape. This could be circumvented by establishing the detector array's responsivity with an integrating sphere to generate uniform illumination. These data can be used to correct the measured fringes before the wavelength is calculated.

11.3.5 Limitations and Variations

The main limit of this basic arrangement of the Fizeau wavemeter, as shown in Fig. 10, is that there is a lateral shear of wavefront at the detector array. Thus, any

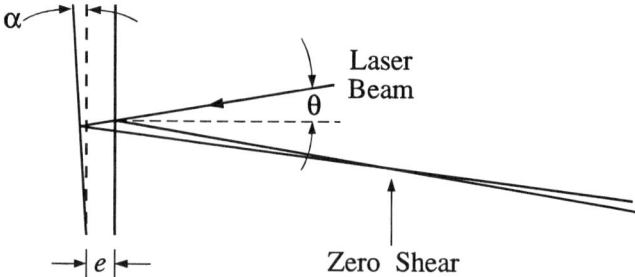

FIG. 11. Shear compensation in Fizeau interferometer. After reference [48].

change of the input beam wavefront curvature induces errors in the measured wavelength. This wavefront curvature change is inevitable because of the chromatic aberration of the optical elements used in the system, especially the laser beam expander. To reduce the chromatic aberration of the beam expander, an off-axis parabola mirror was used for collimation [45, 48]. To reduce the sensitivity of this wavefront curvature dependence, the input beam was tilted in the plane normal to the wedge axis (defined as the intersection of the wedge faces of the interferometer), as shown in Fig. 11 [45, 48]. The detector array then was placed in the zero-shear line to minimize the errors induced by wavefront curvature. This approach also eliminated the beam splitter and improved the efficiency.

Gardner [49] proposed an alternative method. Using a lens or a mirror to image the detector array at the rear surface of the interferometer, where there is also a zero shear position, the input beam is tilted, in this case, in the plane parallel to the wedge axis to eliminate the beam splitter and to create room for the imaging element. In this case, the wedge angle of the optical flat and its orientation should be chosen carefully to avoid the error due to undesired interference. A problem posed by this "solution" is dispersion of the front flat, which leads to an error in the wavelength measurement. A compensation plate could be used to circumvent this problem [50].

In addition, Gardner calculated the wavefront curvature effects on Fizeau wavemeter fringes [49]. The fringe period, X, is

$$X = \frac{\lambda}{2\alpha} \frac{s+d}{s-e}, \qquad (19)$$

where s is the radius of curvature at the front interferometer surface, d is the distance between the detector and the front interferometer surface, and e is the thickness of the interferometer. In order to obtain the required fringe precision to determine the interference order (~3000), d/s should be kept at ~10^{-4}. Thus, it is advantageous to keep the detector close to the interferometer. In the Fizeau wavemeter constructed by Gardner, the total internal reflection of a prism was used to steer the interference fringes to the detector array [51]. This arrangement reduced the incidence angle to 2° and kept the detector close to the front surface of the interferometer. The small incidence angle reduced the dispersion effect of the front optical flat. Gardner also reduced the overall size by sealing his air-spaced interferometer [51] instead of putting it into a vacuum enclosure. Information on air dispersion was used in calculation of the wavelength.

Gray *et al.* calculated the fringe periods when the detector array was rotated in the plane normal to the exiting beam axis [52, 53]. They found that for a given incidence angle and detector-wedge separation a rotation angle could be chosen so that the measured fringe periods did not depend on the curvature of the input beam wavefront.

Other variations include using a solid wedge, for example, a fused silica wedge, to eliminate the vacuum enclosure [54]. The (well-documented) dispersion of the material may be used for the correction. A combination of a single-mode optical fiber and an off-axis parabolic mirror could be used in the beam expander [55]. This arrangement simplified coupling of the laser beam into the interferometer. If the spectral range were broader than the single-mode operation range of one fiber, several fibers with a standard fiber connector could be employed.

Although we discuss only two-beam Fizeau interferometers in this chapter, we have to mention one limit posed by the higher-order beams. This limit will show up as the signal-to-noise ratio improves and other limits are taken care of. The next high-order beam generates two additional interference fringes with fringe periods $\lambda/2\alpha$ and $\lambda/4\alpha$, respectively. The intensity of each of these two fringes is about 4% of the main fringe given in Eq. (7). The additional fringe with period $\lambda/4\alpha$ could cause distortion of the main fringe and induce an additional error. The influence of this additional fringe depends on the algorithm used for wavelength calculation. If the intensity profile of the input beam is well known, this additional fringe could be included in the global fitting routine to reduce its influence.

11.4 Plane-Parallel Interferometers with CCD Readout

A pioneering effort to make useful wavelength measurements by TV imaging of Fabry–Perot rings was made by Byer *et al.* in 1977 [7]. This system made use of four etalons of differing thickness, so that the wavelength could be determined fully without a priori knowledge. The measuring system was based on vidicon cameras, and it was found that deflection gain instability and drift over time impacted the calibration data. With the later availability of CCD detectors in which the detector positions are permanently set by photolithography, one can anticipate a reinvestigation of this system. Sansonetti *et al.* have published several interferometric wavelength determinations based on this technique [56]. To give a flavor of these studies, we recall a few results from a recent JILA investigation.

Eickhoff and Hall have described use of plane-parallel interferometers with cylindrical beam illumination, producing an exiting fringe pattern sampled along a diameter of the rings [57]. After imaging onto a linear detector, the fringes are digitized, averaged, and stored in a PC. The fitting function models a broad Gaussian envelope of the fringes and approaches its final fitting in several stages. At first, the fringes are fit with Synder's algorithm as described above. These output data form the starting point for a full nonlinear least-squares fit of the whole profile. The particular interferometer employed has a finesse of 15 and a measured S/N of about 20–100. Numerical simulation of our algorithm with corresponding synthetic data over a range of phases gives a typical fit uncertainty of 2×10^{-4} fringe orders. When the corresponding experimental data are fit, using a frequency-locked dye laser to establish the accurate fringe phase steps, we find that the fitted fringe phases are about five- to tenfold less certain. This is due to some systematic defect in the shape of the experimental fringe (a very small asymmetry), which we attribute to a small residual wedge of the interferometer over the relevant illuminated zone. Still, a 1-millifringe dependable accuracy leads to a fairly useful bootstrap from the raw order number which we obtain with an air-track λ-meter. The computing time was about 20 seconds with a venerable 10-MHz 80286 machine. A natural direction for improvement is the use of a longer spacer and a higher finesse. Unfortunately, both of these exacerbate the problem of digitization by requiring more detector pixels per fringe order and by making the angular dispersion smaller. There are many opportunities for clever design here!

11.5 Summary and Outlook

The art of precise laser wavelength measurement has evolved over the last two decades to the place where several viable approaches are commercially available and refined laboratory systems can give a routine accuracy of 10^{-8}. The authors hope that this article will help even lower-budget experimental groups to easily prepare their own equipment and operate with useful precision. In principle, a preunderstanding of the errors lurking in interferometry will help all users to obtain better and more reliable results. With the exponential precision wall at ~10^{-9} and the rapid progress in frequency measurement technology, such as optical comb generation [58, 59] and molecular overtone frequency standards [60–62], we expect to see users who really need better accuracy become contributors to and innovators in frequency-domain technology.

References

1. Hocker, L. O., Javan, A., Ramachandra Rao, D., Frenkel, C., and Sullivan, T. (1967). *Appl. Phys. Lett.* **10**, 147.
2. Jennings, D. A., Evenson, K. M., and Knight, D. J. E. (1986). *Proc. IEEE* **74**, 168, and the references therein.
3. McIntyre, D., and Hänsch, T. W. (1988). Paper THG3 in *Digest of the Annual Meeting of the Optical Society of America*, Optical Society of America, Washington, DC.
4. Kowalski, F. V., Hawkins, R. T., and Schawlow, A. L. (1976). *J. Opt. Soc. Am.* **66**, 965.
5. Hall, J. L., and Lee, S. A. (1976). *Appl. Phys. Lett.* **29**, 367.
6. Snyder, J. J. (1977). In *Laser Spectroscopy*, Vol. 3, p. 419, J. L. Hall and J. L. Carlsten (eds.), Springer-Verlag, Berlin.
7. Byer, R. L., Paul, J., and Duncan, M. D. (1977). In *Laser Spectroscopy*, Vol. 3, p. 414, J. L. Hall and J. L. Carlsten (eds.), Springer-Verlag, Berlin.
8. Juncar, P., and Pinard, J. (1975). *Opt. Comm.* **14**, 438.
9. Juncar, P., and Pinard, J. (1982). *Rev. Sci. Instrum.* **53**, 939.
10. Michelson, A. A. (1881). *Am. J. Sci.* **22**, 120.
11. Michelson, A. A. (1881). *Phil. Mag.* **13**, 236.
12. Monchalin, J.-P., Kelly, M. J., Thomas, J. E., Kurnit, N. A., Szoke, A., Zernik, F., Lee, P. H., and Javan, A. (1981). *Appl. Opt.* **20**, 736.
13. Bergquist, J. (1996). "Doppler-Free Spectroscopy," in *Experimental Methods in the Physical Sciences: Atomic, Molecular, and Optical Physics*, Volume 29B, Academic Press, New York, and the references therein.
14. Peck, E. R. (1962). *J. Opt. Soc. Am.* **52**, 253.
15. Steel, W. H. (1985). *Appl. Opt.* **24**, 3433.
16. Ishikawa, J., and Watanabe, H. (1993). *IEEE Trans. Instrum. Meas.* **IM-42**, 423.
17. Monchalin, J.-P., Kelly, M. J., Thomas, J. E., Kurnit, N. A., Szoke, A., Javan, A., Zernik, F., and Lee, P. H. (1976). In *Frontiers in Laser Spectroscopy*, p. 695, Les Houches Summer School 1975, R. Balian, S. Haroche, and S. Liberman (eds.), North-Holland, Amsterdam.
18. Rowley, W. R. C. (1966). *IEEE Trans. Instrum. Meas.* **IM-15**, 146.
19. Sternberg, R. S., and James, J. F. (1964). *J. Sci. Instrum.* **41**, 225.
20. Jasny, J., Jethwa, J., and Schafer, F. P. (1978). *Opt. Comm.* **27**, 426.
21. Docchio, F., Schafer, F. P., Jethwa J., and Jasny, J. (1985). *J. Phys. E: Sci. Instrum.* **18**, 849.
22. Drullinger. R. E., private communication. This design is used widely in a number of institutes.
23. Hall J. L., and Lee, S. A. (1979). U.S. Patent No. 4,165,183; see also reference [5].
24. This was one of the features of the first JILA λ-meter.

25. See, for example, Malacara, D., ed. (1992). *Optical Shop Testing*, 2nd edition, Wiley, New York.
26. See, for example, Neumann, E.-G. (1975). *Single Mode Fibers*, Springer-Verlag, Berlin.
27. Snyder, J. J., Raj, R. K., Bloch, D., and Ducloy, M. (1980). *Opt. Lett.* **5**, 163.
28. Baer, T., Kowalski, F. V., and Hall, J. L. (1980). *Appl. Opt.* **19**, 3173.
29. Niebauer, T. M., Faller, J. E., Godwin, H. M., Hall, J. L., and Barger, R. L. (1988). *Appl. Opt.* **27**, 1285.
30. Nez, F., Biraben, F., Felder, R., and Millerioux, Y. (1993). *Opt. Comm.* **102**, 432.
31. Ye, J., Swartz, S., Jungner, P., and Hall, J. L. (1996). *Opt. Lett.* **21**, 1280.
32. Yanagimachi, T., Oguri, H., Nayyer, J., Ishihara, S., and Minowa, J. (1994). *Appl. Opt.* **33**, 3513.
33. Chu, D. C. (1974). *Hewlett-Packard Journal*, June, p. 12.
34. See, for example, Rohde, U. L. (1983). *Digital PLL Frequency Synthesizers*, Prentice-Hall, Englewood Cliffs.
35. Chu, D. C. (1975). U.S. Patent No. 3,921,095.
36. Kowalski, J., Neumann, R., Noehte, S., Schwarzwald, R., Suhr, H., and zu Putlitz, G. (1985). *Opt. Comm.* **53**, 141.
37. Snyder, J. J. (1981). In *Proc. 35th Freq. Control Symposium*, p. 464, Ft. Monmouth, NJ.
38. Junttila, M.-L., Stahlberg, B., Kyro, E., Veijola, T., and Kauppinen, J. (1987). *Rev. Sci. Instrum.* **58**, 1180.
39. See, for example, Griffiths, P. R., and de Haseth, J. A. (1986). *Fourier Transform Infrared Spectrometry*, Wiley, New York.
40. Davis, D. S., Larson, H. P., Williams, M., Michel, G., and Connes, P. (1980). *Appl. Opt.* **19**, 4138.
41. Fizeau, H. (1862). *Ann. Chim. Phys.* **66**, 429.
42. Rogers, J. R. (1982). *J. Opt. Soc. Am.* **72**, 638.
43. Lichten, W. L. (1985). *J. Opt. Soc. Am. A* **2**, 1869.
44. Lichten, W. L. (1986). *J. Opt. Soc. Am. A* **3**, 909.
45. Morris, M. B., McIlrath, T. J., and Snyder, J. J. (1984). *Appl. Opt.* **23**, 3862.
46. Snyder, J. J. (1980). *Appl. Opt.* **19**, 1223.
47. See Helmcke, J. (1997). "Optical Wavelength Standards," in *Experimental Methods in the Physical Sciences: Atomic, Molecular, and Optical Physics*, Volume 29C (this volume), Academic Press, New York, and the references therein.
48. Snyder, J. J. (1981). *Proc. SPIE* **288**, 258.
49. Gardner, J. L. (1983). *Opt. Lett.* **8**, 91.
50. Miller, C. K. (1982). "Wavelength Meter for Pulsed Laser Applications," Internal Report SAND 81-0310, Sandia National Laboratories, Albuquerque, NM.
51. Gardner, J. L. (1985). *Appl. Opt.* **24**, 3570.

REFERENCES

52. Gray, D. F., Smith, K. A., and Dunning, F. B. (1986). *Appl. Opt.* **25**, 1339.
53. Gardner, J. L. (1986). *Appl. Opt.* **25**, 3799.
54. Reiser, C., and Lopert, R. B. (1988). *Appl. Opt.* **27**, 3656.
55. Kedzierski, W., Berends, R. W., Atkinson, J. B., and Krause, L. (1988). *J. Phys. E* **21**, 796.
56. Sansonetti, C. J., Gillaspy, J. D., and Cromer, C. L. (1990). *Phys. Rev. Lett.* **65**, 2539.
57. Eickhoff, M. L., and Hall, J. L. (1997). *Appl. Opt.*, **36**, 1223.
58. Kourogi, M., Nakagawa, K., and Ohtsu, M. (1993). *IEEE J. Quantum Electronics* **29**, 2693.
59. Brothers, L. R., Lee, D., and Wong, N. C. (1994). *Opt. Lett.* **19**, 245.
60. de Labachelerie, M., Nakagawa, K., Awaji, Y., and Ohtsu, M. (1995). *Opt. Lett.* **20**, 572.
61. Suzumura, K., and Sasada, H. (1995). *Jpn. J. Appl. Phys.* **34**, L1620.
62. Ye, J., Ma, L.-S., and Hall, J. L. (1996). *Opt. Lett.* **21**, 1000.

12. OPTICAL MATERIALS AND DEVICES

Sami T. Hendow

Ditech Corporation
Sunnyvale, California

12.1 Introduction

This chapter is a practical discussion of some components found in today's optics labs. Included are brief discussions of the types of popular glasses, mirrors, polarizers, optical isolators, and other active and passive components. Also highlighted are practical implications of using these components, their advantages, disadvantages, and typical applications.

12.2 Optical Materials and Performance

12.2.1 Types of Glass

There are several glasses in commercial use today. In general, the higher the index homogeneity, the more expensive the glass, with better characteristics for UV transmission and high power damage threshold. Figure 1 and Table I give transmission and optical characteristics for common optical glasses.

12.2.1.1 Crown Glass
BK7 is a common crown glass. It is often used as a low-cost mirror substrate. However, it is thermally not as stable as Pyrex or fused silica, and, due to its contents of platinum and other impurities, is more susceptible to laser damage than fused silica. Typically, BK7 lenses are matched to flint glass lenses to correct for chromatic dispersion as well as other aberrations.

12.2.1.2 Pyrex®
Pyrex is a low-cost borosilicate glass designed for low optical deformation under thermal shock. Pyrex mirrors provide low reflected wavefront distortion and scatter for routine applications and are well-suited for mirror substrates, windows for high-temperature chambers, or harsh chemical environments. The main drawback of Pyrex is that its index is less homogenous and contains more optical striations than such other glasses as BK7 and hence is less applicable to transmission optics. It is commonly used for test plates and optical flats. It is generally less expensive than BK7; however, a specification for high index homogeneity can increase cost beyond that of BK7 or fused silica.

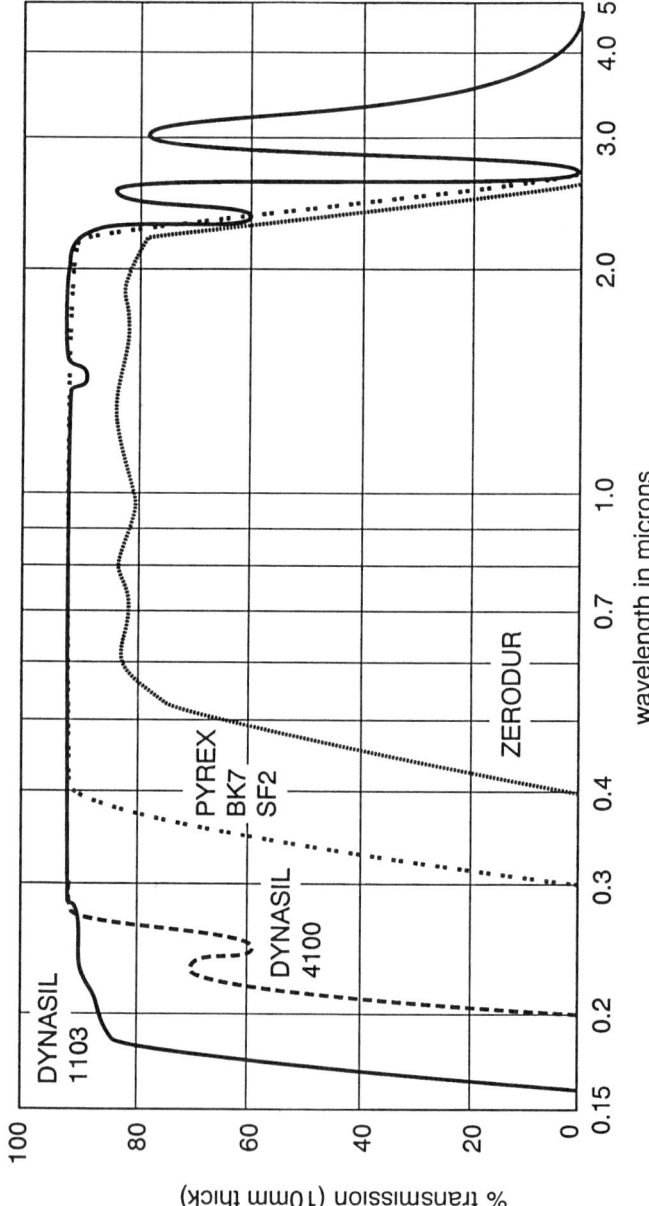

FIG. 1. Transmission characteristics of four types of glasses.

TABLE I. Properties of Optical Materials

	Pyrex	Zerodur	BK-7	SF2	Fused silica
Index n at 1060 nm			1.507	1.628	1.449
Index n at 643.8 nm	1.473		1.515	1.643	1.457
Index n at 546.1 nm	1.477	~1.54	1.517	1.652	1.460
Index n at 486 nm	1.481		1.522	1.661	1.463
Index n at 346.6 nm					1.477
Index n at 248.2 nm					1.508
Abbe number (V_d)			64.2	33.8	67.8
Birefringence (nm/cm)	10	5	6	6	5
Expansivity (10^{-7}/°C)	32.5	−0.2	71	84	5.5
Conductivity (mW/cm °C)	11.3	16.3	11.3	7.3	13.8
Heat capacity (J/g °C)	0.75	0.85	0.85	0.5	0.75
Maximum (°C) temperature	500	600	280	200	950
Density (g/cm^3)	2.23	2.52	2.51	3.86	2.2
Hardness	460	550	510	350	500
Young's modulus (kN/mm^2)	65.5	90.2	81.5	55	70.3

12.2.1.3 Zerodur® Zerodur is a unique glass-ceramic material whose thermal expansion is nominally zero at room temperature. This stability is essential in diffraction-limited systems where the optical figure must not vary under thermal changes. Zerodur is not used for transmissive optics due to bulk scattering that occurs mostly at its crystalline boundaries. The expansion coefficient at room temperature falls in the range of $\pm 0.15 \times 10^{-6}$/°C. By design, this material has a change in the sign of the thermal coefficient at room temperature. Typical application of Zerodur is in interferometry, holography, and space-borne systems.

12.2.1.4 Fused Silica Fused silica (amorphous silicon dioxide, for example Dynasil 1103) is a synthetic optical material that features broadband transmission, thermal stability, and resistance to laser damage. Fused silica mirrors can be polished to extreme accuracy, thereby minimizing wavefront distortion and scatter.

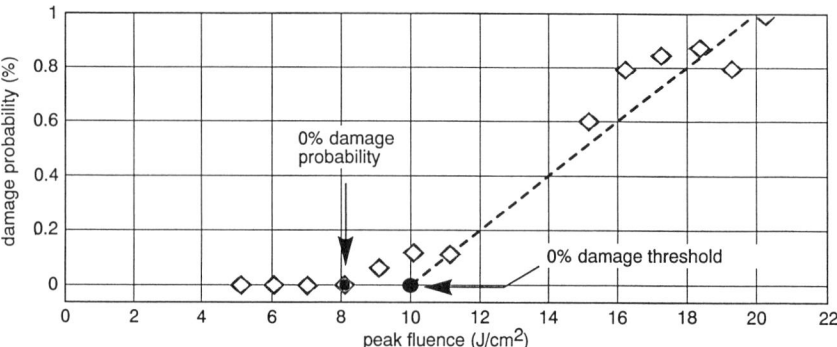

FIG. 2. Damage threshold specification of an optic using the 0% damage threshold point and the 0% damage probability point.

It is far purer than fused quartz and has excellent transmission over a wide spectral range.

Fused silica is the most economical glass if index homogeneity and UV and IR transmission are required. With the exception of an absorption band at 2.7 µm, fused silica is usable from 185 nm to 4.0 µm. The high purity and homogeneity of fused silica also eliminate potential laser damage sites.

Optical systems subjected to thermal stress are prime candidates for fused silica optics. Lenses are typically rated for continuous use from −273 to 900°C. Possible applications include pyrometry and coupling of optical energy into cryogenic systems.

12.2.2 Laser Damage

Damage testing is a method for determining how much laser energy can be applied to an optic before it is physically damaged [1]. There are a number of data interpretation methods used for damage evaluation. For example, most antireflection coatings pit at relatively low fluence levels when illuminated by a pulsed laser. Only at much higher levels will the coating experience macroscopic damage. For most applications, these small pits are of no consequence.

Damage threshold specifications. A popular method for defining the damage threshold is the x-axis intercept of a line fitted to the damage probability (see Fig. 2). It is suitable for comparing data and reduces the chance that a single contaminated site will affect the results. A second and more conservative method is based on the highest fluence for which the damage probability is zero, as shown in Fig. 2.

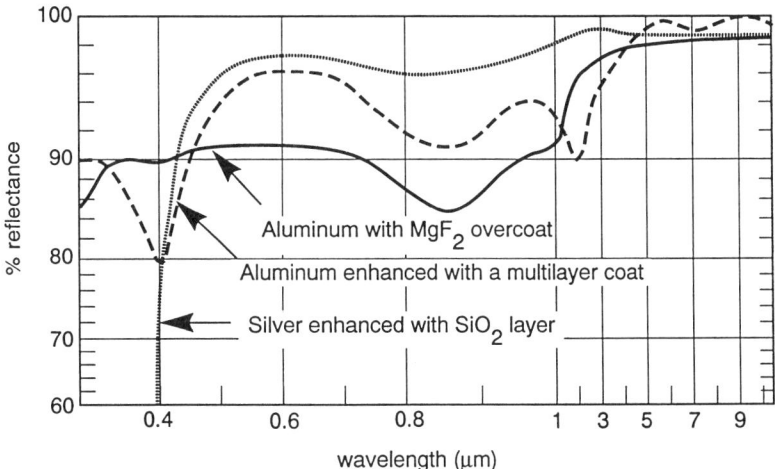

FIG. 3. Reflectance of coated aluminum and silver mirrors.

12.3 Optical Components

12.3.1 Mirror and Coating Types

12.3.1.1 Metallic Mirrors Metallic mirrors represent a good mixture of performance and economy. They are relatively insensitive to the wavelength, angle, or polarization of the incident light wave. To prolong the life of these coatings, high power levels should be avoided and only gentle cleaning should be performed.

Three types of metal are used for coating: aluminum, silver, and gold. Gold has the highest reflectivity but is not commonly used because it is soft, tarnishes easily, and is costly. Silver reflects further into the visible than gold. The most commonly used coating, with the broadest bandwidth, is aluminum. For protection against oxidation and for reflectivity enhancement, these metals are used with overcoats of silicon, silicon dioxide (SiO_2), or magnesium fluoride (MgF_2), as shown in Fig. 3.

12.3.1.2 Dielectric Mirrors Dielectric mirrors can minimize losses in nearly all optical applications [2]. They are capable of near total reflection over a selected bandwidth. Coatings are durable, cleanable, and resistant to laser radiation. The best performance in reflectivity, bandwidth, and damage threshold is obtained when the optic is used with S-polarized light, which is normal to the plane of incidence. P-polarized light has lower reflectivity with a bandwidth that is narrower and shifted toward the UV.

Two main technologies are used today for dielectric coatings: electron-beam (*e*-beam) and ion-beam-sputtering (IBS) deposition. Electron-beam deposition [3–6] evaporates a material by heating it with an electron beam, while IBS uses an ion beam to sputter a target material. In both cases, the material condenses on an optic forming a thin film. Electron-beam optics are more economical and are suitable for most applications. IBS coatings [5, 6] produce high-performance high-reflectivity optics with ultralow scatter (less than 100 parts per million loss). Advanced polishing and state-of-the-art IBS coating technologies combine to provide the highest reflectivities and durability with minimum scatter losses. IBS also has the potential for extending the bandwidth of dielectric mirrors.

Ultralow-Loss Optics Typical applications of ultralow-loss optics are high-reflectivity ultralow-scatter [7] beam delivery and high-performance optical resonators, such as the kind used with low-gain laser media, high-finesse optical spectrum analyzers, and optical parametric oscillators.

Electron beam mirror reflectivities are typically specified at 99.7%. While this may sound adequate for many precision applications, this corresponds to losses of 3000 parts per million (ppm). IBS-coated mirrors, on the other hand, offer loss between 1 and 50 ppm. For products such as ultrahigh-resolution spectrum analyzers, this difference can mean a factor of 100 to 1000 improvement in resolution or finesse.

In a typical *e*-beam evaporative coating chamber, the heating of the target causes the material to boil off the surface. IBS systems, however, use ions rather than electrons. The higher mass of the ions allows them to remove individual molecules of coating material by momentum transfer. The higher energy of the ions translates into greater surface mobility and an ability to fill in small voids in the developing film, producing a coating with a glassy amorphous structure. This makes it possible to produce coatings that are much closer to bulk material and display much lower scatter and absorption, as well as an improved laser damage threshold.

12.3.1.3 Antireflection Coatings

There are three typical types of AR coatings: a single layer of an MgF_2 coat, a V coat, and broadband AR coat. MgF_2 (magnesium fluoride), whose index is halfway between air and glass, reduces the reflectivity of glass from 4% to about 2% over a broad bandwidth. V coating is composed of three to five layers of SiO_2, TiO_2, or MgF_2 that reduces reflectivity to less than 0.25% at a specific wavelength. Beyond that wavelength, the reflectivity increases rapidly and is higher than that of uncoated glass. Broadband AR coating, also referred to as dual wavelength or dual V, can decrease the reflectivity to less than 0.5% over a broad range. A typical two-wavelength design is 1064 and 532 nm.

12.3.2 Types of Beam Splitter

Table II shows common beam splitters, their advantages, limitations, and applications.

TABLE II. Types of Beam Splitters

Types	Design/principle	Application	Comments
Uncoated precision optics	Optical surface samples the beam by Fresnel reflection.	Ideal when minimum disturbance of the transmitted beam is required. Typically, the back surface is AR-coated to eliminate ghost images.	Depending on incident polarization, front surface reflectances vary from 1 to 10% at 45° incidence.
Dielectric beam splitters	Single optical flat, coated for a 50/50 split. The back surface is AR-coated to eliminate ghost images.	General-purpose beam splitting, laser cavity coupling, and interferometry.	Split ratios typically change with incident beam polarizations, wavelengths, and angles of incidence.
Nonpolarizing beam splitter cubes	Permanently cemented prisms after a thin film dielectric beam splitter coating is applied to one of the prisms.	General-purpose beam splitting applications.	Perform well within ±5% of the design wavelength.
Metal beam splitters	Metallic coating on a flat substrate.	General-purpose and economical, e.g., neutral-density filters.	Some metallic coatings also cause transmission losses.
Metal beam splitter cubes	Cemented prisms, one of which has a metallic beam splitter coat.	General-purpose and economical compared to dielectrically coated beam splitters.	Some metal coatings have absorption losses that limit the damage threshold.
Harmonic beam splitter	Dielectrically coated substrate.	Separates one wavelength from its harmonic.	For example, it transmits 1064 nm and reflects its harmonic at 532 nm, typically at 45°.
Pellicle beam splitter	5-μm thick plastic membrane stretched over a flat metal frame. Touching or immersing the membrane is usually not recommended. Cleaning is only performed by gently blowing the dust.	Ultralightweight applications. Losses and multiple reflections, commonly associated with more massive glass optics, are reduced or eliminated.	The Fabry–Perot effect between the surfaces of the membrane causes a ±5% wavelength-dependent change in the reflectivity. Membranes may resonate due to vibrations in the environment.

12.3.3 Optical Filters

Filters are used to control the spectrum of light in such diverse applications as spectral radiometry, medical diagnostics, chemical analysis, and astronomy [8, 9]. Generally, they are grouped in three categories: bandpass filters [9], edge filters, and neutral density filters. Spectral selection is obtained either by colored glass, dielectric coatings, metallic coatings, or their combinations. Metallic coatings give flatter performance as compared to the other two, while colored glass is the most economical.

Bandpass filters are most common. They are defined in terms of their bandwidth (full-width half-maximum, FWHM), rate of attenuation with respect to wavelength change along the slopes of the bandpass, and transmission outside the bandpass region. The attenuation of the signal is frequently expressed in terms of the optical density (O.D.), given by

$$\text{O.D.} = \log_{10}(P_o/P_i), \quad (1)$$

where P_o is the transmitted and P_i the incident optical power. Bandpass and edge filters that are designed with dielectric thin layers will exhibit a shift in the bandpass center toward the UV with tilt. This character is advantageous for tuning the center wavelength by changing the angle of incidence up to $\pm 15°$. For example, a bandpass filter designed for 650 nm shifts its center wavelength to 645 nm at $15°$. Temperature changes have a minor effect on the center wavelength. A typical temperature coefficient is 0.01 nm/°C.

Edge filters are used in such applications as hot/cold mirrors and color separation. Many narrow-band filters (less than 50 nm bandpass) are assembled from two edge filters by bonding a long wave filter to a short wave filter. Interference edge filters exhibit much sharper change from transmittance to rejection than absorption-type filters. Note that the transmission coefficient usually exhibits a 5–10% modulation.

There are other techniques and components used for spectral selection and filtering not included here. For example, monochromators have transmission bandwidths from 10 to 0.001 nm, depending on their slit widths, grating type, and focal length. Monochromators, however, suffer from poor transmission efficiency and slow switching speeds. Acoustooptic filters, which operate on Bragg reflection, have no moving parts and can switch in microseconds, yet their bandwidth is limited to about 1 nm. Much sharper bandwidths can be obtained by using Fabry–Perot interferometers. These interferometers exhibit a periodic bandpass and can be solid (called etalons) or air-spaced. Air-spaced interferometers are usually driven by a piezoelectric transducer with millisecond switching speeds. Section 12.5.4 includes further discussion on the types, applications, and limitations of Fabry–Perot interferometers.

12.3.4 Lenses

12.3.4.1 Lens Shapes In general, the imaging performance of a lens or combination of lenses may be determined by an exact trigonometric ray trace for a variety of ray F-numbers [10]. The plano-convex lens shape is preferred for focusing laser beams to a spot, collimating expanding beams, or imaging objects at great distances. In these cases, the convex surface minimizes spherical aberrations in situations where the object and image are at asymmetrical or unequal distances from the lens. The optimum situation is where the object is placed at infinity (parallel rays entering lens) and the image is the final focused spot.

To obtain the sharpest focus for a plano-convex lens, the curve surface should be oriented towards the distant object. Furthermore, spherical aberrations is inversely dependent on the cube of the F-number, hence only the central portion of the lens should be illuminated.

The symmetric nature of biconvex and biconcave lenses minimizes spherical aberration in situations where the object and images are at symmetrical or near symmetrical distances. When the entire object lens–image optical system is fully symmetric (1:1 magnification), not only is spherical aberration minimized, but coma and distortion are identically canceled. As a guideline, biconvex and biconcave lenses perform with minimum aberrations at magnifications between 5 and 1/5. Outside of this magnification range, plano-convex lenses are usually more suitable.

12.3.4.2 Achromatic Lenses The simplest lens that corrects for chromatic aberration is a Fraunhofer cemented doublet. In this type of lens, a positive crown glass lens is cemented to a concave flint glass element such that the focal length is constant at two wavelengths, typically chosen in the red and blue regions of the spectrum. This simple achromat exhibits minor focal length shifts, termed a secondary spectrum, at intermediate wavelengths. With newer optical glasses, the secondary spectrum can be reduced to about 0.01% of the nominal focal length. Achromatization of a lens can effectively eliminate spherical aberration and coma and give diffraction-limited performance at a central wavelength. Given an optical system of two cemented lenses, no other aberrations can be corrected.

At greater than 2° field-of-view angles, diffraction-limited performance is degraded by astigmatism, distortion, and field curvature. Dispersion effects increase the nominal focal length by 0.75% when operating the doublet outside its wavelength range (still superior to a singlet). It should be noted that these effects are not due to faulty design, but rather to lack of adjustable parameters. An air-spaced optical doublet or a triplet may be used to obtain multiwavelength diffraction limited performance.

12.3.4.3 Graded-Index Lenses Graded-Index lenses, also referred to as GRIN or SELFOC® lenses [11], are lenses that have an index that is highest at the

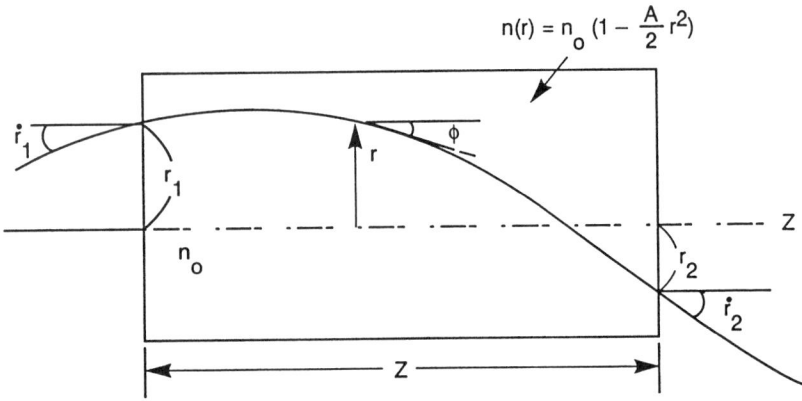

FIG. 4. The locus of a ray passing through a graded index lens. This ray is described by height r and angle $\dot{r}$.

center, and decreases radially. These lenses refract light not by a curved interface but by refraction through a medium of varying index. The index of refraction of these lenses is given by

$$n(r) = n_0(1 - Ar^2/2), \quad (2)$$

where A is the refractive index gradient constant, n_0 is the refractive index on axis, and r is the radial distance from the optical axis. For effective use of this lens it is sufficient to analyze the locus of optical rays passing through it, as shown in Fig. 4. These rays are defined by its height from axis r and its propagation angle $\dot{r}$.

For a lens of known parameters, any ray entering the GRIN lens can be traced by using the following ray equation:

$$\begin{bmatrix} r_1 \\ \dot{r}_1 \end{bmatrix} = \begin{bmatrix} \cos\sqrt{A}\,z & (\sin\sqrt{A}\,z)/n_0\sqrt{A} \\ -n_0\sqrt{A}\sin\sqrt{A}\,z & \cos\sqrt{A}\,z \end{bmatrix} \begin{bmatrix} r_2 \\ \dot{r}_2 \end{bmatrix}. \quad (3)$$

Note that Eq. (3) also contains the refraction effects of the front and exit surfaces. Typically, these lenses are fabricated with several index gradient constants and in diameters of 1 to 3 mm. The flexibility of the GRIN lens comes from its small size and the ability to select the appropriate length z that corresponds to the application. They are used successfully for focusing and collimating laser beams, as well as for imaging and mode matching. However, these lenses exhibit third-order aberrations. Hence, aspheric or achromatic lenses are better suited for diffraction-limited applications. To reduce spherical aberrations, the diameter of the input aperture should be limited to less than 50% of the front surface.

12.4 Polarization-Controlling Components

12.4.1 Types of Polarizing Optics

Table III shows common types of optical polarizers, their advantages, limitations, and applications.

12.4.2 Types of Waveplates

The orderly arrangement of atoms in crystalline structures results in an orientation of the electric field vector that might be different from the crystalline axes. This results in different refractive indices for different polarizations. Birefringence is the change in the index of refraction with input polarization. Unlike dispersion, birefringence is easy to avoid: use amorphous material such as glass, or crystals that have simple symmetries, such as NaCl or GaAs.

Birefringence [12–14] can be used to modify the polarization state of light. By taking just the right slice of a crystal with respect to the crystalline axes, we can arrange it so that the index of refraction exhibited for one polarization orientation of the input wave, say vertical, is different from that exhibited for the opposite polarization, that is, horizontal. These two axes are termed the "fast" and "slow" axes of the crystal depending on the phase velocity of the corresponding propagating wave.

The difference in the indices of refraction between the fast and slow axes translate to phase retardation between waves whose polarizations are oriented along these axes. Retardance units are usually expressed in units of wavelengths. For example, a half-waveplate produces a phase difference of half a wavelength between the waves whose polarizations are along the fast and slow axes. Effectively, the optical length of a crystal along the fast axis is shorter than the slow axis by half a wave.

Since waves repeat themselves every 2π radians, we can have the same phase retardance between crystal plates that have different thicknesses. Accordingly, we see two common types of waveplates: zero-order and multiple-order waveplates. This order refers to the number of wavelengths difference in optical length between the fast and slow axes.

By far the most commonly used phase plates are quarter- and half-waveplates. The half-waveplate can be used to rotate the plane of polarization of plane-polar-

TABLE III. Types of Polarizing Optics

Optic	Polarization contrast ratio	Description	Acceptance angle	Comments/applications
Polarizing sheets	100:1	Laminated plastic	±30°	General-purpose and least expensive, typical transmission of 40%.
Laser line polarizing beam splitter cubes	500:1	Cemented prism pair	±2.5°	Efficient narrow-band polarization for moderate-power lasers. Ideal for focused beams, pulsed lasers, and polarization purification. Prism coating is typically optimized for power handling as opposed to broadband use.
Broadband polarizing beam splitter cubes	1000:1	Cemented prism pair	±2.5°	Meets general polarization requirements for most low-power laser and laser diode applications. 300 nm bandwidth usable range is common.
Polarcor™ polarizers	10,000:1	Dichroic glass	±15°	Compact, high-contrast, large acceptance angle, and low wavefront distortion make it suitable for laser diodes, without the need for collimators or bulky positioners. Typically used for near-IR.
Glan–Thompson polarizers	100,000:1	Cemented calcite prism pair	±7°	Very high polarization purity for low-power laser systems. Large acceptance angle permits use with diverging beams. A single polarizer offers high-purity polarization for broadband or multiple wavelength sources, with transmission >90%
Glan–Laser polarizers	100,000:1	Air-spaced calcite prism pair	Wavelength-dependent, ±0.5 to ±2°	Extreme polarization purity plus high laser damage resistance.

ized light from one plane to any desired plane. Quarter-waveplates are used to turn plane-polarized light into circularly polarized light and vice versa. Note that, if the waveplate is not exactly half-wave (or quarter-wave) at the operating wavelength, it will produce elliptically polarized light at the output. Small errors in retardation of a multiple-order waveplate can be corrected by rotating it a small amount

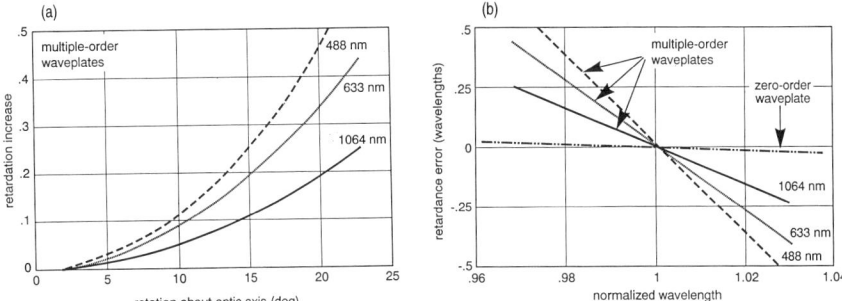

FIG. 5. (a) Change in retardance of multi-order waveplates with rotation about its optic (slow) axis. (b) Change in retardance of multi-order and zero-order waveplates with wavelength changes.

around its fast or slow axes (Fig. 5a). Rotation around any axis increases the retardation.

A waveplate of practical thickness (~1 mm) produces multiples of $\lambda/2$ or $\lambda/4$ retardation. Consequently, higher orders cause retardation to vary dramatically with wavelength. As shown in Fig. 5b, a ±1% departure from the design wavelength causes serious retardation errors. Waveplates are also sensitive to temperature variations, for instance, typical plates have temperature coefficients of about 0.0015 $\lambda/°C$.

Zero-order waveplates are achieved by combining two multiple-order retarders with their optical axes oriented at right angles to each other. By matching thicknesses, the multiwave phase shift developed by one plate is removed by the second except for a residual $\lambda/2$ or $\lambda/4$ retardance. The primary benefit of zero-order plates is moderate insensitivity to wavelength. A ±2% deviation from center wavelength results in a 0.01 wave retardance error (3.6°). For example, a zero-order plate designed for a laser diode of 750 nm will provide useful retardance from 735 to 765 nm. Typical zero-order plates have temperature coefficients of 0.0001 $\lambda/°C$, rendering them insensitive to temperature changes.

12.4.3 Liquid Crystal Devices

These devices provide solid-state control of light without moving parts. The key component is an optical cell containing an anisotropic fluid whose birefringence is proportional to an applied electric field [15]. The assembly of this cell with other optical elements enables voltage controlled adjustment of beam attenuation, polarization, phase, or beam splitting ratio, as well as the processing of light, such as incoherent to coherent conversions and color switching.

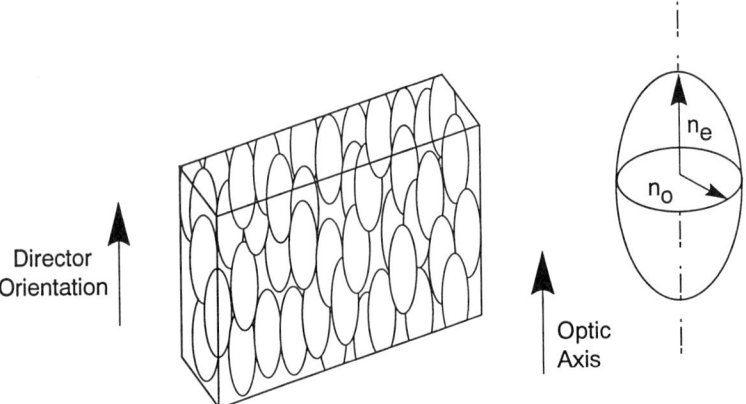

FIG. 6. Spontaneous organization of the structures in a nematic liquid crystal, where birefringence is equal to $n_e - n_0$ and is a function of the applied voltage.

12.4.3.1 Types of Liquid Crystal Nematic liquid crystals represent the basic form of compounds. The elongated nematic molecule is optically anisotropic and is characterized by a uniaxial optical index. The birefringence of the molecule is positive, and its optic axis (extraordinary refractive index) is coincident with the molecular long axis (see Fig. 6).

Other forms of liquid crystal are cholesteric liquid crystals. These are generated by adding chiral twisting agents to the nematic liquid crystal, generating a twist in the nematics, and creating a molecular tilt that follows a helical pattern from one layer to the next. The planes of orientation of the optical director for the liquid crystals changes orientation periodically, as shown in Fig. 7, where P is the pitch of the cycle. Light that has the wavelength and circular polarization that matches this pitch will be strongly scattered by the liquid crystal. The liquid crystal has no effect on waves that have circular polarization handedness opposite to that of the molecular plane orientation. All other wavelengths pass unaffected.

12.4.3.2 Liquid Crystal Cell Optical cells are typically constructed from a thin layer of liquid crystal material, a few microns to a few tens of microns thick, and held between two fused silica optical flats. The inner faces of the flats are coated with conductive indium–tin–oxide (ITO), to which an ac electrical signal is applied. Preparation of the inner surfaces may include rubbing or mechanically shearing the glass substrates, which creates a preferred direction for the liquid crystals. This preferred direction extends to tens of microns into the bulk with an applied external field. A birefringent polymer is sometimes sandwiched by a third optical flat to offset residual birefringence in the liquid crystal cell, enabling a retardance range of 0 to 1/2 wave over the specified bandwidth. For a nematic liquid crystal device, the typical rise and fall times are a few msec and a few tens of msec, respectively.

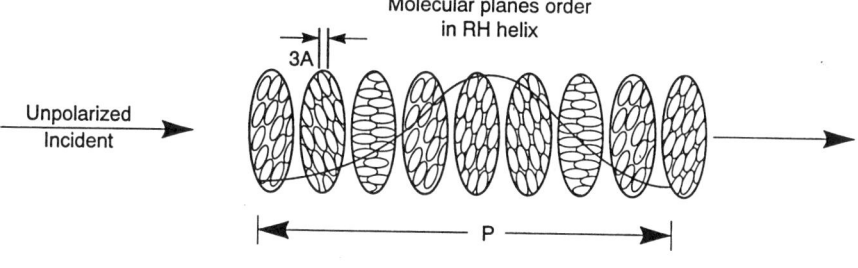

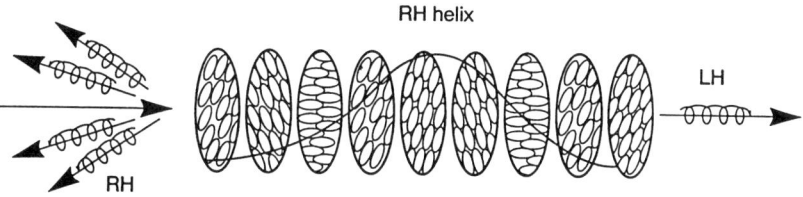

FIG. 7. Spontaneous organization of cholesteric liquid crystal and the corresponding selective reflection.

This device can be used as a variable polarization rotator by cascading a linear polarizer, a nematic liquid crystal cell, and a quarter-waveplate. The liquid crystal converts the linear polarization to elliptical. Upon passing through the quarter-waveplate it is converted back to linear with the same orientation as the elliptical component leaving the cell.

12.4.4 Polarization Meters

The propagation of light through optical or fiberoptic components may lead to a change in the state of polarization of the light, or it may lead to attenuation due to the component's polarization sensitivities [16]. This section illustrates a common technique for describing the polarization of a wave.

The most common method used for characterizing the polarization state of light is by the Stokes parameters, together with its visual representation, the Poincaré sphere. To discuss the Stokes parameters, let us consider an input polarized wave having two orthogonal polarizations given by

$$E_x = e_x \cos(\omega t - kz),$$
$$E_y = e_y \cos(\omega t - kz + \delta), \qquad (4)$$

where δ is the phase difference between the two polarizations. The Stokes parameters of a polarized wave can be measured with the aid of a linear polarizer, a quarter-wave retarder, and calibrated detectors, and are given by

$$S_1 = e_x^2 + e_y^2,$$
$$S_2 = 2e_x e_y \cos \delta,$$
$$S_3 = 2e_x e_y \sin \delta,$$
$$S_0^2 = S_1^2 + S_2^2 + S_3^2, \qquad (5)$$

where S_0 represents the total light intensity. An elliptically polarized beam of ellipticity $e = \tan \chi$ and orientation ψ is thereby defined as

$$\sin(2\chi) = S_3/S_0,$$
$$\tan(2\psi) = S_2/S_1. \qquad (6)$$

Equations (5) and (6) can be used to present the Stokes parameters and the state of polarization pictorially by plotting the S_1, S_2, and S_3 parameters on three orthogonal axes. Any polarization state of light is represented by a unique point on the surface of a sphere, called the Poincaré sphere. A normalized ($S_0 = 1$) Poincaré sphere is shown in Fig. 8. The poles correspond to oppositely handed states of circular polarization, while the equator corresponds to linear polarization, at different orientations. If a wave changes its polarization, its position on the surface of the sphere will change accordingly. Opposite points on the sphere surface are always orthogonal in polarization.

12.5 Passive Optical Devices

12.5.1 Optical Isolators

In simple terms, an isolator is a "one-way valve" for light. Energy from a source is transmitted through the isolator, while returning energy is strongly attenuated. This attenuation is expressed in decibels (dB), where attenuation is equal to −10 log [(transmitted power)/(incident power)].

12.5.1.1 Polarizer/Quarter-Wave Isolator A simple isolator can be constructed from a polarizing beam splitter cube and a quarter-waveplate. Light in a forward direction is polarized by the cube, then circularly polarized by the quarter-waveplate. The returning beam is converted to linear polarization and is blocked by the polarizer. In practice, this type of isolator can attenuate the returning beam by 30 dB. This degree of isolation, however, is reduced if the output beam passes through birefringent material such as an optical fiber or is reflected by optics that alter its state of polarization by other than π. The major drawbacks of this form of isolation are the narrow wavelength coverage and the circularly polarized output.

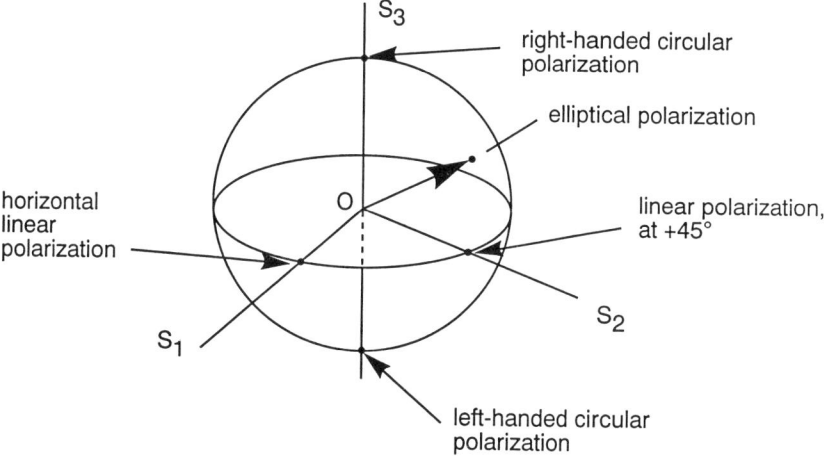

FIG. 8. The Poincaré sphere.

12.5.1.2 Acoustooptic Isolation Another form of isolation that is useful for isolating low-gain lasers, such as helium–neon lasers, is to use an acoustooptic modulator. These devices deflect the input beam by Bragg diffraction off an index gradient formed by an applied acoustic wave. Isolation is due to the nonreciprocal nature of the frequency shift of the diffracted wave. This shift occurs every time the wave is diffracted by the Bragg grating. Hence, the backreflected wave will have a frequency difference of twice the acoustic frequency as compared to the input wave (if the first-order diffraction order is used). In cases where the source is a low-gain laser with high Q, or a high finesse cavity, this backreflected beam will have a frequency that is sufficiently removed from cavity resonance that it will not perturb the laser frequency.

Acoustooptic isolation, however, has its limitations due to the degraded quality of the intensity profile of the diffracted beam, and the limited efficiency of the modulator. Most important, this method cannot be used with high-gain low-cavity finesse lasers such as laser diodes and dye lasers. To eliminate any laser instabilities due to backscatter, the linewidth of the laser's cavity should be much less than the shift in frequency of the backreflected wave.

12.5.1.3 Magnetooptic Isolators A magnetooptical isolator [17–19] can remedy many of the deficiencies of the beam splitter/waveplate isolator and acoustooptic isolation. These devices utilize crystals that rotate the plane of polarization by 45° when they are placed in a magnetic field. The rotation is independent of the direction of propagation (nonreciprocal); therefore, returning light will rotate by 90° with respect to the input. If input and output polarizers are oriented 45° to each other, the return beam will be strongly attenuated (30–40 dB).

The amount of polarization rotation (θ) of an isolator crystal is given by

$$\theta = V(\lambda)Hd\cos(\gamma), \tag{7}$$

where H is the applied magnetic field, d is the thickness of the crystal, γ is the angle between the field and the direction of light travel, and $V(\lambda)$ is the Verdet constant, a wavelength-dependent materials constant. Ideally, materials with large Verdet constants are desirable because only thin wafers are needed, significantly reducing attenuation, distortion, and size.

Isolator crystals whose Verdet constants vary slowly with wavelength are most desirable. This allows an economical device to provide superior isolation within a wavelength range. Several types of Faraday rotator materials have emerged for common use in isolators: YIG (yttrium iron garnet), TGG (terbium gallium garnet), and BIG (bismuth-substituted iron garnet). Both YIG and TGG are bulk crystals, while BIG is a crystalline film ranging in thickness between 200 to 600 μm. The choice of material for an isolator depends on application, environment, and location.

The wavelength sensitivities of YIG and BIG are 0.046 and 0.068 °/nm, respectively, at 1.55 μm. When translated to a bandwidth for –35 dB isolation, the usable ranges of the isolators are ±14 nm for BIG and ±20 nm for YIG. Temperature sensitivities for BIG and YIG are ±16 and ±30°C, respectively. This relative insensitivity to changes in temperature for YIG is one of the principal reasons for its wide appeal.

12.5.2 Binary Optics

Binary (or diffraction) optics are microoptics fabricated as a flat surface with a relief structure that resembles binary steps. These structures are made using high-resolution lithography and integrated circuit techniques. The relief structure must be smaller than or comparable in size to a wavelength of light and of typical height of half a wave. Similar to holographic optical elements, the principle of operation relies mainly on diffraction to coherently alter the reflected or transmitted wavefront. This principle can be applied to fabricate optical elements, such as lenses, multiplexers, splitters, and filters either discretely or as applied to surfaces of bulk optics.

Binary optics are most effectively used with monochromatic light. Example applications of binary optics are beam splitting of coherent optical beams and spherical aberration correction for optical elements such as mirrors and lenses. Figure 9 shows a single binary optical phase grating in a scheme to phase-lock a laser array, as well as to combine the individual laser beams into a single diffraction order [20].

The main drawbacks of diffraction optics are its optical diffraction efficiency and high cost. Although low diffraction losses of 3–5% are common, this is not

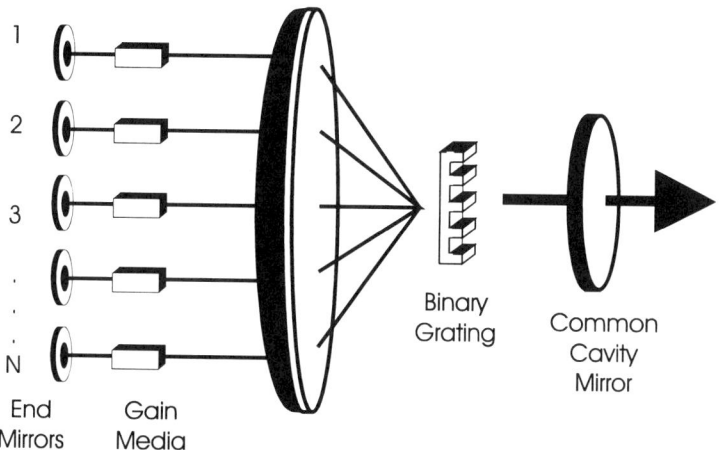

FIG. 9. A binary optical element is used to phase lock and combine outputs of several lasers.

favorable when compared to those of bulk optics. Automation and advances in optics fabrication have combined to reduce the cost of bulk optics significantly. In addition, bulk optics can be coated with dielectric multilayers to enhance their efficiency and damage threshold. Consequently, bulk optics are used for most optical applications, while diffraction optics are used in unique and specialized situations where lightweight, small-sized, and simple optical system configuration are important.

12.5.3 Holographic Optical Elements

Holographic diffraction gratings [21–24] are made by recording the interference pattern formed where two beams from a laser intersect. They are made by exposing a photosensitive layer on top of a glass substrate. Chemical treatment of the film selectively dissolves it to form a relief pattern on the substrate. This pattern can then be used as a transmission grating or is coated with a thin metallic layer to produce a reflection grating. Holographic gratings can be recorded on plane or curved substrates.

One of the main advantages of a holographic optical element (HOE) is that its function is essentially independent of the substrate, unlike conventional optical elements. Also, multiple HOEs can be recorded in one hologram, or, if needed, the recording wavefront can be predistorted to correct for aberrations in the hologram or the optical system. Transmission HOEs are often used in large-aperture systems where it is possible to construct extremely lightweight holographic lenses that have very large apertures. The most common transmission HOE is as a specialized lens that transmits light of specific wavelengths and at a specific angle.

Holographic optical elements are currently being used in heads-up displays, multiple-imaging systems, and other lightweight systems performing complex functions, such as image processing and pattern recognition. They can perform as diffraction gratings, holographic scanners (bar code readers and laser printers), beam splitters, and beam combiners. HOEs can be used as optical filters and can be made in the form of honeycomb microlens arrays with potential uses as solar concentrators or in three-dimensional imaging. While HOEs are not likely to replace conventional optical components due to their limited diffraction losses, they will certainly find increasing use in specialized applications

12.5.4 Fabry–Perot Interferometers

Fabry–Perot interferometers are optical resonant cavities formed by two or more mirrors. These devices have well-defined resonances that allow them to be used as spectral filters in a variety of applications, the most common as spectrum analyzers. Other applications are as external references for laser stabilization, wavelength division multiplexers as communication filters, and intensity multipliers in harmonic and nonlinear wave generation.

A Fabry–Perot interferometer can be as simple as two flat mirrors facing each other, or as complex as multimirror out-of-plane ring cavities [25, 26]. In either case, resonance is established between the cavity and the input beam when the full round-trip cavity length is a multiple of the wavelength of the input beam. At resonance, the cavity transmits the incident light wave (minus losses). Off resonance, all the incident light is reflected back to the source.

The principle of operation is based on internal multireflection of the light ray between the two mirrors. Constructive interference occurs at resonance, allowing the energy of the input ray to penetrate the first mirror and build up in amplitude inside the cavity. Equilibrium is reached when internal losses balance the input energy. The transmitted intensity, which is equal to the input energy for a lossless cavity, is equal to the output mirror transmission T multiplied by the intracavity circulating power.

The transmission curve for a cavity with flat mirrors is shown in Fig. 10. This curve illustrates cavity parameters as defined by cavity bandwidth (γ), free spectral range (FSR), and cavity finesse ($\mathcal{F}$), where:

$$\mathcal{F} = \text{FSR}/\gamma, \qquad (8)$$

and

$$\text{FSR} = c/2L, \qquad (9)$$

where $2L$ is the round-trip cavity length and c the speed of light. The transmission peaks of Fig. 10 are referred to as the longitudinal or fundamental cavity modes. Finesse is a good measure of cavity performance. It is influenced by mirror

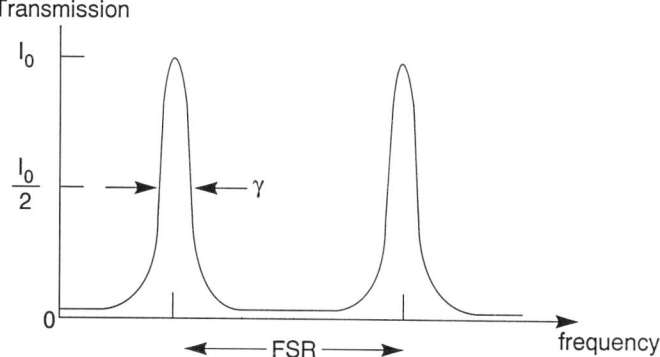

FIG. 10. Transmission characteristics of a flat-mirror Fabry–Perot interferometer.

reflectivity, mirror deformations (or form), cavity losses, and diffraction, and is related to the number of round trips a photon makes inside the cavity before it is absorbed or transmitted.

Finesse is affected by mirror deformations in the following manner. If mirror aberrations are present, an input ray will gradually "walk away" from its intended course. This decreases the number of round trips the photon makes inside before it is lost, thereby decreasing the total finesse. Wavefront diffraction has the same effects. An input collimated beam will diffract if it propagates far enough into the Fraunhofer regime, hence limiting the total finesse of a flat-mirror Fabry–Perot cavity. Diffraction effects can be reduced by keeping the cavity length short or by using spherical mirrors. Typical finesse values for flat mirror interferometers is 10–50 for 20-mm diameter mirrors. Irradiating smaller areas on the mirrors reduces the effects of mirror deformations, allowing for a finesse up to 500. Finesse values of 100–200 are common. For higher finesse, a cavity with curved mirrors is a good alternative.

The following is a list of equations illustrating the relationship between the various parameters together with that for cavity transmission efficiency (η):

$$1/F^2 = 1/F_R^2 + 1/F_F^2, \tag{10}$$

$$F_R = \sqrt{R}\,\pi/(1 - R), \tag{11}$$

$$\eta = (T/(T + A))^2, \tag{12}$$

where R, T, and A are the mirror reflectivity, transmission, and absorption, respectively; $1 - R = T + A$; F_R is the reflectivity finesse, and form finesse is $F_F = M/2$, where λ/M is the flatness of the mirror over the radiated area; F is the total finesse.

12.5.4.1 Flat Fabry–Perot Optical Spectrum Analyzers
An optical spectrum analyzer is one where the resonance frequency of the cavity is scanned with respect to an input laser beam. This is commonly performed by scanning one of the mirrors with a piezoelectric transducer, and detecting the transmitted intensity. The net effect is that the laser spectrum is displayed at the output of the analyzer as a set of transmitted pulses. The width of these pulses is the convolution of the cavity linewidth and the individual laser mode linewidth. To analyze finer details of the laser frequency contents, it is therefore important that the cavity simultaneously have a small bandwidth (γ) and a large free spectral range (FSR), that is, high finesse.

12.5.4.2 Nonconfocal Resonant Cavities with Curved Mirrors
In this case, the mirrors are spherical instead of flat, and the distance between the mirrors is less than the radius of curvature of the mirrors. The introduction of curvature to the mirrors overcomes both the effects of beam diffraction and aberrations by refocusing the propagating wave inside the cavity. Finesse values of more than a million have been reported for ultralow-loss cavities [27].

The introduction of spherical mirror curvature, however, sets up boundary conditions where off-axis rays resonate in the cavity. These resonances are referred to as cavity modes or "eigenmodes," and, depending on their order, they are termed "fundamental" or "transverse." These cavity modes have well-defined resonant frequencies and transverse amplitude distributions. The resonance frequencies for these modes are given below and are illustrated in Fig. 11:

$$\nu_{mnq} = \frac{c}{2L} \left\{ q + \frac{1}{\pi}(m+n+1) \cos^{-1} \sqrt{(1 - L/R_1)(1 - L/R_2)} \right\}, \qquad (13)$$

where $m, n = 0, 1, 2, \ldots$, represent the order of the transverse modes, while q is the longitudinal mode order, which represents the number of half-waves that fit inside the cavity. The shift in resonance frequency of the transverse mode as compared to the fundamental highlights the effect of mirror curvature. This curvature causes an off-axis ray to have a shorter round-trip cavity length. To maintain resonance with the same order, its wavelength is proportionately shorter, and hence its resonance frequency shows up higher than the fundamental mode.

For optical spectrum analyzer cavities, transverse modes introduce the need for mode matching. These cavity eigenmodes have the effect of Fourier decomposing an input beam to a set of inputs that couple proportionately to their respective eigenmodes. An input beam that is misaligned with respect to the cavity optic axis will pump more energy into the higher-order transverse modes. Nonconfocal cavities, therefore, have the disadvantage that the input beam has to be aligned and "mode-matched" to the fundamental cavity mode. Mode-matching is accomplished by matching the input beam parameters, such as divergence, waist, and intensity profile, to that of the fundamental Gaussian cavity modes. This is often

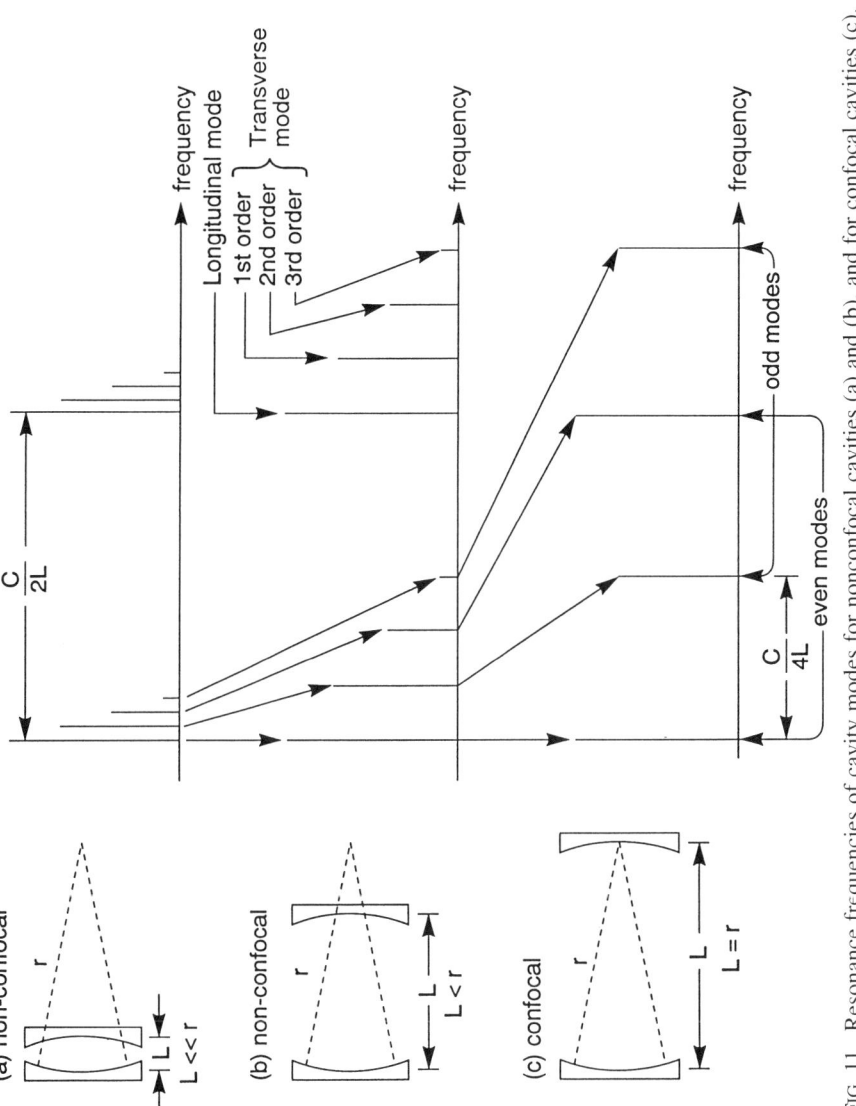

FIG. 11. Resonance frequencies of cavity modes for nonconfocal cavities (a) and (b), and for confocal cavities (c).

performed by one or two lenses between the laser and the optical spectrum analyzer cavity.

12.5.4.3 Confocal Resonant Cavities Confocal resonant cavities [28] are a special case of the curved cavity mirror geometry. The confocal condition is formed when the radii of curvature of the two mirrors is equal to the spacing between them. At this condition, all the even modes and all the odd modes of the cavity have degenerate frequencies, as shown in Fig. 11c.

The main advantage of confocal cavities is in the ease of coupling light to it. The degeneracy of the cavity modes has the effect that any incident ray will resonate in the cavity. Effectively, input rays retrace themselves after four internal mirror bounces, so that minimal mode matching is required to operate the cavity. The disadvantage of such cavities is that the effective free spectral range is half that of nonconfocal cavities. Furthermore, the requirement for short radii of curvatures introduces fabrication difficulties that limit the finesse to approximately 200. Common radii of curvatures are 5 to 50 mm.

References

1. Wood, R. M. (1986). *Laser Damage in Optical Materials*, Adam Hilger, Bristol.
2. Macleod, H. A. (1986). *Thin Film Optical Filters*, McGraw-Hill, New York.
3. Bunshah, R. F., ed. (1982). *Deposition Technologies for Films and Coatings*, Noyes Publications, Park Ridge, NJ.
4. Maissel, L. I., and Glang, R. (1970). *Handbook of Thin Film Technology*, McGraw-Hill, New York.
5. Rossnagel, S. M., Cuomo, J. J., and Westwood, W. D. (1990). *Handbook of Plasma Processing Technology*, Noyes Publications, Park Ridge, NJ.
6. Martin, P. J. (1986). "Ion-Based Methods for Optical Thin Film Deposition," *J. Mat. Sci.* **21**, 1.
7. Bennett, J., and Mattsson, L. (1989). *Surface Roughness and Scattering*, Optical Society of America, Washington, DC.
8. Driscoll, W. G., and Vaughan, W., eds. (1978). *Handbook of Optics,* McGraw-Hill, New York.
9. Barr, E. E. (1974). "Visible and Ultraviolet Bandpass Filters," *Proc. SPIE* **50**, 87.
10. See, for example, Kingslake, R. (1978). *Lens Design Fundamentals*, Academic Press, New York.
11. See, for example, product literature from NSG America Inc., Somerset, NJ.
12. Born, M., and Wolf, E. (1975). *Principles of Optics*, Pergamon, New York.
13. Azzam, R. M. A., and Bashara, N. M. (1987). *Ellipsometry and Polarized Light*, North-Holland, Amsterdam.
14. Hecht, E. (1987). *Optics*, Addison-Wesley, Reading, MA.

15. Jacobs, S. D., Cerqua, K. A., Marshall, K. L., Schmid, A., Guardalben, M. J., and Skeffett, K. J. (1988). "Liquid-Crystal Laser Optics: Design, Fabrication, and Performance," *J. Opt. Soc. Am. B* **5**(9), 1962.
16. Bennett, J. M., and Bennett, H. E. (1978). "Polarization," in *Handbook of Optics*, W. G. Driscoll and W. Vaughan (eds.), pp. 10–76, McGraw-Hill, New York.
17. Samuelson, S., and Amaya, P. (1991). "Faraday Rotators: BIG vs. YIG," *Lasers Optronics*, August, p. 39.
18. Sirasaki, M., and Asama, K. (1982). "Compact Optical Isolator for Fibers Using Birefringent Wedges," *Appl. Opt.* **21**, 4296.
19. Fischer, G. (1987). *The Faraday Optical Isolator*, *J. Opt. Comm.* **8**, 18.
20. Leger, J. R., Swanson, G. J., and Veldkamp, W. B. (1987). "Coherent Laser Addition Using Binary Phase Gratings," *Appl. Opt.* **26**, 4391.
21. Hariharan, P. (1984). *Optical Holography*, Cambridge Univ. Press, Cambridge.
22. Marginos, J. R., and Coleman, D. J. (1985). "Holographic Mirror," *Opt. Eng.* **24**(5).
23. Bennett, S. J. (1976). "Achromatic Combinations of Hologram Optical Elements," *Appl. Opt.* **15**, 542.
24. Sweatt, W. C. (1977). "Achromatic Triplet Using Holographic Optical Elements," *Appl. Opt.* **16**, 1390
25. Dorschner, T. (1983). "Nonplanar Rings for Laser Gyroscopes," *Proc. SPIE* **412**.
26. Martin, G. (1984). "Non-Planar Gyros and Magnetic Biases," *Proc. SPIE* **487**, 94.
27. Rempe, G., Thompson, H. J., Kimble, H. J., and Lalezari, R. (1992). "Measurement of Ultralow Losses in an Optical Interferometer," *Opt. Lett.* **17**(5), 363.
28. Hercher, M. (1968). "The Spherical Mirror Fabry–Perot Interferometer," *Appl. Opt.* **7**(5), 951.

13. GUIDED-WAVE AND INTEGRATED OPTICS

Leon McCaughan

Department of Electrical and Computer Engineering
The University of Wisconsin at Madison

13.1 Introduction

Of all the rich areas of experimental physics, perhaps no field has provided more powerful investigative tools to more areas of science and technology than has optics. Many of the recent contributions are a direct result of the development of the laser, and, as a consequence, these techniques are usually carried out with unguided light beams. Driven by the advantages afforded by light wave communication (e.g., bandwidth, immunity to electromagnetic interference), a new set of capabilities based on optical waveguides has become accessible to the experimentalist. These include remote sensing in hostile environments, very-high-speed data encoding and transmission, programmable optical paths (i.e., optical switching), and the ability to optically sample chemical interactions at surfaces by way of evanescent fields. In this chapter, I survey both guided-wave optics (including fiberoptics) and integrated electrooptic devices, focusing on their capabilities, performance characteristics, and methods for laboratory implementation.

13.2 Optical Waveguides

Waveguides can be operationally divided into three classes: planar, channel, and fiber (i.e., cylindrical). All are composed of a light-carrying central region (called the core or film) surrounded by a material of lower refractive index (the surround, cladding, or substrate). The fundamental properties of all three waveguide forms are most easily elucidated by examining the light propagation characteristics of a symmetric slab waveguide (Fig. 1). For simplicity we assume that both the core index (n_c) and surround index (n_s) are constant. Light propagating along the z-direction (either with its electric field polarized parallel to the waveguide planes, i.e., transverse electric, TE, or polarized perpendicular, the transverse magnetic, TM, polarization) is given by the sum of transverse modes:

$$\mathbf{E}(x,z,t) = \sum_j \mathbf{E}_j(x) e^{i(\beta_j z - \omega t)}, \tag{1}$$

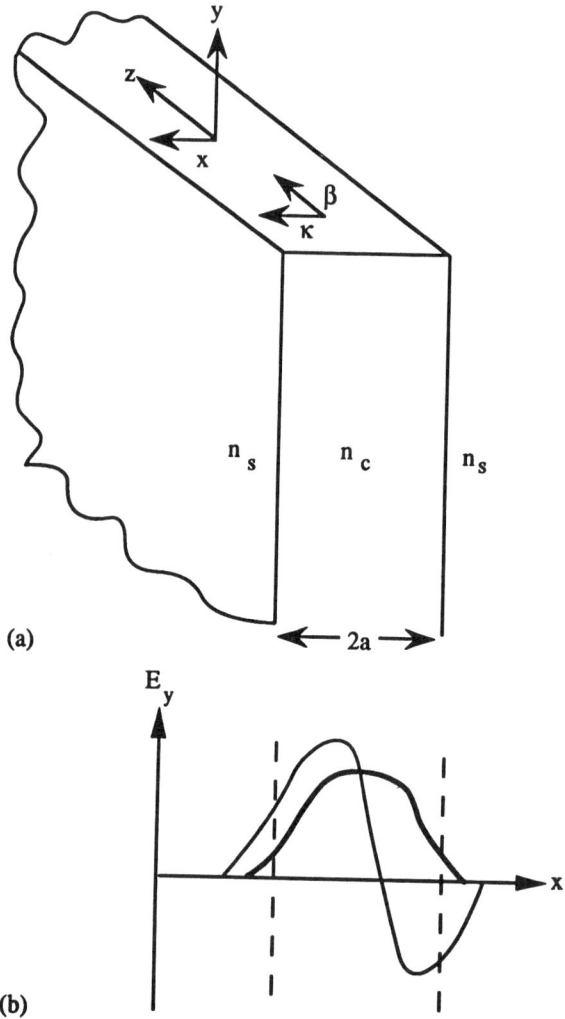

FIG. 1. (a) Planar waveguide with constant core (film) and surround (cladding) indices, n_c and n_s. The axial (β) and transverse (κ) components of the wavevector of a particular mode are related by $(\beta_j^2 + \kappa_j^2) = n_c^2 k_0^2$. (b) Plot of field amplitude vs. position of first- (fundamental) and second-order TE modes.

where $E_j(x)$ are the transverse field profiles of the modes that make up the optical field, and β_j is the propagation constant (i.e., the magnitude of the axial component of the wavevector). As depicted in Fig. 1, mode profiles have a trigonometric solution in the core region (with argument $i\kappa_j x$) and an exponentially decaying solution in the surround (called the evanescent tail, with argument $-\gamma_j x$). The axial, β_j, and transverse, κ_j, components of a mode's wavevector are related by

$$\beta_j^2 + \kappa_j^2 = n_c^2 k_0^2, \tag{2}$$

where k_0 is the magnitude of the free space wavevector ($k_0 = \omega/c = 2\pi/\lambda_0$). Similarly, β_j is related to the decay constant γ_j of the optical field in the surround by

$$\beta_j^2 - \gamma_j^2 = n_s^2 k_0^2. \tag{3}$$

The dispersion relation of a mode $\beta_j(\omega)$ can be derived from the scalar Helmholtz equation [1], and is usually given in terms of a normalized propagation constant,

$$b_j = \frac{\beta_j^2 - n_s^2 k_0^2}{n_c^2 k_0^2 - n_s^2 k_0^2}, \tag{4}$$

and a normalized frequency, sometimes called the V-number,

$$V = 2k_0 a \sqrt{n_c^2 - n_s^2}, \tag{5}$$

where $2a$ is the thickness of the guiding region, and n_c and n_s are the refractive indices of the core (guiding) and surround (evanescent).

A measure of the ability of a waveguide to confine light is given by the normalized propagation constant, $b(V)$ (see Fig. 2). Unconfined light ($n_c \to n_s$) effectively travels as a plane wave in the surround; that is, as $b_j \to 0$, $\gamma_j \to 0$ and $\beta_j \to n_s k_0$. From Fig. 2 we see that, with increasing V, the degree of transverse confinement increases (decreasing κ_j) and the number of modes supported increases. This can be brought about by increasing the core/surround refractive index difference, increasing the core size, or reducing the wavelength of the launched light (see Eq. (5)). In terms of the number of modes supported by a particular structure, we will be primarily concerned with the two extremes: single-mode waveguides that support only the fundamental mode and highly multimode structures ($\gtrsim 100$ modes). As seen in Fig. 2, the requirement that a slab waveguide be single-mode is given by $V < \pi$. For light transmission, it is usually desirable to have strong transverse confinement (i.e., $V \sim \pi$), since waveguide losses, especially those at bends in a guide, decrease with increasing confinement of the light. For sensors that depend on the overlap of the evanescent field with the environment, weak confinement is preferred.

The $b(V)$ curves for orthogonal polarization (the transverse magnetic or TM modes) are very similar to those in Fig. 2. More detailed descriptions of waveguide fundamentals can be found in Kogelnik [1] and Saleh and Teich [2].

13.3 Fibers

Due to the cylindrical symmetry of optical fibers, the exact shape of the dispersion curves $b(V)$ are slightly different from that shown in Fig. 2. (see, e.g.,

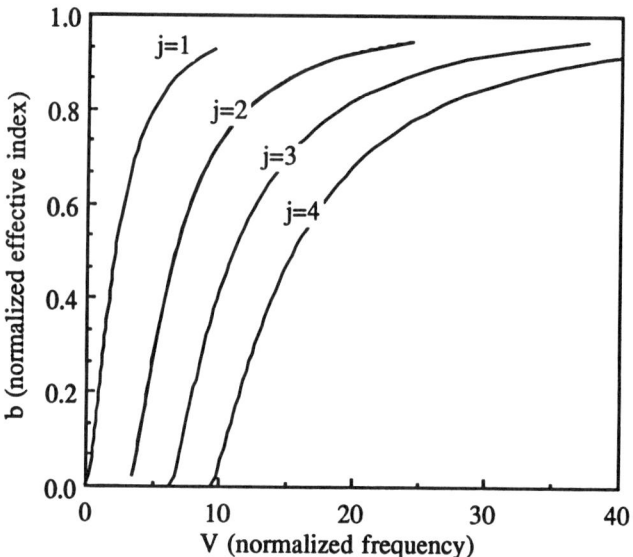

FIG. 2. Normalized dispersion curve for the first four modes of the TE (transverse electric) modes of a symmetric slab waveguide.

Gloge [3]). In particular, the condition for which a step-index fiber supports only one transverse mode is $V \leq 2.405$. (We note that there are two degenerate polarizations for each mode.) Nevertheless, this figure is instructive. As seen from Fig. 2 and Eqs. (4) and (5), a waveguide will support an ever-increasing number of modes as the wavelength of the light becomes shorter, as the core becomes larger, or as the core/cladding index difference becomes greater. For V & 5 we can approximate the number of modes by

$$M \approx 4V^2/\pi^2. \tag{6}$$

Typical multimode telecommunications fiber is designed to support ~500 modes. Because of the larger core diameter (typically ~50 vs. 10 μm for single-mode fiber) and the larger core/cladding index difference ($(n_c - n_s)/n_s \approx 2\%$ for multimode vs. 0.4% for single-mode, $n_s \cong 1.46$ for silica fiber), coupling tolerances between multimode fibers are more relaxed than for single-mode guides (see Section 13.3.3). However, because of the different group velocities of these many modes, light pulses will spread much faster than a single-mode pulse as they travel down the fiber. The consequent degradation of the bandwidth, Δf, with travel distance, L, is given approximately by

$$\Delta f L \approx c/2\Delta n, \tag{7}$$

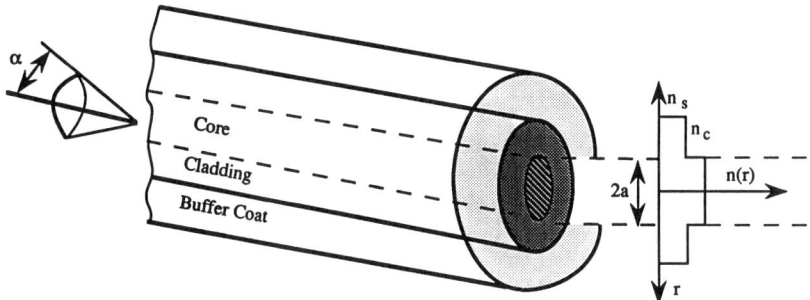

FIG. 3. Step-index optical fiber. n_c and n_s are the core and cladding refractive indices, respectively; α is the maximum light acceptance angle of the fiber.

where $\Delta n = n_c - n_s$. A 20-MHz · km bandwidth–distance product means that a 1 Gb/sec data stream must be restricted to less than 20 m. In order to reduce this modal dispersion, telecommunications fibers (but not most other nonsilica multimode fibers) usually have a refractive index distribution that is graded parabolically from a maximum at the center [4]. Graded index fibers have bandwidth–length products of ~500–1000 MHz · km. If higher bandwidth–length products are required, single-mode fiber must be used.

Since all materials exhibit a wavelength-dependent dispersion, the finite spectral width of the light source must also be taken into account. GaAs/AlGaAs light-emitting diodes (LEDs) and laser diodes emitting at a center frequency ~0.8 µm, for example, have spectral widths of ~25 and ~3 nm, respectively. Silica has a dispersion coefficient of ~100 psec/(km · nm) at this wavelength, limiting the bandwidth–length product to ~400 MHz · km if a GaAs LED is used. With a GaAs laser diode and single-mode fiber, the bandwidth–length product can be raised to ~1.5 GHz · km. This dispersion can be nearly eliminated by operating at a wavelength corresponding to a zero in the material's dispersion coefficient (e.g., ~1.3 µm for silica). (For further details, see Gowar [4] and Saleh and Teich [2].)

13.3.1 Types of Optical Fiber

13.3.1.1 Silica Fiber
The vast majority of optical fiber is manufactured for telecommunications applications. This fiber consists of a Ge-doped silica core (Fig. 3) of circular cross-section (~10 µm for single-mode fiber, ~50–100 µm for multimode fiber) surrounded by an undoped cladding region (~125 µm outer diameter). The fiber exhibits extremely low losses (<0.5 dB/km in the wavelength range between 1.3 and 1.55 µm), is inexpensive (about $0.25 per meter), and has great tensile strength (~10^2 kpsi). Single-mode fiber has a step-index profile for

the core (as in Fig. 3); the multimode version has a parabolically graded index core profile to reduce modal dispersion. The fiber is usually coated with a strippable acrylate buffer layer (final diameter ~250 μm). In the lab, this buffer layer can be easily removed with an organic solvent (e.g., acetone or a commercial varnish stripper) or can be mechanically stripped with a commercially available tool. Removal of the buffer layer will result in slow degradation of the tensile strength of the fiber caused by exposure to atmospheric water vapor and mechanical abrasion [5]. Water corrodes the fiber by way of the following reaction:

$$\text{Si—O—Si} + \text{H}_2\text{O} \leftrightarrow 2\text{SiOH}, \tag{8}$$

which is accelerated at mechanical flaws.

Silica fiber has its minimum attenuation (~0.1 dB/km) at 1.55 μm and zero chromatic dispersion at 1.3 μm (with a slightly higher loss: ~0.5 dB/km). Most fiberoptic components (e.g., laser diodes, LEDs, photodetectors, couplers) are optimized for one or both of these wavelengths. At shorter wavelengths intrinsic Rayleigh scattering, which goes as $\sim\lambda^{-4}$, is the dominant loss mechanism. At longer wavelengths, the infrared absorption tail from Si—O vibrations limit silica's usefulness. Nevertheless, silica fiber will transmit light from 2.5 μm & λ_0 & 0.4 μm with . 1 dB/m loss.

In addition to telecommunications applications, these fibers have the potential for remote sensing in, and data transmission through, such hostile environments as those possessing a high vacuum, intense electromagnetic fields, lethal radiation, chemically reactive species, or high temperatures. The glass transition temperature of silica (~2000°C) suggests a useful operating range in excess of 1000°C for bare fiber. SiO_2 is also reasonably inert, making it a good candidate for remote chemical sensing such as spectroscopic and interferometric monitoring. Fiberoptic vacuum feedthroughs are commercially available (e.g., Pave Technologies, Dayton, OH). Fiber is not, however, immune to ionizing radiation, which can form optically absorbing defect centers in silica fibers [5]. Electrons, protons, neutrons, UV light, X-rays, and gamma rays can all give rise to defects in the glass network. For example, a dose rate of 10^5 rad/h for one minute from a ^{60}Co source (~1.25-MeV gamma rays) produced a 10-dB/km loss in GeO_2/SiO_2 single-mode fiber [6]. The losses have both permanent and transient components. The defect centers sometimes may be bleached away by exposure to UV or even visible light. A more complete review of the effects of ionizing radiation is given by Friebele *et al.* [7].

With the appropriate modifications, fibers themselves can be turned into transducing elements. For example, while optical fiber is generally considered to be immune to conventional electromagnetic interference (EMI), silica fiber does exhibit a small Faraday effect, that is, the rotation of the plane of polarization of light induced by a magnetic field oriented parallel to the propagation direction. The rotation angle is given by

$$\phi = V \int \mathbf{H} \cdot d\mathbf{s}, \qquad (9)$$

where V is the Verdet constant of the material, $\mathbf{H}$ is the magnetic field intensity, and $d\mathbf{s}$ is a line element along the fiber. The Verdet constant for silica fiber is ~4.3 × 10^{-6} rad/A. This rotation can be frustrated by making the fiber birefringent (e.g., coiling the fiber into a loop; see Section 13.3.3). Magnetic field, current, acoustic pressure, rotation, temperature, acceleration, and strain sensors have all been demonstrated with treated silica fiber. A complete review is given by Giallorenzi et al. [8].

13.3.1.2 Nonsilica Fiber Heavy metal fluoride fibers were first developed to permit telecommunication transmission in the mid-infrared (2–4 μm). The most common of the fluoride fibers are made of zirconium–barium–lanthanum–aluminum fluoride, with sodium fluoride used to vary the core-to-cladding refractive index difference (ZBLAN glasses). Commercially available fluorozirconate fibers have transmission windows ~0.5 to ~4 μm (e.g., Infrared Fiber Systems Inc., Silver Spring, MD; Galileo Corporation, Sturbridge, MA). Typical attenuation coefficients in this wavelength range are 0.01 to 0.1 dB/m. Fluoride fibers have also been demonstrated with windows in the 6–8 μm range [9]. Their tensile strength is expected to be about that of silica fiber (~100 kpsi). These fibers open up the possibility of carrying out several types of remote infrared experiments, including IR imaging, temperature sensing, IR absorption spectroscopy, and Raman spectroscopy [10]. As a specific example, the IR vibrational signature of a chemical species (e.g., the C—H stretch bond has an absorption peak at ~3.5 μm) might be remotely monitored via a fluoride fiber. In addition, these fibers, unlike their silica counterparts, do not darken with exposure to radiation. However, fluorozirconate fibers have a much lower glass transition temperature (~250°C) than silica fiber (~2000°C). Finally, these fibers are expensive relative to their silica counterparts (~$75 compared to ~$0.25 per meter for silica fiber).

Chalcogenide glasses are amorphous semiconductors (e.g., Ge/Se/Fe) with transmission windows extending from ~0.5 to ~20 μm [9, 11]. Operating temperatures must be kept low (~150 to 200°C). Commercially available fiber (Infrared Fiber Systems Inc., Silver Spring, MD) is transparent (typically 1 to 5 dB/m) in the 5 to 10 μm wavelength range. Potential applications for this fiber include transmission of long-wavelength radiation (e.g., 10 μm CO_2 laser light) and far-infrared spectroscopy.

Metal oxides (e.g., Al_2O_3, sapphire) offer the highest usable temperature (~2000°C for Al_2O_3) and in principle should transmit from the UV well into the IR (~0.2 μm ≤ λ_0 ≤ 6 μm). In addition, their chemical stability makes them excellent candidates for experiments involving optical monitoring in reactor vessels. Commercial production of oxide fibers is just beginning to get underway.

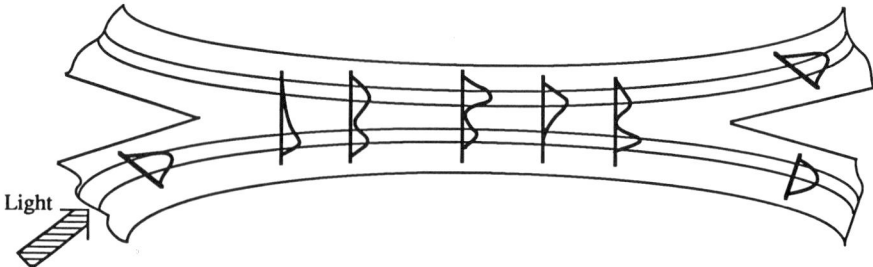

FIG. 4. Diagram of evanescent coupling between two closely spaced fiber cores. Light launched into one waveguide couples periodically between the two guides. The fraction of coupled power is controlled by adjusting the waveguide core separation and the length of interaction.

13.3.2 Fiberoptic Components

13.3.2.1 Fiberoptic Delay Line One of the most straightforward applications of fiberoptics takes advantage of the finite propagation velocity of light and the fact that even a large length of fiber occupies a very small volume. Thus, a kilometer of standard graded index multimode telecommunications fiber (50-μm core) wound on a 15-cm mandrel occupies a cross-section less than 100 mm² and provides a ~4.8 μsec delay with a loss (due primarily to micro- and macrobend losses) less than 3 dB/km [12].

13.3.2.2 Fiber Couplers Light can be made to couple periodically between two single-mode fibers if one core is brought within a few evanescent decay lengths of the other (see Fig. 4). The fraction of coupled power is controlled by the separation between the guides and the interaction length. Fused fiber couplers (the most common type) are fabricated by heating and pulling two fibers that have been stripped of their buffer coat and wrapped around one another. As the two claddings fuse, the fiber is stretched, causing the core diameters to narrow and the optical modes to spread and overlap one another (i.e., V becomes smaller and mode confinement is reduced). The fraction of optical power coupled between the guides is given by

$$\frac{P_x}{P_0} = \frac{\kappa^2}{\kappa^2 + (\Delta\beta/2)^2} \sin^2\left\{(\kappa L)^2 + (\Delta\beta L/2)^2\right\}^{1/2}. \tag{10}$$

The coupling strength, κ, is controlled by the overlap between the modes of each fiber (see Section 13.4.4.2). Note that complete transfer of optical power can only occur if the propagation constant difference, $\Delta\beta$, between the two modes is zero. Since the coupling strength, κ, increases with increasing wavelength (decreasing

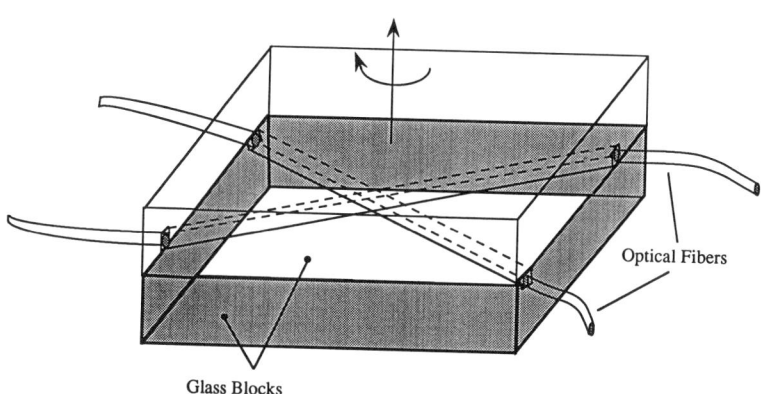

FIG. 5. Diagram of a fiberoptic variable coupler. The fibers are mounted near the surface of the glass blocks with their cores partially exposed. The amount of coupling can be adjusted by varying the intersection angle of the guides.

V), it is possible to make a wavelength divider/combiner by designing the coupling period, $\kappa(\lambda)L$, to be an odd multiple of $\pi/2$ at one wavelength and an even multiple of $\pi/2$ at the other. Packaged 2×2 couplers with a variety of power-splitting ratios are commercially available (e.g., Gould Inc., Glen Burnie, MD).

Fused multimode fiber couplers operate on a different principle since it is not practical to design a coupler with the same coupling period for all of its modes. Instead, the core diameters are decreased by heating and drawing until they can no longer support their modes, and the fused cladding now becomes the new common waveguide. At the output end, the light is coupled back into the individual cores in approximate proportion to their cross-sectional areas. As a consequence of this last statement, light may be distributed to (or from) many fiber ports by fusing the appropriate number of fibers together. A number of different configurations are commercially available.

A tunable fiber coupler was demonstrated by Parriaux *et al.* [13]. The device consists of a pair of glass blocks into which is cut a shallow slot (Fig. 5). Fibers are cemented into the slots and the blocks are polished to expose the fiber cores. Optical coupling is achieved by laying one exposed core over the other at a slight angle (an index-matching fluid may be used). The amount of coupling is increased by decreasing the intersection angle between the two cores.

13.3.2.3 Polarizing Elements

A standard single-mode fiber with a circular cross-section supports two degenerate modes corresponding to the two polarization components. Small off-axis perturbations in the refractive index of the fiber caused by strain can cause light energy to exchange between the (linear or circular) polarization components. The two modes can be decoupled by introducing a large stress-induced birefringence, $\Delta\beta$, between the two principal axes (see Eq. (10)).

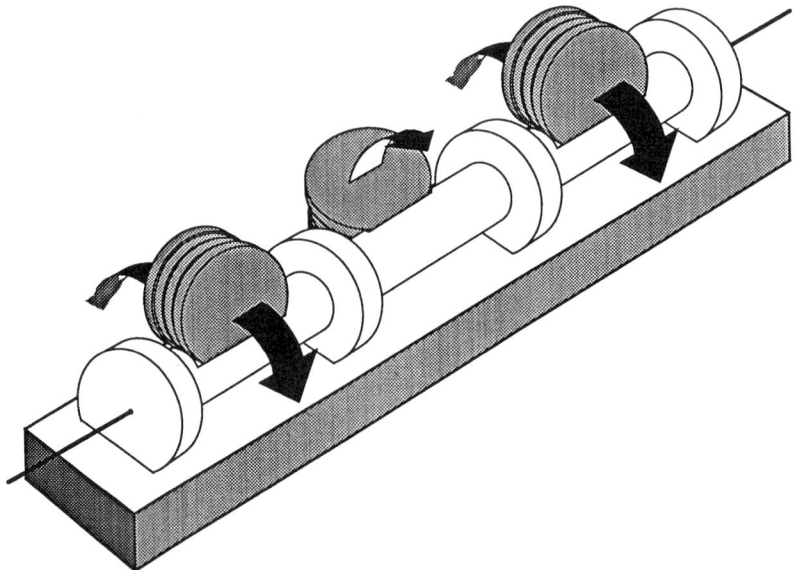

FIG. 6. Fiberoptic polarization controller. The outer coils are $\lambda/4$ retarders; the center coil is a $\lambda/2$ retarder.

As a result, linearly polarized light remains linearly polarized. (Or, equivalently, the crosstalk between the two polarization components is eliminated.) Note that the phase coherence between the two polarization components is rapidly lost (typical beat lengths are $2\pi/\Delta\beta \sim 2$ mm).

Stress birefringence is produced by introducing a material with a different thermal expansion coefficient (e.g., a borosilicate glass) into the cladding layer near the core. Descriptions of the methods for producing birefringence can be found in reviews by Kaminow [14] and Noda *et al.* [15]. Both polarization-preserving fiber and polarizing fiber (only one polarization is guided; the other is a "leaky" wave) are commercially manufactured (e.g., Corning Inc., Corning, NY).

A uniform stress applied to a fiber, such as coiling the fiber, can also be used to produce a birefringence and convert standard single-mode fiber to a polarization-preserving version. A polarization controller based on this principle [16] is shown in Fig. 6. Winding a fiber of radius b into a loop of radius R produces a stress-induced retardation between the orthogonal polarization components that is proportional to $(b/R)^2$. The first and third loops are quarter-wave retarders; the second is a half-wave retarder. By adjusting the planes of the loops, an arbitrary input polarization can be transformed to any desired output polarization.

13.3.3 Source-to-Fiber Coupling

The ends of most glass fibers can be made acceptably smooth simply by removing the buffer coat (usually with acetone or a mechanical stripping tool), gently scoring the surface with a hard edge (such as tungsten carbide or diamond) to initiate cleavage, and pulling on the fiber until it separates. A flat mirror surface should be produced. Alternatively, commercial equipment that performs the scribing and breaking is available (e.g., Newport Corp., Fountain Valley, CA). Holding one end of the fiber near almost any light source will provide sufficient light to illuminate the fiber core. A small magnifier can be used to confirm that the core area is free of imperfections.

There are two principal sources of loss in coupling light from a light source (e.g., thermal, LED, laser): Fresnel reflection at glass/air interfaces, and the finite "collecting aperture" of the fiber. The fraction of optical power lost due to reflection at the interface between the fiber core and the medium in which the light source is immersed (index n_a) is given approximately by

$$R = \left(\frac{n_c - n_a}{n_c + n_a}\right)^2. \tag{11}$$

Each glass/air interface, for example, produces ~3% (~0.15 dB) loss.

The light collecting power of the fiber can be described by its numerical aperture, NA, which is related to the maximum acceptance angle α (Fig. 3) and the core and cladding indices (n_c and n_s, respectively) by

$$\text{NA} = \sin \alpha = \sqrt{n_c^2 - n_s^2}. \tag{12}$$

The V number of a fiber is related to the NA by

$$V = 2\pi a \, \text{NA}/\lambda_0. \tag{13}$$

We define coupling loss as the ratio of light coupled into a waveguide to that emitted from the source:

$$\eta = \frac{P_{wg}}{P_0}. \tag{14}$$

For a single-mode fiber, the coupling efficiency is given by the overlap of the optical field profile of the light source, $E_s(x,y)$, with that of the optical fiber, $E_f(x,y)$:

$$\eta = \frac{\left[\int_{-\infty}^{\infty}\int_{-\infty}^{\infty} E_f(x,y)\, E_s(x,y)\, dx\, dy\right]^2}{\left[\int_{-\infty}^{\infty}\int_{-\infty}^{\infty} E_f^2(x,y)\, dx\, dy\right]\left[\int_{-\infty}^{\infty}\int_{-\infty}^{\infty} E_s^2(x,y)\, dx\, dy\right]}. \tag{15}$$

If it is assumed that the two mode profiles are Gaussian along both the x and y axes, then

$$\eta = \frac{4}{\left(\dfrac{w_x}{d} + \dfrac{d}{w_x}\right)\left(\dfrac{w_y}{d} + \dfrac{d}{w_y}\right)}, \tag{16}$$

where w_x and w_y are the $1/e$ intensity halfwidth and depth of the waveguide mode, and d is the $1/e$ intensity radius of the fiber mode. Direct butt-coupling of a single-mode fiber to a diode laser, for example, produces relatively poor coupling efficiency (~10%), primarily due to the poor match between the elliptical profile of the laser ($w_x \sim 0.8$ μm, $w_y \sim 0.5$ μm) and the circular profile of the fiber ($d \sim 5$ μm). Lenses can be used to raise the field overlap to ~50–60% [17], but alignment tolerances become more severe. For routine research work, a pair of back-to-back 20× microscope objectives mounted on translating stages provides a convenient method of introducing optical elements (e.g., mechanical chopper, filters, beam splitters) between the light source and fiber. Because of the small numerical aperture of single-mode fibers, negligible amounts of light can be coupled from broad area (e.g., LEDs and thermal) sources.

In order to transmit reasonable amounts of power, multimode fiber is usually required for coupling to extended light sources (such as LEDs, lasers, thermal sources, gas discharge sources). Consider a light source of area A_s with radiant intensity $I(\theta)$ lens-coupled to a step-index fiber supporting many modes (V ≳ 10). A lens of focal length f placed at distance s_s from the source and distance s_f from the fiber will produce an image on the fiber end of area $A_i = M^2 A_s$, where $M = s_f/s_s$ is the magnification of the lens system. The maximum emission angle θ_m of light that can be accepted by a fiber with NA = $\sin \alpha$ is given by (see Fig. 7)

$$\tan \theta_m = \frac{R}{s_s} = M \frac{R}{s_f} = M \tan \alpha, \tag{17}$$

where R is the aperture radius of the imaging system. The fraction of light that can be coupled into a step-index multimode fiber with core radius A_c is therefore given by (e.g., [18])

$$\eta_{SI} = \frac{\displaystyle\int_0^{\theta_m} I(\theta) \sin \theta \, d\theta}{\displaystyle\int_0^{\pi/2} I(\theta) \sin \theta \, d\theta} \times \begin{cases} 1 & \text{for } A_s < A_c \\ A_c/A_s & A_s \geq A_c \end{cases}. \tag{18}$$

For an isotropically radiating source, Eq. (18) can be approximated [19, 20] by

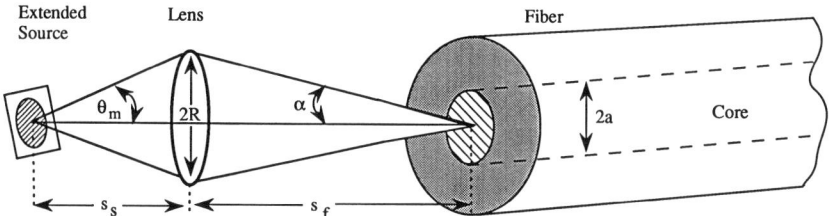

FIG. 7. Diagram of lens coupling from an extended source to multimode fiber.

$$\eta_{SI} = \frac{(NA)^2}{4n_0^2} \times \begin{cases} 1 & \text{for } A_s < A_c \\ A_c/A_s & A_s \geq A_c \end{cases}, \quad (19)$$

where n_0 is the refractive index into which the source is immersed. For a Lambertian source (i.e., $I(\theta) = I_0 \cos \theta$), Eq. (19) should be multiplied by a factor of 2. For a fiber with a parabolically graded core index, the NA is calculated (Eq. (12)) using the on-axis core index for n_c, and the fraction of coupled light (Eq. (19)) is reduced by a factor of 2. As an example, for a typical graded index multimode fiber (NA ~ 0.3) and assuming a Lambertian source, $n_0 = 1$, and $A_c = A_s = 2 \times 10^3$ µm², the coupling efficiency is η_{GI} ~ 0.02 (−16 dB). This is approximately a factor of 2 better than direct butt-coupling of the Lambertian source to the fiber [21].

Finally, it should be noted that a variety of connectors and alignment elements for fiber splices are commercially available (for a review, see [22]).

13.4 Guided-Wave Integrated Optics

Guiding of light is not limited to the cylindrical geometry of a fiber. Both single- and multimode channel waveguides have been fabricated in most transparent materials, including a variety of semiconductors [23], glass [24], and polymers [25, 26]. The distinct advantage of fabricating waveguides on planar substrates is that standard photolithographic processing can be used to define channel waveguide geometries of nearly arbitrary cross-section and configuration. In addition, any combination of guided-wave elements can be integrated on a single substrate with a single set of processes. As outlined later, various functionalities are possible, depending on the substrate used.

Very briefly, the photolithographic process consists of spin-coating a planar surface with a thin (~1 µm) layer of photoresist (a photosensitive polymeric solution; e.g., Shipley Products, Newton, MA). The substrate is exposed to UV light through a glass mask, which has been patterned with the desired features. For a "positive" photoresist, the exposed regions dissolve away, leaving the desired waveguide pattern in the form of protected areas on the substrate surface. The exposed areas can be doped or etched, as the particular waveguide forming technique dictates.

By convention, the two polarization components of an optical mode in a channel waveguide are labeled transverse electric (TE) if the electric field is in the plane of the substrate and transverse magnetic (TM) if perpendicular to the substrate. Unlike optical fiber (in which pairs of modes with orthogonal polarization are degenerate), β_{TE} and β_{TM} are not necessarily identical.

13.4.1 Optical Plumbing

Aside from the straight waveguide, there are four elements that form the basis set for all of integrated optics: the s-bend, the y-branch, the directional coupler, and intersecting waveguides (Fig. 8). S-bends provide a way of redirecting guided light transverse to its propagation direction. Transverse offsets must be gentle in order to avoid curvature-induced coupling of the guided light into radiation modes. In particular, the curvature change should be continuous. For this purpose a modified trigonometric function is often used:

$$y(z) = \frac{hz}{L} - \frac{h}{2\pi} \sin \frac{2\pi z}{L}. \tag{20}$$

With this geometry, a rise/run ~100 μm/3 μm will exhibit losses . 1 dB (see, e.g., [27]).

Branching waveguides (y-branches) can be broadly divided into (a) symmetric geometries [28, 29] that are used as passive power dividers/combiners, and (b) asymmetric geometries [28, 30]. In the latter geometry, the propagation constants of the branches are unequal, so that the coupling of an optical field from the base waveguide to the branch guides depends on the symmetry of the launched mode. Asymmetric branches have been used with the electrooptic effect to demonstrate a 1 × 2 electrooptic switch [31]. As with s-bends, y-branch devices must have small rise/run ratios (branching angles . 2°) to keep radiation losses to reasonable levels (~1 dB).

As with two closely spaced fiber cores (see Section 13.3.2.2), light can be made to couple between a pair of parallel channel waveguides by way of their modes' evanescent fields. By themselves, these directional couplers are used as power dividers and wavelength-selective dividers/combiners. A number of directional coupler-based active elements have been made by fabricating evanescent couplers in electrooptic effect materials (e.g., $LiNbO_3$, see Section 13.4.4). These include optical modulators, switches, polarization transformers, and wavelength demultiplexers.

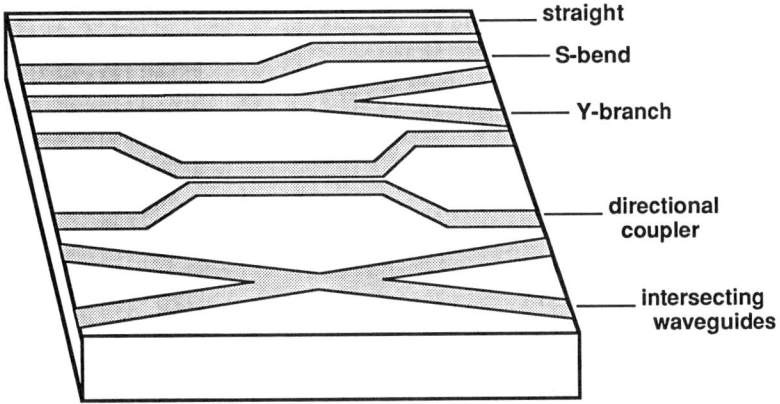

FIG. 8. Waveguide elements of integrated optics.

Intersecting waveguides are very much like directional couplers with nonconstant separation. They are generally used as optical cross-connects and four-port power splatters in switch arrays. This waveguide geometry has the unique property that the coupling between the two guides is quasi-periodic [32] with the intersection angle: at specific angles the coupling between the guides goes to zero. Unlike electronic integrated circuitry, integrated optic paths can therefore be made to intersect without crosstalk. The coupling between these guides can also be varied by changing the refractive index of the intersection region [33]. Optical switching has been demonstrated using the electrooptic effect to alter the effective index of the waveguides [34–36].

While the optical properties of s-bends, y-branches, directional couplers, and intersecting waveguides are reasonably well understood, a successful model (one that begins with the fabrication parameters as input and makes predictions of optical characteristics that agree with experimental results) has yet to be developed. The closest approximation to date is a computational approach, such as the beam propagation method [37].

13.4.2 Channel Waveguides in Glass Substrates

Waveguides are relatively easily fabricated in glass substrates by ion exchange of Na^+ for (an)other cation(s) in a molten salt bath. A metal (e.g., Al) or an oxide (e.g., SiO_2) is first deposited (usually under vacuum by evaporation or sputtering) on the glass substrate, followed by photolithographic patterning of the waveguides into photoresist. Dissolution of the unexposed photoresist and subsequent exposure of the protecting film to an etchant leaves waveguide features on the substrate open to ion exchange. Submersing the patterned substrate in a mixture of $TlNO_3$,

KNO$_3$, and NaNO$_3$ at a temperature of 530°C, for example, will give rise to a large surface index change ($\Delta n_{max} \sim 0.1$) with low propagation losses (~0.1 dB/cm) in a few minutes time [38]. Because Tl$^+$ is toxic, such other ions as Li$^+$, K$^+$, or Ag$^+$ are often substituted [24], although the index change that they can produce is smaller ($\Delta n_{max} \lesssim 0.01$). The ion exchange can be allowed to proceed by diffusion or be augmented with an electric field applied between the bottom of the substrate (via a Cr/Au electrode) and the molten salt (via a Pt wire). This has the advantage of reducing the aspect ratio of the waveguide cross-section by enhancing diffusion normal to the substrate surface, as well as speeding the exchange process [39]. Under sufficiently large electric fields, the index profile becomes step-like. For a more thorough review of ion exchange glass waveguides, see Ramaswamy and Srivastava [24] and Findakly [40].

Light can be end-fire coupled into the glass endfaces after polishing with conventional (e.g., Al$_2$O$_3$) abrasives. Single-mode or multimode fibers can be aligned and butted to the waveguides. A microscope objective (10×–40×) or a prism [41] can be used instead, if polarization considerations are important.

13.4.3 Polymer Waveguides

Optical polymers provide a method for producing waveguide structures on a wide variety of surfaces [42]. Polymers that produce good-quality waveguides include polymethylmethacrylate (PMMA) [43], polydiacetylenes [44], and polystyrene [45]. Thin (~5–10 μm) layers of the polymer are usually obtained by spin-coating. Thickness is controllable to ~±1%. Channel waveguides can be fabricated by wet or dry etching [42], photopolymerization [46], or photobleaching [47]. Waveguide propagation losses are typically 1–5 dB/cm [48]. Because thin polymer layers tend to conformally coat surfaces, optical connections between different levels of a substrate are feasible, given that the vertical offset is not too abrupt. PMMA films as thick as 300 μm with 80:1 height-to-width aspect ratios have been demonstrated using x-ray lithography [49]. Inasmuch as the endfaces of polymer guides are difficult to polish, unless the waveguide endfaces can be cleaved or can be defined as part of photolithographic processing, prism coupling [41] must usually be used.

In addition to their waveguiding properties, organic polymers can produce reasonably large second-order optical nonlinear effects, such as the electrooptic effect [50] and second-harmonic generation [51]. The active moieties consist of electron-donating and attracting groups, connected by conjugated π-bonds. These may be part of the polymer chain, or can be appended (e.g., stilbenes and diazo dyes; see [52]). The polymer must be aligned to produce a net second-order effect. This is usually done by applying a strong electric field (~100 V/μm) while cooling through the polymer's glass transition temperature ($T_g \sim 150°C$). An axial electrooptic coefficient of ~40 pV/m has been observed [53]. Electrooptic devices have been demonstrated (e.g., [54]). A lack of thermal stability at elevated temperatures (T & $100°C$), larger optical losses, and the absence of off-diagonal electrooptic coefficients are the most serious limitations for the experimentalist. Unless guided-wave electrooptic devices must be incorporated on a particular substrate, $LiNbO_3$-based devices are preferred.

13.4.4 $LiNbO_3$ Integrated Electrooptics

A large number of materials are available for bulk electrooptic devices [55], though only a few have shown promise as substrates for guided-wave optics. Of these, lithium niobate ($LiNbO_3$) has proven to be the most versatile. It is transparent from about 0.35 μm to about 5 μm and is uniaxially birefringent ($n_e \cong 2.16$, $n_0 = 2.24$). $LiNbO_3$ is a ferroelectric crystal (Curie temperature ~ 1125°C) that exhibits large piezoelectric, pyroelectric, photoelastic, frequency doubling, and electrooptic effects [56]. These nonlinear properties are due to a large spontaneous polarization in the material. This polarization is usually introduced during crystal growth by poling, that is, drawing a current along the z-axis while cooling the material through its Curie temperature. High-quality poled $LiNbO_3$ is commercially available (e.g., Crystal Technology, San Jose, CA). The property that distinguishes $LiNbO_3$ from all other electrooptic materials is the ability to produce low-loss waveguides without degrading optical nonlinearities. Also, unlike semiconductor optoelectronics, no sophisticated equipment (other than a photolithography station, metal deposition capabilities, and a diffusion furnace) is required to produce these devices.

A variety of thermally diffused metals were examined by Kaminow and Carruthers [57] as a way of changing the refractive index of $LiNbO_3$. Ti has proven to be the most attractive of those studied because: (a) it can be easily patterned into any desired two-dimensional geometry using standard photolithographic techniques; (b) it produces large and nearly equal index changes for both polarizations ($\Delta n \sim 0.6$ per weight fraction Ti); (c) it has negligible absorption in the visible and near-IR; and (d) its diffusion coefficient is sufficiently large that relatively modest diffusion times and temperatures (typically 1000°C, 6–9 h) are needed to make tightly confining waveguides [58]. Diffusion of Ti is carried out under a flowing gas (typically O_2, N_2, or Ar). Water vapor (obtained by bubbling the gas through a water column) is used to suppress Li_2O loss (so-called outdiffusion). During cool-down, O_2 is used to reoxidize the crystal (which otherwise appears yellow in its reduced state). Care must also be taken to remain below the Curie temperature of the substrate. It has been shown that a variety of processes (see, e.g., [59, 60]) can cause the $+z$ surface to depole below its Curie temperature. For this reason,

TABLE I. Ti Strip Dimensions and Optical Losses for Single-Mode Waveguides

Wavelength (mm)	Width (mm)	Thickness[a] (nm)	Diffusion (h)
0.6	3	~40	~5.0 @ 1000°C
0.8	4	40	~6.0 @ 1000°C
1.3[a]	6.5	65	~6.0 @ 1050°C
1.3[c]	6	60	~6.0 @ 980°C
1.55	9	75	~9.0 @ 1050°C

[a]Ti thickness values are given for sputter or electron beam deposition sources. For thermal evaporation, multiply this value by ~1.4.
[b]Minimizing optical loss.
[c]Maximizing overlap of electrode electrical fields with waveguide optical field.

devices are patterned on the $-z$ face, which has a higher depoling temperature. (The "polarity" of the z-axis is determined by the polarity of the charge that develops across the z-face, due to the pyroelectric effect, on cooling the crystal.) Straight waveguides can now be routinely fabricated by Ti diffusion (see Table I) with propagation losses at longer wavelengths as low as ~0.1 to ~0.2 dB/cm and fiber–waveguide butt-coupling loss ~0.6–1 dB (for details, see [61–65]).

Proton exchange [66] is a low-temperature process for making optical waveguides that circumvents the high-temperature processing problems associated with Ti diffusion (Li_2O outdiffusion, depoling of the substrate, and $LiNbO_3$ reduction). A variety of organic and inorganic acids (e.g., [66, 67]) have been shown to produce this index enhancement. For example, immersion into benzoic acid (~200°C, 15 min) causes H^+ to exchange with Li^+ and produces a large (Δn_e ~ 0.12) step-index change. The substrate must subsequently be annealed (~2 h @ ~300°C) to restore the electrooptic effect and reduce waveguide losses [68]. Unlike Ti diffusion, only the index along the extraordinary polarization is increased (the ordinary polarization decreases slightly). Channel waveguides can be patterned by photolithographically masking the substrate (e.g., Ni, Al) before exchange. Because of the absence of any change in surface topography (as occurs with Ti-diffused waveguides), H^+-exchanged channel waveguides are more difficult to image under a microscope.

After fabrication, the endfaces of the waveguides (Ti-diffused or H^+-exchanged) are polished to permit end-fire coupling to the guides with either a microscope objective (~5×–40×) or a fiber. In order to avoid chipping and rounding of the waveguide endfaces, a pair of $LiNbO_3$ blocks are first epoxied over the waveguides (see Fig. 1 in Murphy *et al.* [69]). The ends are cut (usually with a wire saw) perpendicular to the waveguide axis and polished (e.g., 12, 3, and 1 μm Al_2O_3 grits plus a colloidal silica polish).

13.4.4.1 The Electrooptic Effect

The property that distinguishes $LiNbO_3$ integrated optics from passive guided-wave optics (e.g., ion-exchanged waveguides in glass) is the ability to change the material's refractive index by way of an electric field. The electric field is usually produced by applying a voltage to an electrode pair patterned on the substrate surface above a waveguide. This index change in turn alters the propagation constant, β, of a waveguide segment. The index change, $\Delta n_{ij}(E_j)$, is dependent on the directions of both the applied control field, E_j, and the polarization of the light. Using a reduced notation, the index change in $LiNbO_3$ is given by Kaminow [55]

$$\begin{bmatrix} \Delta[1/n^2]_x \\ \Delta[1/n^2]_y \\ \Delta[1/n^2]_z \\ \Delta[1/n^2]_{yz} \\ \Delta[1/n^2]_{xz} \\ \Delta[1/n^2]_{xy} \end{bmatrix} = \begin{bmatrix} 0 & -r_{22} & r_{13} \\ 0 & r_{22} & r_{13} \\ 0 & 0 & r_{13} \\ 0 & r_{51} & 0 \\ r_{51} & 0 & 0 \\ -r_{22} & 0 & 0 \end{bmatrix} \begin{bmatrix} E_x \\ E_y \\ E_z \end{bmatrix}, \quad (21)$$

where $r_{51} = 28.0$, $r_{22} = 3.4 \times 10^{-6}$ μm/V; $r_{33} = 30.8$, $r_{13} = 8.6 \times 10^{-6}$ μm/V; $n_0 \cong 2.237$, $n_e \approx 2.157$; and r_{ij} are the (reduced) electrooptic coefficients. To first order, (Eq. 21) becomes

$$\Delta n_i = -\frac{1}{2} n_{e,0}^3 \sum_j r_{ij} E_j, \quad (22)$$

where $n_{e,0}$ is either the extraordinary or ordinary refractive index, as appropriate. The change in the propagation constant of a waveguide is given by

$$\Delta \beta_i(E) = 2\pi \, \Delta n_i(E)/\lambda_0 . \quad (23)$$

The applied electric field, E, is related to the voltage across the electrodes by $E \approx V\Gamma/g$, where g is the electrode gap and Γ is an electrical/optical overlap coefficient ($\Gamma \sim 0.25$). Figure 9 shows three common *single-mode* guided-wave electrooptic devices: a Mach–Zehnder modulator, a 2 × 2 directional coupler switch, and a polarization rotator. (Good reviews of the principles and performance of $LiNbO_3$ integrated optic devices can be found; see, for example, Alferness [70] and Korotky and Alferness [71].) Electrodes may be positioned at the side of, or above, the waveguide features, depending on the desired electrooptic effect to be activated. For the latter arrangement, it is usually necessary to deposit a buffer layer

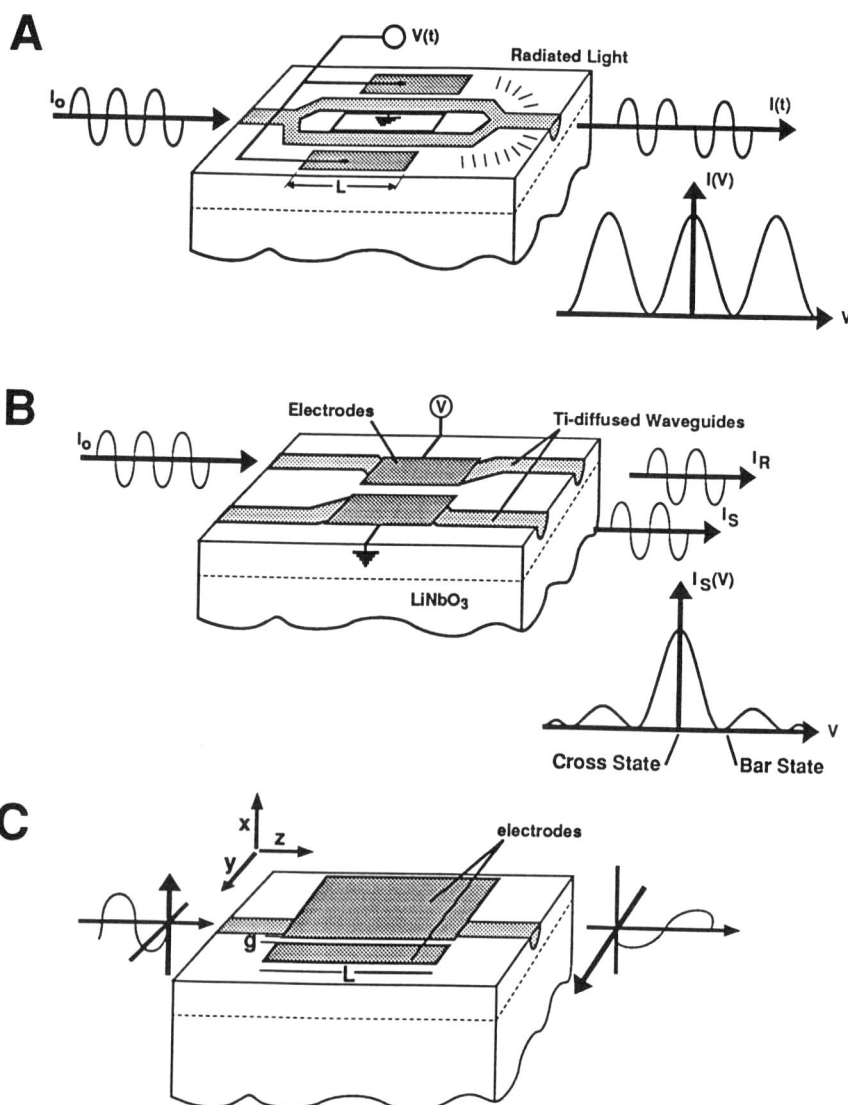

FIG. 9. Diagrams of three basic electrooptic functions: (a) an amplitude modulator consisting of a pair of phase shifters incorporated into an interferometer; (b) a 2 × 2 directional coupler switch; (c) a polarization rotator.

(e.g., ~0.2 μm SiO_2) between the $LiNbO_3$ substrate and the metal electrodes to reduce optical absorption losses. Because these devices rely on changing the propagation constant of a mode by a specific amount, multimode devices are usually not practical.

13.4.4.2 Guided-Wave Electrooptic Devices

The Mach–Zehnder interferometer (Fig. 9a) consists of a pair of y-branch waveguides with electrooptic phase shifters. Because of the electrode arrangement, light in the interconnecting arms experience opposite phase shifts:

$$\Delta\beta L = \pm 2\pi \, \Delta n(V) L / \lambda_0. \tag{24}$$

For a total phase shift of π, the two light fields will recombine antisymmetrically and cannot be guided by the exit waveguide, which supports only the fundamental (symmetric) mode. The light intensity at the exit port is given by

$$I_{out} \propto \cos^2(\Delta\beta L/2). \tag{25}$$

These modulators are usually fabricated on $-z$-cut or x-cut substrates (with y-propagating waveguides) in order to take advantage of the large r_{33} electrooptic coefficient. Fiber device/fiber insertion losses are typically 2–3 dB. Taking into account the optical field/electrical field overlap (~0.25), these devices require an on/off voltage of ~3 V for 1-cm electrodes. Modulation depths in excess of 95% are readily achieved. Devices with center-fed electrodes (as shown in Fig. 9) can achieve modulation rates (bandwidths) up to ~2 GHz for a 1-cm electrode. Modulation rates in excess of 40 GHz have been obtained using a Mach–Zehnder waveguide geometry and replacing the lumped center-fed electrode with a transmission strip line [76]. Note that, unless special design precautions are taken, these modulators will work optimally for only one polarization. Coupling into and out of the device must therefore be done either with polarization-preserving fiber or lenses.

A directional coupler switch/modulator (Fig. 9b) uses the electrooptic effect to mismatch the propagation constant difference, $\Delta\beta(V)$, between two closely spaced (and therefore coupled) waveguides. The fraction of coupled light is given by Eq. (10), where the coupling coefficient is

$$\kappa = \left[\int_0^L dz \, \kappa_1 \, e^{-b(z)/\gamma} \right], \tag{26}$$

where b is the center–center separation between the waveguides. The parameters κ_1 and γ can be experimentally determined from sets of coupled power measurements [73]. Typical waveguide fabrication parameters are listed in Table I. Approximate values of the coupling parameters for a $\lambda = 1.3$ μm wavelength are κ_{1TE}

$\cong 0.02$ μm^{-1}, $\kappa_{\mathrm{ITM}} \approx 0.2$ μm^{-1}, $\gamma_{\mathrm{TE}} \approx 3.4$ μm, and $\gamma_{\mathrm{TM}} = 1.9$ μm [74]. For a single pair of electrodes (as depicted in Fig. 9b), complete coupling (i.e., the crossed state) requires

$$\kappa L = (2m + 1)\, \pi/2, \quad m = 1, 2, \ldots, \qquad (27)$$

and $\Delta\beta L = 0$ (as can be deduced from Eq. (10)). The uncrossed (bar) state is then met for

$$(\kappa L)^2 + (\Delta\beta L/2)^2 = (m\pi)^2, \quad m = 1, 2, \ldots. \qquad (28)$$

Since κL (and therefore the crossed-state condition Eq. (27)) is determined by the waveguide geometry alone, accurate fabrication is required. Both the crossed and uncrossed states can be made electrically tunable by employing two electrode pairs [75]. Typical extinction ratios (crosstalk) for the crossed and uncrossed states are ~30 dB or better. As with the Mach–Zehnder interferometer, these devices are inherently polarization-dependent unless special design considerations are met [77].

The off-diagonal elements of the electrooptic tensor (Eq. (21)) can be used to produce polarization rotation. The most elementary example [78] is shown in Fig. 9c, in which the r_{22} component rotates the principal axes by $\pi/4$. X- or y-polarized light will therefore experience the equivalent of a voltage-tunable waveplate as it traverses the electrode region. Polarization transformers that are also wavelength filters have been demonstrated using a periodic electrode array [79].

At the time of this writing, pigtailed (usually with polarization-preserving fiber at the input end and standard single-mode fiber at the output) and packaged Mach–Zehnder modulators with ~1–20 GHz bandwidth are commercially available from a variety of vendors (e.g., Uniphase Telecomm Products, Hartford, CT).

13.4.4.3 Acoustooptic Devices

LiNbO$_3$ also possesses a piezoelectric effect. Interdigitated electrodes can thus be used to produce a mechanical (acoustic) wave in the material. The photoelastic effect converts this mechanical strain to a change in refractive index. This acoustooptic effect thus results in an index grating, either traveling or standing, that can be used to diffract light. Beam deflectors, switches, and rf spectrum analyzers have been demonstrated using this effect [41].

13.4.4.4 Quasi-Phase Matching Using Domain Inversion

As was mentioned earlier (Section 13.4.4), there are conditions under which the $+z$ surface of LiNbO$_3$ will depole below its Curie temperature. In related experiments, electrodes on x- and y-cut LiNbO$_3$ have been used to deliberately produce periodic domain inversion at relatively modest temperatures (200–300°C) [80, 81]. Periodically patterned regions of inverted domains have also been produced in $+z$

LiNbO$_3$ using a variety of other techniques [82–86]. This controlled-domain inversion has made it possible to produce quasi-phase-matched guided-wave frequency doubling (second-harmonic generation, SHG) in LiNbO$_3$. Periodically inverting the sign of the second-order optical nonlinearity coefficient by inducing domain inversion reestablishes the proper phase relation between the fundamental light wave and the second harmonic (which travel at different phase velocities). The required-phase matching condition is

$$\Delta\beta = 2m\,\pi/\Lambda, \quad m = 1, 2, \ldots, \tag{29}$$

where $\Delta\beta$ is the difference in the propagation constant between the fundamental and second harmonic, and Λ is the inversion period. Note that, while a first-order grating ($m = 1$) is about nine times more efficient than one of second order, the periodicity required for frequency doubling 1-μm wavelength light in LiNbO$_3$, $\Lambda \approx 0.5$ μm, necessitates a holographic or electron beam-written grating.

13.4.4.5 The Photorefractive and Pyroelectric Effects

A change in the refractive index of LiNbO$_3$ due to the presence of visible light was first described by Ashkin *et al.* [87]. The effect is due to the presence of impurities, primarily Fe^{2+} (~5 ppm) in the case of unintentionally doped LiNbO$_3$. In terms of a band model, the Fe^{2+} is photoionized to Fe^{3+} and the conduction band electrons diffuse along the z-axis [88, 89], leaving behind the donor ion, which acts as an electron trap. The space charge resulting from the electron diffusion (or drift) causes a change in the refractive index via the electrooptic effect. Light with a wavelength as long as 1300 nm has been shown [90] to produce changes in the refractive index of LiNbO$_3$ at very large power densities (in excess of 5×10^4 W/cm^2) over long (~100 h) time periods. This corresponds to ~25 mW in a single-mode Ti:LiNbO$_3$ waveguide—higher by one to two orders of magnitude than that carried by typical optical circuits. Visible light, however, will produce measurable index changes at power densities $\&\, 0.2$ W/cm^2 with ~0.5 h lifetimes [91]. The index change caused by this photorefractive effect can be reversed by heating the crystal to ~200°C for 1 h. The photorefractive effect can be greatly reduced by doping LiNbO$_3$ with $\&\, 5$ mol% MgO [92]. MgO-doped LiNbO$_3$ is now commercially available (Crystal Technology, San Jose, CA). With or without MgO doping, the experimentalist must be aware that guided visible light may cause a gradual index change; the operating points of integrated optic devices may therefore drift with time.

As seen from Eq. (22), the use of z-cut material permits access to a large number of electrooptic coefficients, including the largest, r_{33}. It is along the $\hat{z}$-axis, however, that the largest change in polarization with temperature change occurs (the pyroelectric effect). This change in substrate polarization must be compensated by a change in the electric field between the two surfaces (~3×10^5 V/cm/°C). A temperature increase of only 1°C will therefore result in a ~300 V potential difference across a 1-mm substrate; a 100°C increase or decrease will produce an electric field in excess of the material's breakdown field (~10^7 V/m). Such a temperature-induced charging can affect the state of an electrooptic device. The edges of electrodes have been found to exaggerate this effect [93].

13.4.4.6 Remote Sensing As pointed out, guided-wave optical systems can provide a method for remote sensing in hostile environments. $LiNbO_3$ integrated optic circuits can potentially serve as sensor and/or modulator elements for these systems. Modified Mach–Zehnder modulators have been used as electric field sensors [93], for example. Properties like the piezoelectric and pyroelectric effects can also be exploited as optically sampled pressure and temperature sensors, respectively. $LiNbO_3$ and $Ti:LiNbO_3$ waveguides are relatively inert to most chemically reactive agents. (When electrodes are needed, care must be taken to use unreactive metals or protect them.) The large Curie temperature of $LiNbO_3$ makes high-temperature operation of these devices practical. Although exposure of $LiNbO_3$ to gamma-ray radiation produces color centers and lattice defects, increasing the insertion loss of Ti-diffused waveguides [94], the effect is some 1000 times smaller than in silica fiber (e.g., a 0.2-dB increase in insertion loss after 2 h at 3×10^4 rad/min of 1.25-MeV gamma rays).

13.5 Concluding Points

The development of fiberoptics and integrated optics for telecommunications has brought new capabilities to the experimentalist. Low-loss light transmission without line-of-sight restrictions is relatively straightforward. In addition to data transmission, remote spectroscopy and sensing in the visible and infrared are also facilitated by fiber. Channel waveguide devices on planar substrates (e.g., couplers, power splitter, filters) can be used to extend the capability of fiberoptics. Large arrays of waveguide devices can also be integrated on a common substrate. Imagine, for example, an array of intersecting waveguides on a glass substrate. As pointed out earlier, a change in the refractive index of an intersection region changes the coupling between the crossing guides. If the intersection region can be sensitized to specific chemical moieties (e.g., antibodies), and the reaction produces a sufficiently large index change, it should be possible to produce a dense compact chemical or biological sensor array that can be optically interrogated. Finally, integrated electrooptic devices can be used to control the amplitude, phase, and polarization of guided-wave light. With such devices, very-high-speed data encoding and reprogrammable light paths are feasible.

Aside from the references cited herein, Laser Focus magazine (Penwell Publications, Westford, MA) distributes annually a very complete compilation of optical equipment and their manufacturers.

References

1. Kogelnik, H. (1988). Chapter 2 in *Guided Wave Optoelectronics*, T. Tamir (ed.), Springer-Verlag, Berlin.
2. Saleh, B. E. A., and Teich, M. (1991). In *Fundamentals of Photonics*, Wiley Interscience, New York.
3. Gloge, D. (1971). *Appl. Opt.* **10**, 2252.
4. Gowar, J. (1984). Chapter 2 in *Optical Communication Systems*, Prentice-Hall, London.
5. Nagel, S. R. (1988). In *Optical Fiber Telecommunications*, Vol. 2, S. E. Miller and I. P. Kaminow (eds.), Academic Press, New York.
6. Nagel, S. R. (1986). *Proc. SPIE* **717**, 8–20.
7. Friebele, E. J., Long, K. J., Askins, C. G., Geinerich, M. E., Marrone, M. J., and Griscom, D. L. (1985). *Proc. SPIE* **541**, 70–88.
8. Giallorenzi, T. G., Bucaro, J., Dandridge, A., Sigel, G., Cola, J., Rashleigh, S., and Prust, R. (1982). *IEEE Trans. Microwave Theory Techniques* **MTT-30**, 472.
9. Klocek, P., and Sigel, G. H. (1989). *Infrared Fiber Optics*, SPIE Publications, Bellingham, WA.
10. Iqbal, T., Shakriari, M. R., Merberg, G., and Sigel, G. H. (1989). *Proc. SPIE* **970**, 34.
11. Zhang, X. H., Ma, H. L., Blancheterie, C., and Lucas, J. (1992). *J. Non-Crystalline Solids* **146**, 154.
12. McCaughan, L. (1990). Unpublished material.
13. Parriaux, O., Gidon, S., and Kuznetsov, A. A. (1981). *Appl. Opt.* **20**, 2420.
14. Kaminow, I. P. (1981). *IEEE J. Quantum Electronics* **QE-17**, 15.
15. Noda, J., Okamoto, K., and Sasaki, Y. (1986). *J. Lightwave Technol.* **LT-4**, 1071.
16. Bergh, R. A., Leférre, H. D., and Shan, H. J. (1981). *Opt. Lett.* **6**, 502.
17. Presby, H. M., Amitay, N., Scotti, R., and Benner, A. F. (1989). *J. Lighwave Technol.* **1**, 274.
18. Buckman, A. B. (1992). Chapter 6 in *Guided Wave Photonics*, Saunders College Publishing, New York.
19. Gowar, J. (1984). Appendix 5 in *Optical Communication Systems*, Prentice-Hall, London.
20. Barnoski, M. K. (1981). In *Fundamentals of Optical Fiber Communications*, 2nd edition, M. K. Barnoski (ed.), pp. 147–155, Academic Press, New York.
21. Lee, T. P., Burrus, C. A., and Saul, R. H. (1988). Chapter 4 in *Optical Telecommunications*, Vol. 2, S. E. Miller and I. P. Kaminow (eds.), Academic Press, New York.
22. Palais, J. C. (1988). In *Fiber Optic Communications*, 2nd edition, Prentice-Hall, Englewood Cliffs, NJ.
23. Margalit, S., and Yariv, A. (1985). In *Semiconductors and Semimetals*, Vol. 22, W. T. Tsang (ed.), Academic Press, Orlando, FL.

24. Ramaswamy, R. V., and Srivastava, R. (1988). *J. Lightwave Technol.* **6**, 984.
25. Tickor, A. J., Lipscomb, G. F., Lytel, R., Stiller, M. A., and Thackara, J. L. (1989). *Proc. SPIE: Int. Soc. Opt. Eng.* **963**, 165.
26. Booth, B. L., and Marchegiano, J. E. (1988). "Waveguide Properties and Devices in Photopolymers," in *Proc. 14th European Conference on Optical Communications (ECOC)*, Vol. 1, p. 589.
27. Minford, W. J., Korotky, S. K., and Alferness, R. C. (1982). *IEEE J. Quantum Electronics* **QE-18**, 1802.
28. Burns, W. K., and Milton, A. F. (1988). In *Guided Wave Optoelectronics*, T. Tamir (ed.), Springer-Verlag, Berlin.
29. Izutsu, M., Nakai, Y., and Sueta, T. (1982). *Opt. Lett.* **7**, 136.
30. Tol, J., and Learhuis, J. (1991). *J. Lightwave Technol.* **9**, 879.
31. Haruna, M., Hibi, T., and Koyama, J. (1983). *Opt. Lett.* **8**, 534.
32. Bergmann, E. E., McCaughan, L., and Watson, J. E. (1984). *Appl. Opt.* **23**, 3000.
33. Gill, D. M., McCaughan, L., and Agrawal, N. (1991). *IEEE J. Quantum Electronics* **QE-27**, 588.
34. Neyer, A. (1983). *Electron. Lett.* **19**, 553.
35. Silberberg, Y., Perlmutter, P., and Baran, J. E. (1987). *Appl. Phys. Lett.* **51**, 1230.
36. Burns, W. K. (1990). *J. Lightwave Technol.* **8**, 990.
37. Feit, M. D., and Fleck Jr., J. A. (1980). *Appl. Opt.* **19**, 3140.
38. Izawa, T., and Nakogome, H. (1972). *Appl. Phys. Lett.* **21**, 584.
39. Viljanen, J., and Leppihalme, M. (1980). *J. Appl. Phys.* **51**, 3563.
40. Findakly, T. (1985). *Opt. Eng.* **24**, 244.
41. Hunsperger, R. (1984). Chapter 9 in *Integrated Optics: Theory and Technology*, 2nd edition, Springer-Verlag, Berlin.
42. Lytel, R. (1990). *Proc. SPIE* **1216**, 30.
43. Mohr, J., Anderers, B., and Ehrfeld, W. (1991). *Sensors and Actuators* **A25–27**, 571.
44. Townsend, P. D., Baker, G. L, Schlotter, N. E., Klausner, C. F., and Eternad, S. (1988). *Appl. Phys. Lett.* **52**, 1782.
45. Singh, B., Tripathi, K. N., Bishambhu, N. K., Joshi, J. C., Kapoor, S. K, and Dawar, A. L. (1992). *J. Mat. Sci. Lett.* **11**, 382.
46. Ashley, P. R., and Tumolillo, T. A. (1988). *Appl. Phys. Lett.* **52**, 1031.
47. Diemier, M., Suyten, F., Trommel, E., McConach, A., Copeland, J., Jenneskens, L., and Horsthuis, W. (1990). *Electron. Lett.* **26**, 379.
48. McFarland, M. J., Wong, K. K., Wu, C., Nahata, A., Horn, K. A., and Yardley, T. (1988). *Proc. SPIE* **993**, 16.
49. Becker, E. W., Ehrfeld, W., Hagmann, P., Maner, A., and Munchmeyer, D. (1986). *Microelectron. Eng.* **4**, 35.
50. Chen, R. T., Sadovnik, L., Jannson, T., and Jannson, J. (1991). *Appl. Phys. Lett.* **58**, 1.
51. Khanarian, G., Norwood, R. A., Haas, D., Feuer, B., and Karim, D. (1990). *Appl. Phys. Lett.* **57**, 977.

52. Huijts, R. A., and Wesselink, G. L. J. (1989) *Chem. Phys. Lett.* **156**, 209.
53. Horsthius, W. H. G., Heideman, J. P., Koerkamp, M. M. K., Hams, B. H. M., and Mohlmann, G. R. (1993). Paper IMC 1-1 presented at the OSA Symposium on Integrated Photonic Structures, Palm Springs, CA.
54. Girton, D. G., Kwiatkowski, S. L., Lipscomb, G. F., and Lytel, R. S. (1991). *Appl. Phys. Lett.* **58**, 1730.
55. Kaminow, I. P. (1974). *An Introduction to Electrooptic Devices*, pp. 112–117, Academic Press, New York.
56. Rauber, A. (1978). In *Current Topics in Materials Science*, Vol. 1, E. Kaldis (ed.), North-Holland, Amsterdam.
57. Kaminow, I. P., and Carruthers, J. R. (1973). *Appl. Phys. Lett.* **22**, 326.
58. Minakata, M., Saito, S., Shibata, M., and Miyazawa, S. (1978). *J. Appl. Phys.* **49**, 4677.
59. Miyazawa, S. (1979). *J. Appl. Phys.* **50**, 4599.
60. Nakamura, K., Ando, H., and Shimizu, H. (1987). *Appl. Phys. Lett.* **50**, 1413.
61. Alferness, R. C., Ramaswamy, V. R., Korotky, S. K., Divino, M. D., and Buhl, L. L. (1982). *J. Quantum Electronics* **18**, 1807.
62. McCaughan, L., and Murphy, E. J. (1983). *J. Quantum Electronics* **19**, 131.
63. Veselka, J. J., and Korotky, S. K. (1986). *J. Quantum Electronics* **22**, 933.
64. Fukuma, M., Noda, J., and Iwasaki, H. (1978). *J. Appl. Phys.* **49**, 3693.
65. Fukuma, M., and Noda, J. (1980). *Appl. Opt.* **19**, 591.
66. Jackel, J. L., Rice, C. E., and Veselka, J. J. (1982). *Appl. Phys. Lett.* **41**, 607.
67. Bogert, G. A., and Moser, D. T. (1990). *IEEE Photonics Technol. Lett.* **2**, 632.
68. Wong, K. K., DeLaRue, R. M., and Wright, S. (1987). *Opt. Lett.* **23**, 265.
69. Murphy, E. J., Rice, T. C., McCaughan, L., Harvey, G. T., and Read, P. (1985). *J. Lightwave Technol.* **3**, 795.
70. Alferness, R. C. (1988). In *Guided Wave Optoelectronics*, T. Tamir (ed.), Springer-Verlag, Berlin.
71. Korotky, S. K., and Alferness, R. C. (1987). Chapter 2 in *Integrated Optical Circuits and Components*, L. Hutcheson (ed.), Dekker, New York.
72. Gee, C. M., Thurmond, G., and Yen, H. (1983). *Appl. Phys. Lett.* **43**, 998.
73. McCaughan, L. (1984). *IEEE J. Lightwave Tech.* **2**, 51.
74. McCaughan, L. (1993). In *Integrated Optics and Optoelectronics*, pp. 15–43, M. Razeghi and K. K. Wong (eds.), SPIE Optical Engineering Press, Bellingham, WA.
75. Kogelnik, H., and Schmidt, R. V. (1976) *IEEE J. Quantum Electronics* **QE-12**, 396.
76. Korotky, S. K., Eisenstein, G., Tucker, R. S., Veselka, J. J., and Raybon, G. (1987). *Appl. Phys. Lett.* **50**, 1631.
77. Alferness, R. C. (1979). *Appl. Phys. Lett.* **35**, 748.
78. Thaniyavarn, S. (1985). *Appl. Phys. Lett.* **47**, 674.
79. Alferness, R. C., and Buhl, L. L. (1982). *Appl. Phys. Lett.* **40**, 861.

80. Janzen, G., Seibert, H., and Sohler, W. (1992). Paper TuD5 presented at OSA Symposium on Integrated Photonics Research, April, New Orleans.
81. Seibert, H., and Sohler, W. (1991). In *Proc. Int. Conf. on Physical Concepts of Materials for Novel Optoelectronic Device Applications*, Vol. 2: *Device Physics and Applications*, Aachen, Germany, 28 October–2 November 1990; see also *Proc. SPIE* **1362**, 370.
82. Armani, F., Delacourt, D., Lallier, E., Papuchon, M., He, Q., de Micheli, M., and Ostrowsky, D. B. (1993). Submitted for publication.
83. Lim, E. J., Fejer, M. M., and Byer, R. L. (1989). *Electron. Lett.* **25**, 175.
84. Webjorn, J., Laurell, F., and Arvidsson, G. (1989). *IEEE Photonics Technol. Lett.* **1**, 16.
85. Nutt, A. C., Gopalan, V., and Gupta, M. (1992). *Appl. Phys. Lett.* **60**, 2828.
86. Fujimura, M., Suhara, T., and Nishihara, H. (1991). *Electron. Lett.* **27**, 1207.
87. Ashkin, A., Boyd, G. D., Dziedzie, J. M., Smith, R. G., Bellman, A. A., and Nassau, K. (1966). *Appl. Phys. Lett.* **9**, 72.
88. Chen, F. S. (1969). *J. Appl Phys.* **40**, 3389.
89. Glass, A. M. (1978). *Opt. Eng.* **17**, 470.
90. Harvey, G. T., Astfalk, G., Feldblum, A. Y., and Kassahun, B. (1986). *IEEE J. Quantum Electronics* **QE-22**, 939.
91. Schmidt, R. V., Cross, R. S., and Glass, A. M. (1980). *J. Appl. Physics* **51**, 90
92. Byran, D. A., Gerson, R., and Tomaschke, H. (1984). *Appl. Phys Lett.* **44**, 847.
93. Bulmer, C. H., Burns, W. K., and Hiser, S. C. (1986). *Appl. Phys. Lett.* **48**, 1036.
94. Jack, C. A., and Kanofsky, A. S. (1989). *Proc. SPIE* **1177**, 274–279.

Index

A

Achromatic lenses, 351
Acoustooptic devices, lithium niobate, 390
Acoustooptic isolation, 359
Acoustooptic modulator (AOM)
   AOM/EOM for dye lasers, 129
   free spectral range using, 109
   frequency stabilization with, 300–301, 359
   principles, 123–124
   tuning with, 113–114
ac Stark effect, phase shifts generated by, 302
Active modelocking, 175–180
$AgGaSe_2$, as OPO material, 257–259
$AgGaS_2$, as OPO material, 257, 258
Air-spaced interferometer, 350
AlGaAs laser
   coatings, 95
   properties, 80, 81
   tuning, 89
AlGaInP laser, 80
Aluminum, as mirror coating, 347
Amplified spontaneous emission, 154–162
Amplitude modulation, modelocking, 176–179
Angled physical contact connectors (*see* APC connectors)
Antireflective coatings, 346, 348
AOM (*see* Acoustooptic modulator)
APC connectors, 172
Argon laser, 68, 69
Astigmatism, CW dye lasers, 58–59
Asymmetric beam waist, 196
Asymmetric wiggler, 32, 33
Asymmetry, 196
Atomic structure, synchrotron radiation studies, 40
Atomic transitions, laser frequency stabilization, 120
Atom interference, by separated-field excitation, 295–299

Autocorrelation, 187–188
   frequency resolved optical gating, 215, 216–220
   higher-order autocorrelation, 213–215
   intensity autocorrelation, 205, 209–215
   interferometric autocorrelation, 215–216
   multishot autocorrelation, 206
   phase-sensitive autocorrelation, 215–220
   principles, 203–208
   second-order autocorrelation, 205, 206, 208, 209–213
   single-shot autocorrelation, 208

B

Bandpass filters, 350
BBO, as OPO material, 257–259, 261, 262, 263
Beam splitters, 348–349
Beam waist, 196
Bending magnet radiation, accelerating electrons, 24–28
Bethune cell, 162
Biconcave lens, 351
Biconvex lens, 351
BIG (bismuth-substituted iron garnet), 360
Binary optics, 360–361
Birefringence, 353
Birefringent filter, dye lasers, 63
BK7, 343–345
Black-body radiators, 6–7
Borates, BBO, 257, 258
Bordé atom interferometer, 295–298
Broadband polarizing beam splitter cubes, 354
Bulk laser, 172
Bulk optics, 361

C

Calcium-stabilized lasers, 294–295
Carbon dioxide lasers, 4, 148
CCD array, as short-pulse detector, 199, 212
CCL (*see* Color-center laser)
Chalcogenide glasses, 375
Channel waveguides
  fabrication, 384, 386
  on glass substrates, 383–384
Cholesteric liquid crystals, 356, 357
Circular polarization, synchrotron radiation, 30–33
Coatings, 347–348
Collisional absorption, 9, 10
Color-center laser (CCL), 304
Confocal resonant cavities, 366
Connectors, 172–173
Continuous wave (CW) lasers
  dye lasers, 45–46, 138
    alignment, 71–73
    astigmatism in, 58–59
    dyes, 66–71
    flowing dye cell, 59–60
    laser resonator in, 56–58
    mirror transmission, 60–66
    theory, 46–56
    triplet quenching, 60
  error sources, 198
  single spatial-mode semiconductor lasers, 78, 79
Coupling efficiency, single-mode fiber, 379–380
Cross-correlation, 198–203, 220–223, 224
Crossed field undulator, 32, 33
Crown glass, 343
Crystal monochromator, 38–39
CW lasers (*see* Continuous wave lasers)

D

Damage threshold, 346
DBR (*see* Distributed Bragg reflector)
DFB (*see* Distributed feedback)
DFG (*see* Difference-frequency generation)
Dielectric beam splitters, 349
Dielectric mirrors, 347–348
Dielectric susceptibility, 231
Difference-frequency generation (DFG), 248–252, 254
Diffraction gratings, 34, 164–167, 361
Diffraction optics, 360–361
Diode laser array, 90
Diode lasers, 77–78, 149–150, 260
  amplitude noise, 83
  beam quality, 82–83
  current sources, 90–93
  electronics, 89–95
  extended-cavity diode lasers, 83, 84–89, 93, 95, 99
  extended wavelength coverage, 99–100
  frequency conversion, 232
  frequency noise and stabilization, 97–99
  laser degradation, 84
  optical feedback, 83–84, 85
  output power, 80–82
  temperature controllers, 93–95
  tuning, 78–80
Directional couplers, 382, 383
Directional coupler switch/modulator, 389
Distributed Bragg reflector (DBR), 78
Distributed feedback (DFB), 78
Double-passed AOM, 114
DRO, 264
Dye lasers, 146, 151
  advantages of, 146
  amplified spontaneous emission, 154–162
  continuous wave dye lasers, 45–46, 50–73
  disadvantages of, 146
  frequency and phase locking, 128–134
  frequency stabilization, 299
  gain, 52–53
  grazing-incidence dye laser, 164–167
  intracavity spectral filters, 164–169
  multiple passes through the gain medium, 162–164
  output power, 54–56
  principles, 46–50
  safety, 157–158
  single longitudinal mode tunable dye laser, 167–168
  threshold pump power, 53–54
  traveling-wave dye laser (TWDL), 251

INDEX 399

E

ECDL (*see* Extended-cavity diode laser)
Edge filters, 350
Electron-beam deposition, 348
Electron density, of a plasma, 8–10, 11
Electronic structure, synchrotron radiation studies, 40
Electrooptic effect, lithium niobate, 387–389
Electrooptic modulator (EOM), 109, 122–124, 129
Energy levels, dye lasers, 46–50, 60
Energy meter, as short-pulse detector, 199
EOM (*see* Electrooptic modulator)
Er:fiber amplifier (EDFA), 175
Error signal, 106, 115–120, 281–286
Excimer lasers, 4, 147–148, 151, 259–260
Extended-cavity diode laser (ECDL), 83, 84–89
  control, 99
  optical coating, 95
  temperature stability, 93
Extended-cavity laser, 83, 84
Extended x-ray absorption fine structure (EXAFS), 40
External enhancement cavity, 239–243

F

Fabry–Perot cavity, 107
  error signal deviation, 115–120
  mechanical stress, 110–113
  optical properties, 107–110
  phase locking, 131
  tunability vs. stability, 113–115
Fabry–Perot etalons, 64–66, 145, 350
Fabry–Perot interferometer, 350, 362–366
Faraday mirrors, 173–174
Fast Fourier transform (FFT), power spectrum measurement, 330
Fiber couplers, 376–377
Fiber lasers, 171
  cavity building, 171–175
  diagnostics, 187–190
  modelocking, 175–184
  operating parameters, 184–187
Fiberoptic delay line, 376
Fiberoptics, 392
  components, 376–381
  optical fibers, 371–376
Filters (*see* Optical filters)
Finesse, 362, 363
First-order autocorrelation, 205
Fizeau wavemeter
  calibration, 335
  limitations and variations, 335–337
  principles, 331–332
  structure, 332–333
  wavelength calculation with, 333–335
Flappers, 173
Flowing dye cell, 59–60
Fluoride fibers, 375
Fluorozirconate fibers, 375
FM saturation spectroscopy, 120
Fourier spectrometer, scanning Michelson interferometer and, 330
Four-wave DFG, 254
Four-wave mixing interactions, 253–254
Four-wave SFG, 254
Fraunhofer cemented doublet, 351
Free spectral range (FSR), 107, 108
Frequency chain, 304
Frequency conversion, nonlinear (*see* Nonlinear optical) frequency conversion
Frequency doubling
  intracavity, 237–239
  synchronously pumped, 243
Frequency noise, diode lasers, 97–99
Frequency resolved optical gating (FROG), 215, 216–220
Frequency stabilization, 103–104, 299–301
  of diode lasers, 97–99
  of dye lasers, 299
  of tunable lasers, 106–107
    design examples, 126–134
    loop filter, 124–126
    optical frequency references, 107–121
    transducers, 121–124
Frequency uncertainty, 301–303
FROG (*see* Frequency resolved optical gating)
FSR (*see* Free spectral range)
Fused fiber couplers, 376–377
Fused silica, 344, 345–346

G

GaAlAs diode, 150
GaInPAs laser, 80, 149
Gas lasers, 147–148, 151
Glan laser polarizers, 354
Glan–Thompson polarizers, 354
Glasses, 343–346, 351, 375
Gold, as mirror coating, 347
Graded-index lenses, 351–352
Grating monochromator, 34–35
Grazing-incidence dye laser, 164–167
GRIN lenses, 351–352
Guided-wave integrated optics, 381–382, 392
    channel waveguides in glass substrates, 383–384, 392
    lithium niobate, 385–392
    optical plumbing, 382–383
    polymer waveguides, 383–384
    quasi-phase-matching in, 245

H

Half-waveplate, 353–354
Harmonic beam splitter, 349
Helical undulator, 32
Helium–neon lasers
    with dye lasers, 167
    frequency stabilization, 126–128
    iodine-stabilized, 289–294, 307
    as reference laser, 325
    uses, 260
Higher-order autocorrelators, 213–215
High-repetition-rate lasers, 146–147, 150
Holographic optical element (HOE), 361–362

I

IBS (*see* Ion-beam-sputtering deposition)
Incoherent optical sources, laser-produced plasmas and, 1–18
InGaAsP laser, 79

Injection seeding, 145–146
Insertion devices, synchrotron radiation generation, 28–30, 32–33
Integrated optics, 382
Intensity autocorrelation function, 205
Intensity autocorrelators, 209–215
Interference edge filters, 350
Interferometers
    air-spaced interferometers, 350
    Bordé atom interferometer, 295–298
    Fabry–Perot interferometer, 350, 362–366
    Mach–Zehnder interferometer, 389
    plane–parallel interferometer, 337–338
    Sagnac interferometer, 180
    scanning Michelson interferometer, 315–331
Interferometric autocorrelators, 215–216
Intersecting waveguides, 382–383
Intracavity frequency doubling, 237–239
Intracavity spectral filters, 164–169
Inverse bremsstrahlung, 9
Iodine-stabilized He–Ne lasers, optical wavelength standards, 288–294, 307
Ion-beam-sputtering (IBS) deposition, 348
Isolators, 174–175, 358–361

K

Krypton laser, 68, 69
KTP, as OPO material, 24, 257–259, 262
Kuizenga–Siegman equation, 177–178

L

λ-meter (*see* Scanning Michelson interferometer)
Laser-cooled absorbers, wavelength standards utilizing narrow resonances, 294–303, 307
Laser degradation, 84
Laser diodes, 373
Laser dyes, 66–71, 155, 159–160
Laser goggles, 157–158
Laser line polarizing beam splitter cubes, 354

INDEX

Laser-produced plasma, 1–5
　black-body radiators, 6–7
　debris, 17–18
　focusing, 15
　lasers used, 4–5
　physics of, 7–13
　short-pulse-length pumping, 10, 15–17
　targets, 13–15
Laser resonator, 56–58
Lasers (*see also* individual lasers)
　damage testing, 346
　frequency stabilization (*see* Frequency stabilization)
　population inversions, 138
LBO, as OPO material, 257
Lead–salt laser, 78, 149
Lenses, 343, 346, 351–353
Light-emitting diodes (LEDs), 373
Light monochromatization (*see* Monochromatization)
Light polarization (*see* Polarization)
Linear chirp parameter, 195
Linear undulator, 33
Liquid crystal devices, 355–357
Liquid crystals, 356, 357
Lithium niobate ($LiNbO_3$)
　acoustooptic devices, 390
　electrooptic effect, 387–389
　guided-wave electrooptic devices, 389–390
　as OPO material, 257
　photorefractive effect, 391
　properties, 385–387
　pyroelectric effect, 391–392
　quasi-phase-matching, 390–391
　remote sensing, 392
Littman–Metcalf configuration, 87, 260
Littrow configuration, 86, 165, 260
Loop filter, laser frequency stabilization, 124–126
Lyot filter, 63, 178

Magnetic x-ray scattering, 40
Magnetooptic isolators, 359–360
MCP (*see* Microchannel plate)
McPherson monochromator, 35
Metal beam splitters, 349
Metal–insulator–metal (MIM) diode, 251
Metallic mirrors, 347
Metal oxides, 375
"Mickey Mouse" ears, 173
Microchannel plate (MCP), 199, 200
Microscopy, synchrotron radiation studies, 41
MIM diode (*see* Metal–insulator–metal (MIM) diode)
Mirrors, 343, 345–348
Mirror transmission, CW dye lasers, 60–61
Modelocking
　active modelocking, 175–180
　amplitude modulation, 176–179
　nonlinear loop mirror modelocking, 180–182
　nonlinear polarization rotation modelocking, 182–183
　passive modelocking, 180–182
　phase modulation, 179
　semiconductor saturable absorbers, 180
　"soliton" modelocking, 180
　stretched pulse modelocking, 183–184
Monochromatization (light), synchrotron radiation, 33–39
Monochromators, 34–35, 38–39, 350
Monolithic cavity design, 242–243
MOPA laser (monolithic master-oscillator/power-amplifier laser), 80, 82, 90, 162
Multimode fiber, 377, 380
Multiple-order waveplates, 353
Multiple-pass laser, 162–164
Multishot autocorrelation, 206

M

Mach–Zehnder interferometer, 389
Mach–Zehnder modulator, 178

N

NALM (*see* Nonlinear loop mirror modelocking)
Nanosecond OPO systems, 261–262
Nd (neodymium) lasers, laser-produced plasmas, 4, 5
Nd:glass laser, 144

Nd:YAG laser, 172, 259
  energy spectrum, 142
  intercavity frequency doubling, 237–239
  pumping of, 150
  Q-switching, 143–144, 146–147, 260
  wavelength, 151
Nd:YLF laser, 262
Near-edge x-ray absorption fine structure (NEXAFS), 40
Nematic liquid crystals, 356, 357
NIM (*see* Normal incidence monochromator)
Nonconfocal cavities, 110, 364–366
Nonlinear loop mirror modelocking (NALM), 180
Nonlinear optical frequency conversion, 231–233
  difference-frequency generation, 248–252, 254
  four-wave mixing, 253–254
  optical parametric amplifiers and oscillators, 255–266
  Raman shifters, 233, 266–268
  second-harmonic generation, 233–237
    external enhancement cavity, 239–243
    intracavity frequency doubling, 237–239
    quasi-phase-matching, 233, 234, 243–247
    synchronously pumped frequency doubling, 243
  sum-frequency generation, 247–248, 254
  third-harmonic generation, 252
  up-conversion lasers, 233, 268–270
Nonlinear polarization rotation modelocking, 182–183
Nonpolarizing beam splitter cubes, 349
Nonsilica optical fiber, 375–376
Normal incidence monochromator (NIM), 34, 35–36
NYAB, 238

O

OPAs (*see* Optical parametric amplifiers)
OPOs (*see* Optical parametric oscillators)
Optical components
  antireflective coatings, 346, 348
  beam splitters, 348–349
  lenses, 343, 346, 351–353
  mirrors, 343, 345–348
  optical filters, 350
  polarization-controlling components, 353–358
Optical diode, 63
Optical feedback, diode lasers, 83–84, 85
Optical fibers
  properties, 371–373
  types, 373–376
Optical filters, 350
  bandpass filters, 350
  birefringent filter, 63
  edge filters, 350
  interference edge filters, 350
  intracavity spectral filters, 164–169
  loop filter, 124–126
  Lyot filter, 63, 178
Optical frequency measurement, 303–306, 311
Optical frequency references, 106–121, 325
  atomic transistors, 120
  Fabry–Perot cavity, 107–120
  optical phase locking, 121
Optical Kerr shutter, 214
Optical locking, 98–99, 121
Optical materials
  glasses, 343–346
  OPAs and OPOs, 257–259
Optical parametric amplifiers (OPAs), 255–266
Optical parametric oscillators (OPOs), 233, 255–266
Optical plumbing, 382–383
Optical spectrum analyzer, 188, 364
Optical wavelength measurement, 311–312, 338
  Fizeau wavemeter, 331–337
  plane–parallel interferometer, 337–338
  scanning Michelson interferometer, 312–331
Optical wavelength standards, 279–280, 307
  calcium-stabilized laser, 294–295, 307
  iodine-stabilized He–Ne lasers, 288–294, 307
  laser-cooled absorbers, 294–303, 307
  optical frequency measurement, 303–306
  theory, 280–281
    electronic frequency servo control, 286–288
    error signal generation, 281–286
    reproducibility, 288
    stability, 288
    uncertainty, 288
Oscilloscope, diagnostics with, 189–190
Output power, dye lasers, 54–56

Oxazine IR dyes, 155
Oxide fibers, 375

P

Passive modelocking, 180–182
Passive optical devices
    Fabry–Perot interferometer, 350, 362–366
    holographic optical elements, 361–362
    isolators, 174–175, 358–361
PC connectors, 172
Pellicle beam splitter, 349
PGM (see Plane grating monochromator)
Phase-modulated lasers, 179
Phase modulation, modelocking, 179
Phase-sensitive autocorrelators, 215–220
Photoabsorption spectroscopy, 40
Photocathode, 200–201
Photodiode, as short-pulse detector, 199
Photoemission, 40
Photomultiplier, as short-pulse detector, 199
Photon flux, synchrotron radiation, 25–26
Photon optics, monochromatization, 33
Photorefractive effect, lithium niobate, 391
Physical contact connectors (see PC connectors)
P–I curve, CW diode laser, 80
Piezoelectric transducer (PZT), 113, 122
PI isolators, 174
Pitting, laser damage to coatings, 346
Planar waveguide, 369, 370
Planck's law, 6
Plane grating monochromator (PGM), 37
Plane–parallel interferometers, with CCD readout, 337–338
Plano-convex lens, 351
Plasma electron temperature, 11
Plasmas
    atomic number dependence of spectra, 13
    laser-produced (see Laser-produced plasma)
Poincaré sphere, 357–359
Polarcor polarizers, 354
Polarization (light), synchrotron radiation, 30–33
Polarization controller, 173, 353–358, 378
Polarization independent isolators (see PI isolators)

Polarization meters, 357–358
Polarizer/quarter-wave isolator, 358–359
Polarizing elements, 377–378
Polarizing optics, 353–355
Polarizing sheets, 354
Population inversions, 138
Potassium niobate ($KNbO_3$), as OPO material, 257
Pound–Drever–Hall scheme, 115, 116, 119, 124
Power
    synchrotron radiation, 24–25
    undulator, 29–30
Power spectral density, 104–106
Power spectrum measurement, fast Fourier transform, 330
Proton exchange, 386
Pulsed lasers, 137–138
    builder's guide, 153–169
    buyer's guide, 150–152
    cavity mode, 144–146
    dye lasers, 146
    error sources, 198
    gas lasers, 147–148, 151
    high-repetition-rate lasers, 146–147, 150
    injection seeding, 144–146
    Nd:YAG laser, 143–144, 150, 151, 172
    plasma production using, 1–18
    Q-switching, 139, 141–143, 146
    ruby laser, 139, 140
    safety, 157–158
    semiconductor lasers, 149–150
    theory, 138–139
    tunable solid-state lasers, 148–149
Pulsewidth
    autocorrelator, 187–188
    fiber laser, 184, 187–188
Pump lasers, 151, 259–260
Pyrex®, 343–345
Pyroelectric effects, lithium niobate, 391–392
PZT (see Piezoelectric transducer)

Q

QPM (see Quasi-phase-matching)
Q-switching, 139, 141–143, 146
Quadrant diode, as short-pulse detector, 199

Quadratic chirp, 195
Quarter-wave isolator, 358–359
Quarter-waveplate, 354
Quasi-phase-matching (QPM), 233, 234, 243–247, 390–391

R

R6G laser (*see* Rhodamine laser)
Raman shifters, 233, 266–268
Real-time oscilloscope, diagnostics with, 189–190
Reference wavelengths, 279 (*see also* Optical wavelength standards)
Remote sensing, lithium niobate, 392
Retardance units, 353
RF spectrum analyzer, 188–189
Rhodamine 6G (R6G) laser, 48, 56, 66–67, 68, 146, 155
Ring dye laser, 62–63
Ruby laser, 139, 140

S

Sagnac interferometer, 180
Saturable absorbers, modelocking with, 180
Saturation FM spectroscopy, 120
S-bends, 382
Scanning Michelson interferometer
 accuracy and resolution, 315–316
 alignment, 322–325
 data acquisition, 326–331
 principles, 312–315
 reference laser, 325
 structure, 316–322
Schawlow–Townes limit, 104
Schmitter trigger, 327
Second-harmonic generation, 233–237
 external enhancement cavity, 239–243
 intracavity frequency doubling, 237–239
 quasi-phase-matching, 233, 234, 243–247
 synchronously pumped frequency doubling, 243

Second-order autocorrelation, 205, 206, 208
Second-order autocorrelators, 209–213
Second-order parametric interaction, 252
SELFOC® lenses, 351–352
Semiconductor lasers, 77–78, 80, 149–150
 diode lasers, 77–100, 232
Semiconductor photodiode, as short-pulse detector, 199
Semiconductor saturable absorbers, modelocking with, 180
Separated-field excitation, of laser-cooled and trapped atoms, 298–299
Servo control, electronic frequency, 103–104, 106, 286–288
Seya–Namioka monochromator, 35
SFG (*see* Sum-frequency generation)
SGM (*see* Spherical grating monochromator)
SHG, 235–236, 240–242
Short laser pulses, 193–194
 by difference frequency generation (DFG), 250–251
 detectors, 198–199
  autocorrelation, 203–208
  cross-correlation, 198–203, 220–223
  intensity autocorrelators, 209–215
  phase-sensitive autocorrelators, 215–220
  streak camera, 199–203
 error sources, 198
 laser-produced plasma, 10, 15–17
 spatial characterization and focusing, 196–198
 theory, 194–196
 VUV and x-ray regions
  cross-correlation, 224, 226–227
  ultra short pulses, 223
  x-ray streak camera, 223–224
Short-pulse OPOs, 261–264
Side-pumped pulsed amplifiers, 162
Silica optical fiber, 373–375
Silver, as mirror coating, 347
Single longitudinal mode tunable dye laser, 167–168
Single-shot autocorrelator, 208
Soft x-ray monochromator, 34, 36–38
Solid-state lasers, 151, 232
Solitary laser, 83
Soliton energy, 178–179
"Soliton" modelocking, 180
Soliton supporting fiber lasers, 184–187
SOP (*see* State of polarization)

INDEX 405

Source-to-fiber coupling, 379–381
Spherical grating monochromator (SGM), 37–38
Splicing, 172
SROs, 265
SRS (see Stimulated Raman scattering)
SSHG (see Surface second-harmonic generation)
State of polarization (SOP), 173
Static electric field, phase shifts generated by, 302–303
Stefan–Boltzmann law, 7
Stimulated Raman scattering (SRS), 266
Stokes parameters, 357–358
Storage ring
   dipole or bending magnet radiation, 24–28
   light polarization, 30–33
Streak camera, 193, 199–203, 223–224
Streak tube, 201–202
Stress birefringence, 378
Stretched pulse modelocking, 183–184
Sum-frequency generation (SFG), 246–247, 254
Surface domain gratings, quasi-phase-matching in, 245
Surface second-harmonic generation (SSHG), 213
Synchronously pumped frequency doubling, 243
Synchrotron radiation, 23–24
   applications, 40–41
   characteristics, 24–33
   history, 23
   light monochromatization, 33–39

T

Technical noise, 103
TGG (terbium gallium garnet), 360
TGM (see Toroidal grating monochromator)
Third-harmonic generation (THG), 252–253
Third-order parametric processes, 253
Ti:Al$_2$O$_3$ laser, 264
Ti:sapphire crystal, 148, 151
Ti:sapphire laser, 226, 260
Toroidal grating monochromator (TGM), 37
Touchek effect, 26–27

Transducers, laser frequency stabilization, 121–124
Traveling-wave dye laser (TWDL), 251
Triplet quenching, CW dye lasers, 60
Tunable fiber coupler, 377
Tunable lasers
   continuous wave dye laser, 45–73, 138
   frequency stabilization, 103–134
   precise wavelength measurement, 311–312, 338
      Fizeau wavemeter, 331–337
      plane–parallel interferometer, 337–338
      scanning Michelson interferometer, 312–331
   solid state lasers, 148–149
TWDL (see Traveling-wave dye laser)
Two-photon fluorescence, in second-order autocorrelation, 212
Two-photon ionization, in second-order autocorrelation, 212–213

U

Ultralow-loss optics, 348
Uncertainty
   frequency, 301–303
   wavelength, 288
Undulators, synchrotron radiation generation, 28–30, 32–33
Up-conversion lasers, 233, 268–270

V

Vacuum deposition, by laser-produced plasmas, 17
Vacuum photodiode, as short-pulse detector, 199
Vacuum ultraviolet radiation, ultrashort pulses, 224–227

W

Wadsworth monochromator, 35, 36
Waveguides, 369–371
   channel waveguides, 383–384, 386
   fabrication, 384, 386
   intersecting waveguides, 382–383
   polymer waveguides, 383–384
   quasi-phase-matching in, 245–247
Wavelength tuning, 61–66
Wave length uncertainty, 288
Wave-vector mismatch, 234
Wide-strip laser, 90
Wigglers, synchrotron radiation generation, 28–30, 32–33

X

X-rays, ultrashort pulses, 224–227

X-ray scattering, 40
X-ray streak camera, 223–224

Y

Y-branches, 382
YIG (yttrium iron garnet), 360

Z

ZBLAN glass, 375
Zerodur®, 344, 345
Zero-order waveplates, 353, 355
ZnGBeP$_2$, as OPO material, 257, 258

ISBN 0-12-475977-7